Handbook of Food Safety and Security

Handbook of Food Safety and Security

Edited by Jack Pittman

SYRAWOOD
PUBLISHING HOUSE
New York

Published by Syrawood Publishing House,
750 Third Avenue, 9th Floor,
New York, NY 10017, USA
www.syrawoodpublishinghouse.com

Handbook of Food Safety and Security
Edited by Jack Pittman

International Standard Book Number: 978-1-64740-072-9 (Hardback)

Cataloging-in-Publication Data

Handbook of food safety and security / edited by Jack Pittman.
p. cm.
Includes bibliographical references and index.
ISBN 978-1-64740-072-9
1. Food--Safety measures. 2. Food security. 3. Food adulteration and inspection.
4. Food contamination. I. Pittman, Jack.

TX531 .H36 2022
363.192--dc23

TABLE OF CONTENTS

PREFACE

In my initial years as a student, I used to run to the library at every possible instance to grab a book and learn something new. Books were my primary source of knowledge and I would not have come such a long way without all that I learnt from them. Thus, when I was approached to edit this book; I became understandably nostalgic. It was an absolute honor to be considered worthy of guiding the current generation as well as those to come. I put all my knowledge and hard work into making this book most beneficial for its readers.

Food can be contaminated during the processes of production, packaging, transportation, etc. Food safety is concerned with such practices in order to avoid food-borne illnesses. This includes food hygiene, pesticide residue, food labeling, food certification, etc. The measurement of the availability of food along with its accessibility to consumers is known as food security. Some of the indexes to measure food security are Household Dietary Diversity Scale (HDDS), Household Hunger Scale (HDS), Household Food Insecurity Access Scale (HFIAS), etc. The four pillars of food security are availability, accessibility, utilization and stability. This book covers in detail some existing theories and innovative concepts revolving around food safety and security. Different approaches, evaluations, methodologies and advanced studies on these fields have been included herein. This book will serve as a resource guide for food scientists, policy makers, researchers and students involved in these fields.

I wish to thank my publisher for supporting me at every step. I would also like to thank all the authors who have contributed their researches in this book. I hope this book will be a valuable contribution to the progress of the field.

Editor

1

The law pertaining to food issues and natural resources exploitation and trade

François Collart Dutilleul*

Abstract

Background: The questions of the access to natural resources, their use, repartition or commercialization concern both the areas of food safety, protection of the environment and economic development. But whether we look at what the land offers or produces, the repartition of natural resources around the world coincide neither with the map of food needs nor the map of development. This is bitterly obvious in light of the latest failures, or at least partial failures, of the three main international negotiations: the FAO Summit on food safety, the WTO Doha Round on international agricultural trade and the negotiations of the Intergovernmental Panel on climate change (IPCC). In fact, these three negotiations, which are part of a global strategy of sustainable development, will only succeed if we not only take into account the economic, scientific and political aspects but also try to comprehend the judicial causes of the problems.

Results: We wish to determine the appropriate way to determine a law on natural resources that preserves on every continent the food safety. This safety supposes that the agricultural and feeding purposes of the land be maintained, that farmers and their families can access the land and live decently from it and that access to healthy food in sufficient supplying be guaranteed to everyone.

Conclusion: None of the objectives will be achieved without the law conveying the values that society will choose for itself. Between the law and the contract, collective security and individual freedom or between State sovereignty and globalization of trade, one must think of a humanist law to help the Earth to feed Humanity.

Introduction

The link between natural resources and food is too obvious to have to be proven right. Every one of us can experience it in everything we eat that comes from the earth's resources. It follows that food issues concerning trade in natural resources also seem obvious. What is not obvious, however, is the way the connection may be made between the various forms of natural resources exploitation and trade on the one hand and the worldwide consequences in terms of food supplies on the other hand.

Needless to say it partly depends on the way the scope of natural resources is defined. According to the WTO, natural resources consist of 'the material stores that can be found in their natural environment which are both rare and economically useful either in the raw state or after a minimal transformation' [1]. The scope is thus limited and mainly covers fishing and forestry products, fuels, ores, and other ferrous or non-ferrous minerals. We would prefer to consider a broader meaning which is legally more useful regarding food issues, including land itself, agricultural raw materials, water, and biodiversity resources. The criterion of inclusion is more a political than an economic one. The definition comprises land and the resources taken from the natural environment by extraction, capture, or culture which can be traded and have a significant impact on the planet or its peoples, justifying the implementation of public policies of protection, management, and import quotas.

Indeed, the most important point is to know how natural resources exploitation and trade should be organized and thus managed for food security to be optimal on all continents.

The response is partly ideological, dogmatic, or religious. But it is partly also strictly political. More often than not, the advocates of such-and-such a policy dress up their choices in a pseudo-legal form to make it more convincing and logical. When dealing with natural resources exploitation and trade, the laws of economics,

Correspondence: francois.collart-dutilleul@univ-nantes.fr
MSH Ange Guépin, 5 allée Jacques Berque, Nantes cedex 1, 44021, France

the market, nature, science, ethics, or indeed God (depending on one's secular or religious inspiration) are commonly referred to.

For this reason there is much debate to know whether, in order to feed those who starve and ultimately to feed the world, international trade and free trade in agricultural raw materials have to be furthered; whether technical, biological, and chemical innovations should be developed by promoting nanotechnologies or productive farming methods; or whether human and social values should prevail by referring to ethics, morals, religion, cultural and historical diversity, and by encouraging operators and consumers alike to reconsider their habits and criteria by themselves. Should economics, science, or ethics be trusted for the natural resources to meet our basic food needs?

Economic liberalism has enabled a great increase in the generation of wealth, but it has not shown an ability to reduce hunger, poverty, and under-development while preserving our resources. The scientific and technical breakthroughs which have revolutionized agriculture, food, and health have undeniably allowed us to improve overall worldwide food security, but at the cost of degrading our environment and resources as well as opening up a wide gap between the rich North and the poorer South. Should we then turn to ethics? Will promoting ethical consumption and production be enough on its own to neutralize the addiction to wealth shown in *Stupid Cupid*'s behavior[a] and the vagaries of history, geography, and climate?

In fact, all of human experience shows that thinking globally about the exploitation and trade of resources, particularly natural resources, without a framework of law comes down to accepting the law of the strongest, whether the strength be economic, scientific, moral, or religious.

What is missing in the admittedly truly real assets contributed by economics, science, or ethics is the force of law whose first function is, to use Kant's famous distinction, to build on the difference between what has a price and what has dignity[b] and to draw conclusions in terms of social rules. But that is not enough, precisely because some commodities with a price are essential to the respect of people's dignity. Such is the case of commodities enabling people's vital needs to be met - first among them natural resources, made up of land and its output. That is why the missing element is a law whose goal is the respect of people's dignity, starting with the respect of their right to live, and therefore with food security [2].

Thus determining the role of Law regarding food issues and natural resources exploitation and trade is to step in a direction that will pave the way for the respect of life and dignity. Yet questions concerning food cannot be set apart from those concerning freedom either. As Jeanne Hersch aptly put it, 'what is required by the declaration (universal declaration of Human Rights) is that the urge for vital needs (food, housing, etc.) be tempered for humans and their closest fellows so as to increase opportunities for freedom' [3]. Approximately 17,000 or 18,000 years ago, human beings painted beautiful frescoes on the walls of the Lascaux caves, bearing witness to a dramatic cultural development on which freedom had been gradually built. They would never have been able to achieve that much if they had had to spend most of their time nomadically hunting and gathering, day after day, to feed themselves and their families.

The Lascaux frescoes thus resulted from a basic social organization and task-sharing, permanent settlement on a territory and finding ways to share available resources and feed those who lived there. Lascaux encapsulates the origins of the three elements which enabled a social group to develop a common culture on which to found a society and achieve freedom: Law-Food-Land [4].

But how can we conceive and draft the body of law that is needed for the Earth and what it possesses and produces to be able to serve everyone's development, and first of all the goal of food security?

To write this law, it is certainly necessary to start with a diagnosis of the forces that are currently acting - mainly market forces - in order to identify the levers the forthcoming law should act on. Once the acting forces have been identified, it becomes possible to summon up the 'imagining forces of law' according to Mireille Delmas-Marty's well-found expression[c]. This presupposes determining the values on which the law must be conceived in order for humankind to be a reality and not merely a concept and, consequently, to sort out what should be determined by price from what should be governed by dignity.

Acting market forces

To identify the acting forces, we will rely on Karl Polanyi's thesis that he developed in his master work *The Great Transformation* [5]. This choice is justified first because the economic model that emerges includes the food issues resulting from the development of a free trade economy leading to a global self-regulating market. It is also justified by the explanatory relevance of the market model throughout modern history, both in cases of attempts to control the market and in times of crisis. However, we will try to consider this model in an empirical way, with an inductive method adapted from Francis Bacon's *Novum organum* [6]. It has indeed to be remembered that there is no way to scientifically establish if the easing of restrictions on trade is the source of the problem or, conversely, if it must be furthered to lead to a solution. Pure logic does not enable us to state whether speculation is or is not the cause of price volatility and food crises. There is no overall agreement on the benefits and detriments of nanotechnologies to nature and people. In the face of the many controversies – often colored by ideology - the analysis rather requires an empirical, inductive method like that of Bacon's 'Four Idols' doctrine and tables to identify the various sources of error and to compare the cases in which an observed phenomenon does or does not recur. To a legal professional - even though law is not a science - this 'New Tool' offered by Bacon is interesting since it is an alternative to the mainstream syllogism inherited from Aristotle's *Organon.*

The great transformation

The question of food issues in natural resources exploitation and trade is deeply rooted in a broader conception of a global self-regulating market that the economist Karl Polanyi studied in depth in his master work *The Great Transformation.*

In his work, Polanyi built his thesis [7] particularly on the history of 18th- and 19th-century England, a time and place where, according to the author, the reality of such a global self-regulating market was born. It is indeed very enlightening to step back two or three centuries and observe how the three commodities consisting of land and the natural resources it holds or produces, money and the financial resources it provides, and labor and the related human resources, enabled the development of the idea and the reality of such a market, on which we have continued to rely for our living until today. Indeed, for Karl Polanyi, free trade is built upon two pillars.

The first one is the autonomy of the economy with regard to politics. If the economy is subordinated to politics, the market is politically oriented thus legally restricted and as a consequence it cannot self-regulate. So, self-regulation implies an easing of restrictions on trade and a political, legal withdrawal. Besides, it can be seen that in various times and especially in our own, the very right of existence of 'international political economy' as a discipline has been contested [8]. One can also notice how today's economic policies aim at buffering the impact of free trade rather than restricting the play of the free market.

This autonomy of the economy with regard to politics (and thus law) supports the idea developed by Michel Foucault in his analysis of economic liberalism [9] stating that legal and economic goals cannot be pursued at the same time. For Michel Foucault, indeed, *homo oeconomicus* is capable of pursuing his personal interest in an unlimited way. There is no limit to the wealth a person can try and accumulate. One's personal economic interest is certainly not restricted by that of other people. The best, the cleverest, or the luckiest can thus become rich at the expense of others, without having to answer for their actions legally. But, from a legal point of view, each person's interest is restricted by the others' and by the general interest. Nobody can live his personal, family, or social life without taking others' lives into account, or to the detriment of others, unless he is held responsible; thus the law-economy antinomy and the difficulty in designing legal restrictions for a free trade economy.

The second pillar of the great self-regulating market, according to Polanyi, consists of three imaginary commodities: land, money, and labour. These commodities are imaginary since they do not satisfy 'the postulate according to which everything that is bought and sold must have been produced to be sold' [10]. This postulate could well be one of Bacon's 'idols'. Neither labor (human beings' vital force) nor land (nature) nor money (representing purchasing power) is 'produced' to be sold. But they are all at the heart of the economic machinery enabling all trade in goods and services. Indeed, these three imaginary commodities have to be subjected to the market themselves for the economic system - via the organization of the production and distribution of goods and services - to be self-regulated. According to Polanyi, 'humans are expected to behave in such a way as to earn as much money as possible: such is the origin of this kind of economy. It assumes there are markets on which the availability of commodities (including services) at a given price matches demand at the same price (...). Consequently, there are markets for each and every element of industry, not only for goods (including services), but also for labor, land, and money, their prices being respectively called foodstuff prices, wages, land revenue or "rent" and interest' [11]. In this analysis, the self-regulating 'Great Single Market' Polanyi describes was able to develop when land (together with what it holds and produces), money, and labor were themselves subjected to self-regulating markets.

As far as natural resources are concerned, the author demonstrates the birth of a market for land starting with the 'enclosures' movement [12]. This movement, which had already begun in the 13th century, consisted in turning the 'Commons' - the lands which were shared by a community's farmers - into private plots which were regrouped then 'enclosed' and attributed. The phenomenon is partly due to the need to rationalize plots in order to increase yields. In the 18th century, it also aimed to boost breeding to provide the emerging English textile industry with wool. The development of enclosures thus produced wealth linked to the increase in farming production and productivity. But at the same time it led to the expropriation of a whole class of farmers. Some of them would be employed by the new landowners, contributing to the development of farming in enclosures. The others would be expropriated, left on their own without a living, and they would leave the countryside to increase the population of the cities, a phenomenon at the origin of almost all megalopolises, particularly in the developing world. In England, riots slowed the enclosures movement. But enclosures, over the decades, would eventually spread, forced upon the rural economy by hundreds of specific laws passed from 1750 onwards [13]. The enclosures movement thus transformed land and natural resources into goods.

This 'great transformation' had devastating effects, only slightly mitigated by laws introducing a measure of charity for the poor[d].

Legally, the market won the day, with a change from a legal system based on common goods to one based on the private ownership of land, together with charitable laws to maintain social peace in spite of the increase in poverty.

Economic development[e] was thus acquired at a dramatic social cost, so that 'what has a price' clearly won over 'what has dignity'.

Attempts to control the market

Market controls lead to intermediate stages between a fully liberal economy and a fully regulated one.

In their strictest form, these controls can consist in regulating prices to avoid excessive increases - which generate food crises - or too high volatility. In this case, the market no longer self-regulates, it is quite simply neutralized. There was such an attempt in France after Turgot's edict which eased regulation in the grain trade, particularly for the price of wheat, at least until a late stage of the French Revolution [14]. There was at that time a very enlightening series of changes back and forth between imposing and then easing controls on prices. Yet each period of easing triggered a rise in prices, engendering riots which were dealt with by martial law. Conversely, each period of controls was accompanied by the end of martial law. This movement between free prices (during times of easing of restrictions) and free people (during times of price control) provoked over the whole period a social debate between the advocates of free trade and their opponents, a debate whose economic terms can still be used almost word for word today [15] In particular it focused on the fact that grain (or farming products as a whole) is not like any other commodity, being renewable, vital, and so on, and also on the part played by middlemen and speculators.

It is a recurrent debate, and it was particularly relevant, politically speaking, in the 1940s when the Allies, while still at war, sought to set up the bases of a new, prosperous, peaceful world. Initiated by President Roosevelt who wanted to establish the basis of a new worldwide organization for a sustainable post-war peace, 10 or so international conferences met successively in the Allied countries [16]. Among them, those hosted in Hot Springs in May 1943, Philadelphia in May 1944, and Bretton Woods in July 1944 are particularly relevant here.

The Hot Springs conference was set up with the idea that the eradication of starvation and poverty was a criterion for peace. This conference had two major effects. For the first time, it bound together natural resources, farming, and food in the public policies to be implemented at national and international levels. It also gave birth to an international organization that would become the FAO 2 years later. The final Act of Hot Springs referred to food as a right of humankind. The setting up of a policy of plenty was favored, and thus the development of the exploitation of agricultural resources and increased productivity of food resources. International trade in foodstuffs and international cooperation to eradicate starvation were encouraged. States were asked to commit to taking all necessary measures to secure their populations' food security. The specific needs of fragile populations were also noted. Fair prices for farmers were favored. The necessity for states to directly intervene 'from spade to spoon' as one might say now was recognized. The goal was set to limit the fluctuation of the prices of food and farming products. Basically, Hot Springs favored a kind of economic interventionism in the exploitation, management, and trade of food-linked natural resources. This is a long way from thinking of foods as conventional commodities, even further from instituting a self-regulating market. Much closer indeed to numerous NGOs' demands today. But everybody has forgotten Hot Springs, only to remember Bretton Woods.

Nonetheless, the conclusions of the Bretton Woods conference were partly along the same lines since, while promoting international trade, particularly with a set of arrangements later formalized as GATT, it set up an international monetary system which, being organized around the U.S. dollar but tied to gold, shielded currencies from pure market regulation. Bretton Woods also created the international bank and the International Monetary Fund (IMF).

As for the Philadelphia conference which reorganized the International Labor Organization, it started by postulating, among other key principles, that labor is not a commodity.

It can thus be assumed the international community has learnt from history and 'what has dignity' is at least as valuable as 'what has a price'. After these conferences, the economy would lose its autonomy and be subordinated to politics. Secondly and above all, land and the food resources it holds and produces (Hot Springs), labor (Philadelphia), and money (Bretton Woods) would no longer be considered as conventional commodities subjected to self-regulating markets.

But as everybody knows and has noted, the spirit of Philadelphia, to quote the phrase (and analysis) of one author, has given way to the global market [17]. The determination witnessed at Hot Springs has not been strong enough to prevent the trade in food natural resources from being governed by GATT rather than the FAO. The intelligence shown at Bretton Woods was not able to prevent the damage done to the international monetary system in the 1970s. The self-regulating market reasserted itself in its unrestricted, ultra-liberal form which Hayek calls 'the spontaneous market order' [18] and which could well be one of the 'idols' that the method Bacon favoued should enable us to identify.

The crises of the self-regulating market

The pendulum has now swung back, so that a self-regulating market is now clearly seen as 'a crazy idea' [19] One can well remember the sub-primes crisis in the second semester of 2006, which broke on the world in February 2007. One can also remember the first financial crisis which resulted, in summer 2007, and reached a climax in the fall of 2008 when the first effects of the economic recession and unemployment started to be felt worldwide.

On the other hand, several factors have been paving the way for the risk of a food crisis scenario since 2005–2006: the destruction of arable lands through population drain or urban development, the overuse of inputs, climate change, the rise in demand for food, the oil price rise, the increase in bio-fuel crops, and so on. Foodstuff prices are both volatile and bound to increase in the long run. This is the scenario in which the sub-primes crisis and the resulting financial crisis were set. On the one hand, central banks poured hundreds of billions of liquidities into banks which invested in blue-chip stock like agricultural raw materials. On the other hand the hedge funds, torn to pieces by the financial crisis, tried to make up their losses on the agricultural raw materials settlement market. It all led to a huge speculation which unsettled the already fragile prices of agricultural raw materials. This is why from February 2007 onwards, when the sub-primes crises became public and the financial catastrophe began, foodstuff prices rocketed to reach a peak in February 2008, provoking a food crisis in a number of countries[f].

This recent history of crises can help us to understand the phenomena that the issue of food is linked to. Indeed, the sub-prime crisis is related to land-commodity (and to that other basic need, housing); the financial crisis is related to money-commodity; the economic crisis triggering unemployment is related to labor-commodity. It shows that behind the crises the world has experienced, which has sent the least developed countries into a food crisis, there are in fact, in Polanyi's terms, three commodities in crisis: land and its uses, money, and labor.

It is truly a crisis of resources that the world has had to face: the degradation of natural, financial, and human resources. And land and natural resources are in deep crisis. This has become dramatically evident, especially since the end of 2009.

In late 2009, the future of natural resources was decided in three international conferences directly or indirectly dealing with food natural resources: the WTO negotiations on agricultural products trade in Geneva in December, FAO's on food security in Rome in November, and the conference on global warming in Copenhagen in December. Yet, these three meetings failed and in none of these areas has there been any successful outcome so far. What can be learnt from this?

Indeed, these three negotiations recall the main issue of the Hot Springs conference in 1943. They addressed the question of the link between the trade in natural resources and food issues. But the Hot Springs declaration did not provided for any sanctions or other constraints. And the 2009 negotiations did not succeed in forcing anything upon states. Given this, should we not appeal to the imagining forces of law?

The 'imagining forces of law'

To write law, vocabulary and grammar are needed. The vocabulary designates concepts conveying the values we choose to promote and the grammar is the way to connect these concepts to one another, whether they be 'economic liberty', 'political liberty', 'people's dignity', 'individual responsibility', 'sustainable development', 'food security', 'private property', sovereignty, or common goods. To apprehend this vocabulary and grammar, we would prefer to speak in terms of 'law' (or rules) rather than 'regulation' or 'controls', in order to avoid the confusion often made (and which have resulted in the worship of some of Bacon's 'idols'). So let us assume that 'law' differs from 'controls' in the same way as the Highway Code differs from ways to keep the traffic flowing. Prices or speculation can be 'controlled' the way temperature is, that is by setting limits, upper and lower. Controls aim to steady or balance a complex system. Though necessary, they are far from being enough. Law (like 'controls') sets behavioral principles and rules. Through law, social values will be promoted to guide the economy the way society wants it to be, somewhere between maximizing profits and giving priority to people's dignity.

Setting the goal of food security

International law currently governing natural resources exploitation and trade does not aim at securing food security worldwide. Its goals are economic and it relies on two principles: the sovereignty of States over their natural resources and free international trade [20]. The laws currently in force partly match the WTO's, without any specificity regarding agricultural natural resources, subject to the successful outcome of the Doha Round of negotiations[g]. The laws are also partly the result of international agreements and treaties, as well as an increasing number of bilateral treaties [21]. Concerning food resources, it is the organization of fishing and the sharing of halieutic resources that are mainly concerned, precisely because when dealing with the seas and oceans, the allocation of fishing grounds cannot be achieved simply by applying the principle of State sovereignty. Basically, international market law is sufficient when applied to what has a price but not to what has dignity, to quote Kant once more.

We should not hesitate to qualify food security as a priority and a non-negotiable goal. This is justified simply if starving to death is to be outlawed and if it is a condition for world peace, as it was at the Hot Springs conference in 1943. It has been at the core of the language of the FAO ever since, but it has not broken through as law. FAO defined this priority in the worldwide summit on food in 1996: 'Food security is provided when all people, at all times, can economically, socially and physically have access to enough safe, nutritive food to satisfy their nutritional needs and food preferences to enable them to live an active, healthy life' [22].

On the other hand, food security also consists in preventing food crises. In this respect, the billion or so people who do not have sufficient access to adequate food should be seen as the victims of a permanent food crisis. Given this, food security should appear in two forms. One of these would be devoted to all such victims, just as social protection is organized in the form of 'social security'. The other one should consist in designing laws governing the production and trade of food agricultural raw materials that take into account the need to feed 9 billion people by 2050.

Thus conceived, the goal of food security has direct repercussions on the legal treatment of natural resources. Land is a limited, non-extensible commodity. What it produces, though renewable, can become exhausted. Fisheries are the same. So are forests and water. Limited and exhaustible natural resources should be considered as a patrimony or capital and not as goods, especially not as conventional ones. More precisely, their exploitation and trade should be governed by public policies varying from state to state, recognized by international law, thus removing them from sheer market forces. These rules should take into account the fact that the vital character of these food resources gives them a general interest and common dimension at the outset.

Defining a sustainable development strategy

The Geneva, Rome, and Copenhagen negotiations that took place in late 2009 linked together the destinies of an economic issue - the development of international trade and commerce, an environmental issue - global warming, and a social issue - poverty and its links with famine, malnutrition, and food crises. Now these three issues precisely match the three pillars of sustainable development: economic dynamism, environmental care and social progress: Profit, Planet, People.

It will not be a surprise to find here the three imaginary commodities whose history has been highlighted by Polanyi: money as a way to measure profits, land as an issue in protecting the planet and labor as a means for people to secure an income and thus food for themselves and their families. While the global self-regulating market consists in merchandising land, money, and labor, global sustainable development for its part relies on considering the economic, environmental, and social dimensions as indissociable. This means each political action or decision should take into account the three dimensions together, and should be responsible for the consequences of doing so.

Yet history shows that such an 'indissociation' cannot result from a great self-regulating market. Sustainability of development cannot result from a purely economic, scientific, or ethical approach. It can only result from a politically expressed, legally organized will [23].

However, the goal of legislation should not be to put an end to either the great market or free trade. In the wake of Max Weber's doctrine, it is obvious that there is not one single legal system suited to free trade [24]. We simply have to abandon the 'idols' of self-regulation and deregulation which have prevailed in almost all economic sectors since the 1970s. The problem with the three post-war conferences (Hot Springs, Bretton Woods, and Philadelphia) is that they were dissociated from one another. The same problem occurred in November and December 2009 with the separate Rome, Geneva, and Copenhagen negotiations - they eventually failed. This was inevitable: it makes no sense to restrain the unbridled exploitation of natural resources in Copenhagen if their free trade is promoted in Geneva at the same time. And such promotion itself is pointless if we are to work towards the goal of food security in the long run. Food security has no future if what is discussed separately in Rome is neglected in Copenhagen.

To reach its objectives, the law we have to imagine must forge links between economic laws and dynamic trade, the laws of science and technical progress, moral laws, and human values.

Therefore our first problem is the current segmentation of institutions, conferences, negotiations, policies, and decisions. The great market model can only work if it reaches a balance between supply and demand for all goods and services. To do so, we should let the forces of supply and demand act for land and its resources, for money, and for labor. But, with the constant interplay of the various markets and the successive crises, history has shown that this is ineffective. The *homo oeconomicus* is only rational to a certain extent and on condition he has enough food and a safe environment to live in. This is the 'indissociation' that the sustainable development concept allows us to implement. Given this, in these conditions and within the framework established by the United Nations General Assembly [25], what legal resources or instruments could restore 'what has dignity' to its rightful status?

Structuring the legal resources to be implemented

There is a fairly wide range of legal resources available to enable us to come as close as possible to the goal of food security while implementing a sustainable development strategy.

To define them and choose among them, it must first be stated that a sustainable development strategy provides us with at least three categories of values that have been shared by the international community since the Rio declaration of 1992. They constitute a strong basis which avoids immediately confronting the question of the universalism or relativism of values [26]. The legal resources we decide upon and structure should then enable economic dynamism, health and environmental care, and social progress to be achieved. Besides, the instruments we design must be based on legal concepts capable of 'sounding the conscience' [27] as well as appealing to reason.

Up to now, freedom has been favored: free access to resources, free exploitation, free trade, and consequently freedom to do business, contractual freedom, free competition, free movement of goods, capital, and people[h]. Some say failures, poverty, and famine can be explained by restrictions to this freedom[i]. Others, however, think that free trade in natural resources accounts for the problem and the food crises. The main point surely is to lend a legal, constraining form to the limits on freedom, particularly by exploring means of maintaining public policy which that legal form or instrument should target. We can thus try to design an internationally-dimensioned concept of 'food public policy' (or food 'ordre public'), on the model of the WTO's TRIPS agreement for patents[j]. Regarding the freedom of imports and exports, of economic price-setting mechanisms, of speculation on commodities, an 'order' has to be imagined and made 'public'. Economic liberties and legal responsibility also have to be concretely and conceptually re-associated, these values having been largely dissociated since the series of world crises that started in 2005–2006.

Another resource, which has been extensively tried and tested, is basic rights. There are already a considerable number of texts which lay them down but they pose the problem of a lack of efficiency in terms of positive law. Things tend to be different, relatively speaking, when a judge or a court is instituted in parallel to enforce the rights which have been laid down[k]. The basic rights resource is nevertheless very symptomatic of the vital need to ensure each individual's food security. That is why it is promoted by the UN's special rapporteur on food rights to fight the excesses of free trade[l]. But an effective legal system remains to be defined to defend farmers' access to land and people's access to food [28].

A legal framework for the market can also be imagined based on the concept of 'sovereignty' in the form of both access to land (land sovereignty) and access to food (food sovereignty).

Land sovereignty determines the means States can use to keep natural resources, particularly agricultural lands, under control, particularly seeking to prevent their capture by public or foreign powers. This form of sovereignty is a reaction to a new kind of colonization consisting of foreign states or companies buying up developing countries' productive lands. FAO has made proposals supporting the idea [29]. However, these proposals need to be clarified, particularly by designing model contracts preserving the rights of local populations and the interests of host states [30]. The food security of the countries involved requires all lands to be available to populations living on the states' territories. The corollary of this State land sovereignty is the obligation on these states to provide their inhabitants with the means of subsistence. Besides, experience shows that grabbing these lands is often an opaque process, which mainly benefits the capturing investor: a minimal price is paid for the land, legally or not, before or after taking it, and it is at best unclear who the final payee is. Furthermore, the investor's commitments regarding local employment, the species of crops to grow and the destination of the produce are often vague. Eventually, once the contract comes to an end, the investor often leaves a land which has been impoverished by overexploitation without being environmentally restored.

Food sovereignty, on the other hand, determines a state's ability to define public policies designed to provide the population with enough healthy food. This is the means favored by *Via Campesina* [31] which has developed a thorough, demanding food sovereignty concept which extends to a full set of public policies to be implemented to provide the population with food: farmers' access to land, water, seeds, and loans; priority given to local produce to feed the population; the right of farmers and consumers to choose the methods of production and the products consumed; the right of states to protect themselves from low-cost imports by banning or taxing them; populations having their say in political decisions concerning agriculture; protection of women's rights, and so on. In addition to these proposals, which merit thorough debate, it is at least necessary to define the legal tools necessary for the implementation of an agricultural protectionism limiting the power of the WTO and the globalization of trade. In this respect, the concept of food autonomy, compatible with maintaining a system of multilateral trade, would enable states to keep a measure of legal discretion to adapt their agricultural and food policies to their own particular needs. So in the context of sovereignty and autonomy, there is a means to legally regulate exploitation and trade in food natural resources while avoiding excesses. However, a population cannot be confined within a country's borders without the necessary food. Consequently, a state cannot be deprived of the right to make the necessary decisions to feed its population, and both national and international laws should take this into account.

In compatibility with the concepts of sovereignty and autonomy, ownership and the qualification of 'common goods' should be explored as ways to access resources. Several definitions of common goods have been proposed in a thriving, mainly economic, literature [32]. Yet the concept is often assimilated simplistically, as Garett Hardin has done [33], to merely being in competition with private property [34]. In reality, deforestation, excessive inputs, wastage of water, or overexploitation of resources can occur under conditions of common goods as well as private ownership. Ownership can be private or public, individual or collective, absolute or relative, without competition or open to coexisting uses. It is therefore possible to optimize a right to ownership which regulates the owners' power in favor of the common, and others', interests [35], a right which defines a way to manage ownership that respects the specificity and scarcity of the natural resources concerned[m], or even a right that bestows a social function on ownership [36]. Much in the same way, the management of common goods can be legally organized to guarantee a chosen yield, by preserving the quality and quantity of resources. The main point is the values which are promoted rather than the legal means used.

The choice between ownership and common goods also has to be made when turning our attention to incorporeal goods in the field of 'intellectual property'. Patents on living organisms, biodiversity or natural resources [37], plant breeders' rights, brands, and quality signs are legal instruments whose power is underestimated and whose legitimacy is rarely questioned as though it were obvious. In fact, not everything should be eligible for ownership, particularly when goods are necessary for people to live. In this case, it should be forbidden on principle for an owner, whoever he or she might be, to have a power of monopoly on goods which people depend on for their very lives. But intellectual property is like any other property: it can be absolute or relative. International law currently provides few limits to intellectual property's absolute scope. States can rule out an invention's patentability so as to protect public order, health, and animals or people's lives or to avoid serious environmental problems[n] .They can more generally rule out patentability for plants, animals, and their breeding processes, providing they organize an alternative way to protect plant varieties[o]. Exceptions in the International Convention for the Protection of New Varieties of Plants (UPOV) also limit the breeder's rights [38]. Besides, the Convention on Biological Diversity provides for the fair, equitable sharing of benefits from the use of genetic resources [39].

Yet all these limits, exceptions and expectations which could set limits to the 'inventor's' monopoly remain ineffective or too vague to have a full, legal reach, or can be neutralized in contracts between the owners of these monopolies and farmers. There is thus considerable progress to be made in deciding what should be 'common goods' as well as in regulating the powers and rights of the 'owner' of the variety or characteristics of the new animal or plant.

The intellectual property resource is all the more promising as it enables us to imagine an analogical transposition of the so-called right of cultural exception to the issue of food. Cultural exception is an international legal concept aimed at making culture an exception in international treaties and particularly in the World Trade Organization. This exception thus allows States the means to make their sovereignty prevail and to limit free trade in cultural products to the benefit of their national culture and cultural diversity. A food exception based on the same model is not impossible to imagine.

To conclude, regarding the food issues of trade in the land and what it holds, we have to go back to Hot Springs and supply law that would enable the implementation of the values that this quickly forgotten conference sought to enforce. If we do not, we cannot be sure that populations will accept for much longer that 'what has a price' triumphs over 'what has dignity'. Better Hot Springs than spring riots.

Endnotes

[a]See the second section in *Groundwork of the Metaphysics of Morals.*

[b]So far Delmas-Marty M has devoted her Collège de France lectures to this subject and has published four volumes in the Seuil publishings (Paris): *Le relatif et l'universel* (t. 1), *Le pluralisme ordonné* (t. 2), *La refondation des pouvoirs* (t. 3), *Vers une communauté de valeurs* (t. 4).

[c]See particularly the first Poor Law (1601) and the Speenhamland law. The latter, in force from 1795 to 1834, set a minimum income depending on the number of dependent people and the price of bread. It was granted to those without an income and those with an income below the minimum one were granted the difference. It was abrogated in 1834 because it encouraged the textile industry to underpay its employees. Polanyi K, chap. 7, p. 128; Rodriguez J, "*De la charité publique à la mise au travail. Autour du Speenhamland Act*", http://www.laviedesidees.fr/IMG/pdf/20080923_rodriguez.pdf

[d]Reasoned, more productive exploitation of land, sheep breeding, and textile industry development.

[e]Many demonstrations, even riots took place in late 2007 to early 2008 in Africa (Burkina Faso, Cameroon, Senegal, Mauritania, Ivory Coast, Egypt, Morocco) but also in Mexico, Bolivia, Yemen, and in Bangladesh, Pakistan, Sri Lanka, and South Africa.

[f]Let us remember the Doha Round is a negotiations cycle launched in 2001 under WTO's authority. These negotiations deal with agricultural products international trade and the improvement of the conditions for developing countries agricultural products to reach wealthy countries.

[g]One may however observe that in spite of a self-regulating economic pattern, only capitals and goods can circulate without restrictions. The situation is quite different for people.

[h]See Geneva debate (11 May 2009) between Pascal Lamy, WTO's general director, and Olivier de Schutter, UNO's special rapporteur on food law.

[i]See *infra*.

[j]Especially the European Court for Human Rights, Supreme Courts in the USA and Canada, inter-American Court of Human rights.

[k]*Eod. loc.*

[l]See Ostrom, E., *op. préc.*

[m]See art. 27.2 of WTO's agreement on the aspects of intellectual properties dealing with trade (TRIPS).

[n]See art. 27.3-b of TRIPS/WTO agreement.

[o]See art. 27.3-b of TRIPS/WTO agreement.

Author's information

Professor at the University of Nantes, Member of the Institut Universitaire de France, Principal investigator of Lascaux (www.droit-aliments-terre.eu). The Lascaux program (2009–2014) is linked to the 7th Framework Programme of the European Research Council ('IDEAS'). The research leading to these results has received funding from the European Research Council under the European Union's Seventh Framework Programme (FP7/2007-2013)/ERC grant agreement no. 230400.

References

1. World Trade Organization: *Report on 2010 World Trade*. Geneva: WTO; 2010.
2. Collart Dutilleul F, Garcia F: **Dans le domaine de l'alimentation, quels "droits à" dans le "droit de"?** In *In Droit économique et Droits de l'Homme*. Edited by Boy L, Racine J-B, Siiriainen F. Brussels: Larcier; 2009.
3. Hersch J: **Les droits de l'homme d'un point de vue philosophique.** In *In La philosophie en Europe*. Edited by Klibansky R, Pears D. Paris: Folio essais; 1993.
4. [http://www.droit-aliments-terre.eu]
5. Malamoud C, Angeno M, Dumont L: *Aux origines politiques et économiques de notre temps, trad*. Paris: Editions Gallimard; 2009.
6. PUF: *coll*. Paris: Epithémée; 2010.
7. Polanyi K: *chap.1*.38.
8. Tooze R: *"Susan Strange et l'économie politique internationale", L'Économie politique, 2001/2 n° 10, p.101. Also Cox, R. W., "Au-delà de l'empire de la terreur: réflexions sur l'économie politique de l'ordre mondial"*. Vol 2nd edition. H: A contrario; 2004:http://www.cairn.info/revue-a-contrario-2004-2-page-167.htm.
9. Foucault M: **l'économie politique et le libéralisme, Annales Histoire, Sciences sociales.** In *In Naissance de la biopolitique, Cours au Collège de France, 1978–1979, Jan. 10, 1979 lecture*. Edited by Ewald F, Fontana A, Senellart M, Grenier J-Y, Orléan A, Foucault M. Paris: Gallimard-Le Seuil; 2004:279–1155. Vol.62 (5/2007).
10. Polanyi K: *122. chap.6*.
11. Polanyi K: *118. loc. prec.*
12. Neeson JM: *Commoners: Common Right, Enclosure and Social Change in England 1700–1820*. Cambridge: Cambridge University Press; 1993.
13. Chambers JD, Mingay GE: *For lack of amicable agreement, enclosures had to be imposed by law*, In The Agricultural Revolution 1750–1850.: Batsford; 1982.
14. Gauthier F: *De Mably à Robespierre: un programme économique égalitaire 1775–1793*, In Annales historiques de la Révolution française, n° 261, 1985.1792:265. available online: http://www.persee.fr/web/revues/home/prescript/article/ahrf_0003-4436_1985_num_261_1_1116: Also see Robespierre's analysis, Opinion sur les subsistances, speech given at the Convention.
15. Clément A: **La spécificité du fait alimentaire dans la théorie économique. Les fondements historiques et les enjeux.** *Ruralia* 2000, Varia, [http://ruralia.revues.org/178]
16. Deblock C, Hamel B: *Bretton Woods et l'ordre économique international d'après guerre"*, In Interventions économiques, Pour une alternative sociale. 1994/1995, n° 26, Dossier spécial « De l'ordre des nations à l'ordre des marchés – Bretton Woods, cinquante ans plus tard »,12 [http://classiques.uqac.ca/contemporains/deblock_christian/bretton_woods_ordre_econo/Bretton_Woods_ordre_eco.pdf]
17. Supiot A: *L'esprit de Philadelphie: la justice sociale face au marché total*. Paris: Seuil; 2010.
18. Hayek F: *Droit, législation et liberté, t.1, Règles et ordres, chap.2, PUF*. Paris: Quadrige; 2007.
19. Bernard A: *Le marché autorégulé, "une idée folle"?* Paris: Rec. Dalloz; 2009.
20. World Trade Organization: *WTO Report on 2010 World Trade, Natural Resources Trade, part II, section E*. Geneva: WTO; 2010.
21. Le commerce international entre bi et multilatéralisme: In Edited by Remiche B, Ruiz-Fabri H. avril: RIDE; 2010.
22. [http://www.fao.org/cfs/fr]
23. Collart Dutilleul F: **La régulation juridique du développement durable et le droit agroalimentaire européen.** In *In Production et consommation durables: de la gouvernance au consommateur-citoyen, CEDE International Talks Acts*. Edited by Parent G. Québec: Yvon Blais; 2008:393.
24. Coutu M: **Max Weber et les rationalités du droit.** *LGDJ* 1995, :123.
25. *UNO's general assembly first debate on 'harmony with nature'*. [http://www.un.org/News/frpress/docs/2011/AG11075.doc.htm]
26. Delmas-Marty M: *Les forces imaginantes du droit, t.1, Le relatif et l'universel*. Paris: Seuil; 2004.
27. **La réforme du droit des contrats: projet et perspectives.** *Rev des Contrats (RDC)* 2006, :22.
28. *Acts of Lascaux programme talks*. 2010 [http://www.droit-aliments-terre.eu/pages/menu2/forum_lascaux_2.html]
29. *FAO draft report on Directives volontaires sur la gouvernance responsable de la tenure des terres et des autres ressources naturelles*. 2012 [http://www.fao.org/nr/tenure/voluntary-guidelines/e-consultation/fr/]
30. Collart Dutilleul F: *Quel regard sur le phénomène de l'accaparement des terres?* [http://www.droit-aliments-terre.eu/documents/sources_lascaux/editos/edito_FCD/edito_062011_FR.pdf]
31. *On this concept presented in 1996 in FAO's world summit on food in Rome.*: Bamako (Mali); 2007. see http://www.viacampesina.org/. See Nyeleni's statement in the international meeting on food security: http://www.nyeleni.org/.
32. Ostrom E: **Gouvernance des biens communs - Pour une nouvelle approche des ressources naturelles.** In *In Révision scient*. Edited by Baechler L.: de Boeck; 2010.
33. **The tragedy of the Commons.** *Science* 1968, **162**:1243.
34. Coban A: *Entre les droits de souveraineté des États et les droits de propriété: la régulation de la biodiversité, A contrario 2004/2, Vol. 2*. 2004:138 [http://www.cairn.info/resume.php?ID_ARTICLE=ACO_022_0138]
35. *See Robespierre's speech: Propositions d'articles additionnels à la déclaration des Droits de l'Homme et du citoyen*. 1793.
36. Duguit L: **Le droit social, le droit individuel et la transformation de l'État, conférences faites à l'École des Hautes Études Sociales.** In *In La Mémoire du Droit: Les transformations générales du droit privé depuis le code Napoléon*. 2nd edition. Edited by Alcan P. 1999:147.

2

Household food insecurity access: a predictor of overweight and underweight among Kenyan women

Susan Keino[1,2*], Guy Plasqui[3] and Bart van den Borne[1]

Abstract

Objectives: There has been a rise in the number of overweight women across the globe and sub-Saharan Africa has not been spared. Our objective was to describe the performance of household food security as a factor in determining overweight and underweight among Kenyan women of child-bearing age.

Design: A cross sectional survey using a structured questionnaire was carried out among 656 households with women of child-bearing age from both rural and urban settings in the Rift Valley, Kenya.

Results: Of all households involved, 70.1% were categorized as severely food insecure, 21.9% were moderately food insecure, and 3.7% were mildly food insecure, whereas 4.3% were food secure. Urban women had higher mean BMI (M = 24.53, SD = 4.36), compared to rural women (M = 22.52, SD = 3.71; P <0.000). Households with more children had decreased dietary diversity (r = −0.154; P <0.01). Logistical regression indicated that more food-secure households (OR = 0.97; 95% CI 9.4, 1.0), urban residence (OR = 1.82; 95% CI 1.1, 3.0), older age (OR = 1.10; 95% CI 1.1, 1.5), and high standard of living (OR = 1.79; 95% CI 1.2, 1.5) were strong predictors of overweight while being underweight was predicted by younger age (OR = 0.91; 95% CI, 0.83, 1.0).

Conclusions: A food-secure household does not predict underweight but predicts overweight among women.

Keywords: Dietary diversity and women, Food insecurity access, Overweight, Underweight

Introduction

Overweight and obesity and their risk factors are in the increase in sub-Saharan Africa with higher increase among urban-poor dwellers [1-4]. The current estimates are that as much as 20% to 50% of urban populations in Africa are classified as either overweight or obese [5,6], and that by 2025 three quarters of the obese population worldwide will be in non-industrialized countries [7]. Reasons for the increase in overweight and obesity are not fully understood, although physical inactivity and increased consumption of high fat and refined carbohydrates are thought to be contributing factors [8-10]. However, it has also been suggested that overweight and obesity may occur due to limited availability of food and this occurs if food insecurity leads to consumption of obesity-promoting foods [11], overeating when food becomes available, or metabolic changes that hinder the efficient use of energy [12].

Food security exists when all people at all times have access to sufficient, safe, nutritious food for a healthy and active life [13]. Food insecurity is defined as limited or uncertain availability of nutritionally, adequate and safe foods, or limited or uncertain ability to acquire acceptable foods in socially acceptable ways [14]. Food insecurity and malnutrition result in serious health problems and loss of human potential in developing countries [15]. Increasing food insecurity in the developing world paradoxically has resulted in increasing numbers of overweight. Past studies cite increased availability of high sugar and energy, low fibre foods as a major cause for overweight and obesity [16,17]. However, some studies conducted in developed countries have consistently

* Correspondence: s.keino@maastrichtuniversity.nl
[1]Department of Health Education & Promotion, Faculty of Medicine & Life Sciences, Maastricht University, P.O. BOX 616, 6200 MD Maastricht, The Netherlands
[2]Department of Epidemiology & Nutrition, School for Public Health, Moi University, P.O. Box 4606, 30100 Eldoret, Kenya

shown that mild but not severe food insecurity is associated with obesity among women [18-20]. This association between food insecurity and obesity has not been confirmed in studies performed in developing countries, where poverty rates are much higher [21]. In spite of the increasing numbers of overweight people, the underweight population still persists in these food insecure groups, and both may coexist within the same households or neighbourhoods [22,23]. Although the prevalence of overweight exceeds that of underweight in most developing countries [24], the consequence of both are damaging to health. A study among adults and children from low- and middle-income families in Bogota, Columbia, indicated that food insecurity was related to maternal underweight but not to maternal overweight [25]. The association of food insecurity and overweight is not consistent as found by two studies [26,27]. The adverse health effects resulting from overweight and underweight are far reaching and sometimes permanent, and may result in death [28,29]. Women and children are vulnerable to malnutrition and food insecurity. Further, women globally have higher rates of obesity than men [9]. Therefore, this study targets women of childbearing age in Kenya to identify the relationship between food insecurity, overweight, and underweight among this group of women and to highlight the predictors of overweight and underweight.

Methods

Access Scale (HFIAS) Study, setting, design, and sampling

This was a cross-sectional study that took place in Narok County, Kenya. The Narok County is made up of an agricultural and agro-pastrolists community in the South Rift Valley. The study was carried out in the Narok District Hospital, where we purposively selected women of childbearing age (15–45 years old) to participate in this prevalence study. The women were randomly selected from the Mother and Child Health clinics where they were bringing their children for immunization and weighing. The women were approached and assigned a number between 1 and 2 and all the women assigned number 1 were asked if they were willing to participate. A total of 649 participants from rural and urban areas agreed to the study protocol and gave an informed consent. The inclusion criterion was that the women must have one child below five years of age at the time of the study.

All study procedures were approved by the Institute for Research and Ethics Committee of Moi Teaching and Referral Hospital, Eldoret, Kenya. The subjects gave written and verbal consent to both the study procedure and participation. The data were collected by the principal author and three trained research assistants in July and August 2011.

Measurements

Data were collected through interviews and anthropometric measurements. The interview consisted of a structured questionnaire and included questions on household socioeconomic and demographic characteristics, 24-hour dietary recall, and a Household Food Insecurity measurement. After the interview, the respondents presented themselves for the anthropometric measurements. Data on standard of living (SOL), age, household size, marital status, and number of children were collected.

To measure SOL, we considered asset ownership as well as income and expenditure. Income and expenditure were categorized into 1 for high and 0 for low income as well as 1 for high expenditure and 0 for low expenditure. The assets included owning a radio, cell phone, refrigerator, mode of cooking (gas, firewood, or electricity), and type of house as well as having access to a toilet facility. We recorded the type of energy used for cooking into advanced and old methods of cooking: gas and electricity were included in the advanced list whereas charcoal and firewood, and other energy sources, were in the old methods of cooking. The type of house was the last factor, where type of house was defined by the roofing and the building materials. Iron sheets and tiled roofing as well as stone were categorized into 1 for higher levels whereas grass-thatched and mud/iron-sheet roofing were categorized into 0 for the lower level. Principle component analysis and calculation of Cronbach's alpha resulted in an 8-item SOL scale.

Food insecurity was measured by the HFIAS, which is a 9-item scale consisting of two types of questions; an occurrence question followed by a frequency of occurrence question which asked the participant how often the condition reported in the previous occurrences question happened during the previous month [30]. The occurrence questions asked whether the respondent or other household members either felt a certain way or performed a particular behaviour over the previous month. Before inclusion in the questionnaire, the questions were reviewed with key informants as suggested in the protocol. A standard scoring procedure was used with 1 point for occurrence and 0 for non-occurrence. The frequency scores ranged from 0 to 3, where 0 was the score for non-occurrence, 1 for rarely (once or twice in the past four weeks), 2 for sometimes (three to ten times in the past four weeks), and 3 for often (more than ten times in the past month). For the purpose of this paper, we used the total score (9-items based on the frequency score). A total score of 27 represents the most food-insecure household whereas a lower score represents a more food-secure household.

Data on household dietary diversity was collected using a 24-hour recall method and information was entered into the Household Dietary Diversity Score (HDDS) sheet [13].

The HDDS captures dietary diversity in a normal 24-hour period by the household as a whole and not a single member. Food consumed outside the home that was not prepared in the home was not included. A set of 12 food groups was used to guide the scoring as per the food items consumed, with 1 being the minimum score and 12 as the maximum [13].

Anthropometric measurements of weight and height were taken from the woman who was interviewed. Weight was measured to the nearest 0.5 kg using a mechanical scale (SECA 762). Height was measured to the nearest millimetre also using a transportable stadiometer (SECA 214); the scales were calibrated after each session of measurements. Waist and hip measurements were measured using a tape. The waist was measured around the narrowest point between the navel and the ribcage and the hip was measured around the widest part of the hip to the nearest cm. All measurements were taken with light clothing on and without shoes. A total of two measurements were taken and an average calculated.

Table 1 General characteristics of women in rural and urban households in Narok County, Kenya

	Population mean (SD) n = 649	Rural mean n = 162	Urban mean n = 487	*P* value
Weight (kg)	61.8 (12.3)	58.2 (10.6)	63.0 (12.6)	0.000
Height (m)	1.6 (0.07)	1.61 (0.07)	1.60 (0.06)	NS
Hip circumference (cm)	96.9 (10.0)	94.2 (9.4)	98.0 (9.9)	0.000
Waist circumference (cm)	79.0 (10.5)	77.3 (11.2)	79.3 (11.1)	0.05
SOL	4.97 (1.92)	3.62 (1.82)	5.42 (1.73)	0.000
Household size	4	5	4	NS
Age (years)	26	26	26	NS
Post primary education (%)	49.2	12.3	36.9	0.000
Marital status (%)				
• Single	9.5	6.2	10.7	NS
• Married	88.6	92.0	87.5	NS
• Other[#]	1.8	1.9	1.8	NS

SD: Standard deviation, cm: centimetres, SOL:- Standard of living, NS: Not significant, [#]separated/widowed/divorced.

Data management and analysis

The data were entered using Microsoft® Excel 2003 and rechecked for accuracy to minimize errors. Analysis was done using Statistical Package for Social Sciences (SPPS) Version 17.1. Frequency distribution of all the data was done for descriptive purposes. Internal consistency of the measurements (Cronbach's alpha) was calculated for the HFIAS and SOL. Pearson's correlation analysis was performed to compare relationship between variables. Logistic regression analysis was used to examine factors related to underweight (body mass index (BMI) ≤18.5) and overweight (BMI ≥25), and linear regression analysis for predicting the total BMI score.

Results

Anthropometrics measurements and general characteristics

Table 1 presents data on the women's age, marital status, education, and anthropometric measurements. Among the participants, 88.6% were married. At least 5.5% of the women had no formal or informal education and 49.2% had a post-primary level of education in the study group; primary education was reported by 45.3% of the women. Education beyond primary level was reported by 54.4% and 33.3% of the urban and rural participants, respectively. The main source of income, according to 38.6% of the respondents, was from formal employment. Running of a small holdings business, such as small shops or kiosks, as the main source of income was reported by 24.5%, whereas farming was the main source of income for 19.6% of participants. Crop production (28%) was the main agricultural activity followed by mixed farming at 11.6% (livestock and crop).

The percentage of women with more than one child under five years of age was 40%. The average household size was 5 and 4 in the rural and urban areas, respectively. At the interview date, 73.7% of the women had not moved residence within the past six months and 66.6% lived in rental houses. A total of 33.5% of women had access to their own tapped water, while 25.5% of the women used communal tapped-water points, 16% bought water from vendors, and 25% obtained the water from a borehole, a stream/river, or harvested rain water; 7.9% of women had no access to a toilet facility. Relief food, which is common in the arid and semi-arid areas, has not been accessed by 88.6% of the women in the community. Meals were shared at the same time by all family members in 66.3% of the households, whereas 16.9% served children first and 16.2% served husband or male figures in the households first.

Nutritional status and household food security

The BMI of women was higher in the urban (24.5) compared to the rural (22.5) areas. Combined overweight and obesity was higher in the urban (39.6%) compared to that of rural participants (21%). However, underweight was 11.2% and 3.7% among the rural and urban women, respectively. The percentage of the population that had a waist-to-hip ratio (WHR) ≥0.85 was 26.2%, whereas those who had a waist circumference ≥80 cm were 38.9% (Table 2). The HDDS had a mean score of 6.56 for the total population and did not significantly differ between urban and rural settings.

Table 2 Descriptive statistics of nutritional status, dietary diversity, and household food security among rural and urban women in Narok County, Kenya (n = 649)

Variables	Population mean (SD) n (649)	Rural mean (SD) n (162)	Urban mean (SD) n (487)	*P* value
Average meals per day	3.0 (0.51)	2.97 (0.47)	3.03 (0.52)	0.15
HDDS	6.56 (1.3)	6.41 (1.27)	6.61 (1.30)	0.093
HFIAS score	11.5 (6.26)	12.61 (6.80)	11.20 (6.03)	0.013
BMI (kg/m^2)	24.0 (4.3)	22.4 (3.6)	24.5 (4.4)	0.000
- Underweight (%)	5.6	11.2	3.7	0.000
- Normal (%)	59.4	67.7	56.6	0.000
- Overweight (%)	26.0	18.6	28.4	0.000
- Obese (%)	9.1	2.5	11.3	0.000
Waist circumference ≥80 cm (%)	38.9	41.2	32.1	0.04
WHR (%)	26.2	29.0	25.3	0.34

BMI: Body mass index (kg/m^2), WHR: Waist to hip ratio, HFIAS: Household food insecurity access scale, HDDS: Household dietary diversity score.

However, the HFIAS score differed slightly ($P = 0.013$) with rural settings being more insecure (12.61 ± 6.8) than the urban (11.20 ± 6.03). A total of 4.3% of the population were food secure, whereas 70.1% reported severe food insecure access. Affirmative responses to the 9 items in the scale ranged from 53.5% to 88.0%. The item least frequently receiving an affirmative response was question 8, 'Did you or any household member go to sleep at night hungry?' Whereas the item receiving the highest number of affirmative responses was question 1, 'Did you worry that your household would not have enough food?'

Pearson product–moment correlations were computed to assess the relation between SOL, HDDS, education, residence, and number of children (Table 3). There was a positive association between HDDS and SOL ($r = 0.27$, $P < 0.01$) and a negative association between HDDS and number of children ($r = -0.15$, $P < 0.01$), whereas SOL was negatively associated with number of children ($r = -0.31$, $P < 0.01$). When comparing education and BMI, there was a significant positive relationship between education and underweight ($r = 0.15$, $P < 0.01$). Looking at the numbers of those with no education and underweight, the numbers were higher in the rural (66.7%) areas compared to the urban (45.6%) areas. Overweight had a positive association with number of children ($r = 0.082$, $P < 0.05$).

Logistic regression analyses were performed on overweight and underweight as separate outcomes and 8 predictor variables: marital status, education, SOL, number of children, HFIAS score, HDDS, age, and residence.

Tables 4 and 5 show the regression coefficients, odds ratios (ORs) and 95% confidence intervals (CIs) for each of the 8 predictor variables. Predictors of overweight that were significant were SOL, age, number of children in the household, urban residence, and lower HFIAS scores as indicated in Table 4. Women who had higher SOL had higher odds of being overweight (OR = 1.79, 95% CI 1.19, 1.53), and older women had higher odds of being overweight (OR = 1.10, 95% CI, 1.05, 1.15). A negative association was reported between overweight and HFIAS (OR = 0.97, 95% CI 0.94, 1.0), indicating that the lower the HFIAS score the higher overweight

Table 3 Inter-correlation of variables associated with nutritional status among women in Narok County, Kenya (n = 649)

Variables	1	2	3	4	5	6	7	8	9
1. Marital status$^{\infty}$	--								
2. Education$^{\alpha}$	0.054	--							
3. HFIAS	−0.004	0.022	--						
4. SOL	0.004	−0.002	−0.280**	--					
5. HDDS	−0.020	0.048	−0.127**	0.266**	--				
6. Residence$^{\beta}$	0.047	0.137**	0.098*	−0.408*	−0.068	--			
7. Underweight$^{\mu}$	0.044	0.146**	0.078*	−0.162**	−0.037	0.140**	--		
8. Overweight$^{\lambda}$	0.038	−0.052	−0.133**	0.257**	0.042	−0.169**	−0.178**	--	
9. No of children	0.12**	0.138**	0.167*	−0.305**	−0.154**	0.253**	−0.055	0.082*	--

Note: *$P < 0.05$; **$P < 0.01$, HFIAS: Household food insecurity access scale, SOL: Standard of living, HDDS: Household dietary diversity score.
Score. $^{\infty}$1 = married, 0 = single, divorced/separated; $^{\alpha}$1 = post-primary level, 0 = primary and lower level; $^{\beta}$1 = urban, 2 = rural; $^{\mu}$1 = underweight, 0 = not underweight; $^{\lambda}$1 = overweight, 0 = not overweight.

Table 4 Logistic regression analysis of predictors of overweight among women in Narok County, Kenya (n = 648, BMI ≥25)

		95% CI for odds ratio		
	B (SE)	Lower	Odds	Upper
Overweight				
Marital status	−0.510 (0.32)	0.32	0.60	1.12
Education	0.202 (0.21)	0.81	1.22	1.86
SOL	0.298** (0.07)	1.19	1.79	1.53
No. of children	0.117* (0.11)	0.92	1.13	1.38
HFIAS	−0.029* (0.016)	0.94	0.97	1.00
Dietary diversity	−0.009 (0.075)	0.86	0.99	1.15
Residence (urban)	0.597* (0.255)	1.10	1.82	3.00
Age	0.093** (0.023)	1.05	1.10	1.15
Constant	−4.977 (0.831)			

Model x^2 (1) = 94.00; *P <0.05, **P <0.001; SOL: Standard of living, HFIAS: Household food insecurity access scale, SE: Standard error.

and vice versa. Regarding the prediction of underweight, younger women had higher odds of being underweight (OR = 0.91, 95% CI 0.83, 1.0) (Table 5). Although not significant, being underweight reduced with increased SOL.

Linear regression analysis was computed to assess the association between BMI and all 8 predictor variables. There was a positive significant relation between BMI and the predictor variables (R^2 = 0.17, P <0.001) as seen in Table 6. Only age (Beta = 0.28, P <0.001) and SOL (Beta = 0.29, P <0.001) were positively associated with BMI.

Discussion

In this study, urban women had higher numbers of overweight compared to their rural counterparts, and although underweight existed in both groups, the prevalence was lower compared to overweight in the study group. This is in agreement with many studies in developing countries that report that urban dwellers are frequently better nourished than their rural counterparts [31-33]. However, the fact that the urban areas in our study mainly consisted of peri-urban dwellers with a poor infrastructure and low economic status may explain the prevalence of underweight persisting among both the rural and urban dwellers, although the prevalence was high among rural dwellers. In addition, overweight is an emerging problem among the rural dwellers as observed in our study. The predictors of overweight among our participants were SOL, age, HFIAS score, residence, and number of children while age was a significant predictor of underweight.

Table 5 Logistic regression analysis of predictors of underweight among women (BMI ≤18.5, n = 649)

		95% CI for odds ratio		
	B (SE)	Lower	Odds	Upper
Underweight				
Marital status	0.249 (0.53)	0.45	1.28	3.64
Education	0.343 (0.46)	0.57	1.41	3.50
SOL	−0.186 (0.13)	0.65	0.83	1.07
No. of children	0.198 (0.19)	0.84	1.22	1.78
HFIAS	0.013 (0.03)	0.96	1.01	1.07
Dietary diversity	−0.020 (0.15)	0.73	0.98	1.32
Age	−0.093* (.05)	0.83	0.91	1.004
Residence (urban)	−0.568 (0.42)	0.25	0.57	1.295
Constant	−0.093 (1.67)			

Model x^2 (1) = 21.59; HFIAS: Household food insecurity access scale, SOL: Standard of living, No.: Number, SE: Standard error: *P=0.05.

Table 6 Linear regression analysis of BMI and standardized beta of predictors variables among women in Narok County, Kenya (n = 649)

	B	SE B	β
Marital status	0.19	0.50	0.01
Education	−0.08	0.38	−0.01
SOL	0.63	0.10	0.28*
No. of children	0.07	0.18	0.02
HFIAS	−0.02	0.03	−0.04
HDDS	−0.02	0.13	−0.01
Age	0.23	0.04	0.29*
Residence (urban)	0.08	0.41	0.01
Constant	14.95	1.46	

R^2 = 0.17 (P <0.001). *P <0.001, SOL: Standard of living, HFIAS: Household food insecurity access scale, HDDS: Household dietary diversity score.

In our study, the urban dwellers reported a high number of overweight and underweight, respectively, and overweight was related to a high SOL. This was the case in a study by Ziraba et al. that indicated that urban dwellers had the highest cases of obesity [2]. High SOL using wealth index may be translated to mean wealthy households; thus, our study agrees with the findings from a study among Ghanian women which indicated that obesity was common among wealthy households as opposed to the poor [34], and a study by Shayo and Mugusi [35] which indicated that participants with a high socioeconomic status had a statistically significant increased risk for obesity compared to participants with low socioeconomic status. It is worth noting that the study by Ziraba et al. [2] reports an increase in obesity among higher socioeconomic groups as well as an increase in the number of obese among the lower socioeconomic group. In our study, SOL was a measure of income, expenditure, access to toilet facility, advanced cooking methods, owning a cell phone, and type of housing material, thus, it cannot be used interchangeably with socioeconomic status in the reported studies.

Education levels within the study area were still low compared to other counties in Kenya [36]. Education is important in terms of making choices when it comes to resource allocation. Studies indicated that women who receive even a minimal education are generally more aware than those who have no education of how to utilize available resources for the improvement of their own nutritional status and that of their families [37-39]. A cultural practice that is common among the study population is that of marrying-off girls at a young age; thus, within this community one is likely to find young married women. The youngest participant among the study population was 15 years of age. According to a government document released in 2008, secondary education enrolment rate for girls in the study area was 15% and that of boys was 20%, while primary education enrolment rate was 73% and 83% for girls and boys, respectively [36]. Using education as a proxy for socio-economic status the study by Ziraba et al. [2], which included data from Kenya, indicated that the prevalence of obesity was higher among the non-educated than educated group; however, in our study, it is worth noting that the percentage of women who had attained post-primary education was higher among rural dwellers compared to urban dwellers. Thus, the non-educated and the primary level-educated women were mainly found in the urban areas, where overweight was higher. In addition, overweight women were low among households where the respondent had no education but the numbers increased among the primary and secondary level of education respondents. This may relate to income or job opportunities, whereby the non-educated women had lower opportunities of obtaining higher paying employment while those with primary or secondary level of education had higher incomes and thus an increase in access to food; however, this may not always be the case.

Age was an important predictor of BMI among women; the older the women the higher the chances of them being overweight and the younger the women the higher the probability of underweight and thinness [40]. Women tend to gain weight with increasing age and parity [41-44]. However, some studies indicate greatest weight gain among young adults (20–29 y), and in particular women [45]. During pregnancy women gain weight which is difficult to lose after pregnancy and in the event that they become pregnant again additional weight is gained. This concurs with other studies that indicated that if excessive pregnancy weight is not lost soon after breast-feeding, overweight and obesity may occur [46]. However, younger age was an important predictor of underweight as also reported by Griffiths et al. [40]. In our study, underweight prevalence was also higher among rural women compared to urban women, which may be explained by the longer distances women travelled to access social amenities such as shops, health centres, searching for water, etc. Thus, activity levels may be high among this group leading to underweight and thinness. However, this is beyond the scope of this paper and will be discussed elsewhere.

Most households were predominantly male-headed households as indicated by the larger percentage of married women among the participants. According to a study in Ethiopia, marital status of women is associated with household headship and other social and economic status which affect women's nutritional status [37]. When women control resources, children and household members are more well-nourished [47]. However, when it comes to nutritional status of children, conflicting results have also been reported in studies that show children of working mothers to have a lower nutritional status compared to children of mothers who work at home [48]. This brings into focus the issue of care, whereby women who leave their children in the care of others may compromise the nutritional status of their children as opposed to the mothers who work at home and care for their children themselves. In our study, the married participants had higher prevalence of over-weight compared to the single, divorced, or separated; however, underweight was more prevalent among the younger group (15–24 years old).

Household food insecurity arises when families do not have enough to eat due to food being neither available nor accessible. In our study area, food insecurity was predominant in many households as reported in government documents [49]. Seasonal hunger, *per se*, is most common during planting seasons and just before harvest. According to a study by Vaitla et al., most of the world's hunger and under-nutrition occurs during the pre-harvest season [50]. However, as much as this is the case in this population, overweight is still reported among some of the participants in our study, indicating that overweight is not only a disease of the rich but also of the poor. The results from this study indicated that household food insecurity access is a predictor of overweight, as has also been reported in other studies [2,19,21,35]. In this study, household food insecurity access is not a statistically significant predictor of underweight. This is in contradiction with a study which showed that food insecurity predicted underweight but not overweight among adults and children [25]. The increasing overweight and underweight among respondents within food insecure household brings into question the quality of the food consumed, and perhaps also the quantity, particularly of fried foods which are cheap and easily available, predominantly in urban areas.

Parity was also a predictor of overweight in this study, given that when a woman gives birth to many children without losing the pregnancy weight between births, her weight progressively increases. Parity, according to a study done in Ghana, was significant in predicting obesity [34]. However, in our study, women who had one or two children were more likely to be underweight. The increased number of children per household means scarcer resources and increased physical activity for the mother in providing and caring for the children, and thus a lower BMI. Sanitation is equally important in determining nutrition. In the present study, most of the households used communal water points and in the rural area women travelled long distances to get water, whereas within the urban area water was sold at the communal water-points by water vendors. Several studies have looked at the relationship between sources of water and availability of toilet facility in that inadequate water and sanitation increase the probability of infectious disease and indirectly cause some types of malnutrition [38,39]. The relationship between sanitation and nutritional status of women was reported in another unpublished article [51].

Our study had some limitations. Firstly, we did not study cultural diversity; different cultures have different food consumption patterns and dietary habits and thus nutritional status of the different communities (tribes) would differ as a result of varied dietary habits. Second, nutrient security is also important in explaining the overall nutritional status of an individual; our study did not look at the different nutrients in the diet, which would predispose one to either overweight or underweight. Finally, we did not take into consideration physical activity to explain overweight and underweight among women but included this in another ongoing study.

Conclusions

Overweight is a major nutritional problem among women in the study area although food insecurity is also a problem confirming the paradox of over-nutrition occurring where food insecurity exists. In our study, food secure households have higher odds of being overweight and not underweight. Other predictors of overweight included being in an urban settlement, high SOL, increasing number of children, and age, whereas, dietary diversity reduced with increasing household food insecurity.

Abbreviations

BMI: Body mass index; HFIAS: Household food insecurity access scale; HDDS: Household dietary diversity score; SOL: Standard of living; WHR: Waist-to-hip ratio.

Competing interests

The authors agree that they have no conflicting interests to declare.

Authors' contributions

SK participated in the data collection, analysis, literature review and writing up the manuscript, GP and BVB also participated in analysis and interpreting results. All authors read and approved the manuscript.

Acknowledgements

The authors are grateful to the Netherlands Fellowship Programme of NUFFIC for financing the study. They also would like to thank the staff of Narok District Hospital Mother and Child Health department for their cooperation during the data collection period. We also appreciate the effort of the research assistants.

Author details

[1]Department of Health Education & Promotion, Faculty of Medicine & Life Sciences, Maastricht University, P.O. BOX 616, 6200 MD Maastricht, The Netherlands. [2]Department of Epidemiology & Nutrition, School for Public Health, Moi University, P.O. Box 4606, 30100 Eldoret, Kenya. [3]Department of Human Biology, School for Nutrition, Toxicology & Metabolism, Maastricht University Medical Centre, P.O. BOX 616, 6200 MD Maastricht, The Netherlands.

References

1. Villamor E, Msamanga G, Urassa W, Petraro P, Spiegelman D, Hunter DJ, Fawzi WW: **Trends in obesity, underweight and wasting among women attending prenatal clinics in urban Tanzania 1995–2004.** *Am J Clin Nutr* 2006, **83**(6):1387–1394.
2. Ziraba AK, Fotso JC, Ochako R: **Overweight and obesity in urban Africa: a problem of the rich or the poor?** *BMC Public Health* 2009, **9**:465.
3. Abubakari AR, Lauder W, Agyemang C, Jones M, Kirk A, Bhopal RS: **Prevalence and time trends in obesity among adult West African populations: a meta-analysis.** *Obes Rev* 2008, **9**(4):297–311.
4. Martorell R, Khan LK, Hughes ML, Grummer-Strawn LM: **Obesity in women in developing countries.** *Eur J Clin Nutr* 2000, **54**(3):247–252.
5. Kamadjeu RM, Edwards R, Atanga JS, Kiawi EC, Unwin N, Mbanya JC: **Anthropometry measures and prevalence of obesity in the urban adult population of Cameroon: an update from Cameroon burden of diabetes baseline survey.** *BMC Public Health* 2006, **6**:228.
6. Sodjinou R, Agueh V, Fayomi B, Delisle H: **Obesity and cardio-metabolic risk factors in urban adult of Benin: relationship with socio-economic status, urbanization, and lifestyle patterns.** *BMC Public Health* 2008, **8**:84.
7. WHO: *Preventing Chronic Diseases, A Vital Investment.* Geneva: World Health Organization; 2005.
8. Gulliford MC, Mahabir D, Rocke B: **Food insecurity, food choices, and body mass index in adults: nutrition transition in Trinidad and Tobago.** *Int J Epidemiol* 2003, **32**:508–516.
9. WHO: *Obesity: Preventing and Managing Global Epidemic in Report of a WHO Consultation WHO Technical Report.* Geneva: World Health Organization; 2000.
10. Hannack LJ, Jeffrey RW, Boutelle KN: **Temporal trends in energy intake in the US: an ecological perspective.** *Am J Clin Nutr* 2002, **71**:1478–1484.
11. Dietz WH: **Does hunger cause obesity?** *Pediatrics* 1995, **95**:766–767.
12. Alaimo K, Olson CM, Frongilo EA Jr: **Low family income & food insufficiency in relationship to overweight in US children: is there a paradox.** *Arch Pediatrics Adolescents Med* 2001, **155**:1161–1167.
13. Swindale A, Bilinsky P: *Household Dietary Diversity Score (HDDS) for Measurement of Household Food Access: Indicator Guide V. 2.* Washington DC: Food and Nutrition Technical Assistance Project; 2006.
14. Carlson SJ, Andrews MS, Biekel GW: **Measuring food insecurity and hunger in the United Stats: development of a national benchmark measure and prevalence estimate.** *J Nutr* 1999, **129**:510S–06S.
15. Pinstrup-Andersen P, Cohen MJ: **Modern biotechnology for food agriculture: risks and opportunity for the poor.** *Int J Biotechnology* 2000, **2**(1/2/3):145–163.
16. Popkin BM: **An overview on the nutrition transition and its health implications: the Bellagio meeting.** *Public Health Nutr* 2002, **5**(1A):93–103.
17. Clugston GA, Smith TE: **Global nutrition problems and novel foods.** *Asia Pac J Clin Nutri* 2002, **11**(S6):S100–S111.

18. Sarlio-Lahteenkorva S, Lahelma E: **Food insecurity is associated with past and present economic disadvantage and body mass index.** *J Nutr* 2001, **131**:2880–2884.
19. Adams EJ, Grummer-Strawn L, Chavez G: **Food insecurity is associated with increased risk of obesity in California women.** *J Nutr* 2003, **133**:1070–1074.
20. Townsend MS, Peerson J, Love B, Achterberg C, Murphy SP: **Food insecurity is positively related to overweight in women.** *J Nutr* 2001, **131**:1738–1745.
21. Velasquez-Melendez G, Schlüssel MM, Brito AS, Silva AA, Lopes-Filho JD, Kac G: **Mild but not light or severe food insecurity is associated with obesity among Brazilian women.** *J Nutri* 2011, **141**(5):898–902.
22. Doak CM, Adair LS, Monteiro C, Popkin BM: **Overweight and underweight coexist within households in Brazil, China and Russia.** *J Nutr* 2000, **130**:2965–2971.
23. Caballero B: **A nutrition paradox – underweight and obesity in developing countries.** *N Engl J Med* 2005, **352**(15):1514–1516.
24. Mendez MA, Monteiro CA, Popkin MB: **Overweight exceeds underweight among women in most developing countries.** *Am J Clin Nutr* 2005, **81**:714–721.
25. Isanaka S, Mora-Plazas M, Lopez-Arana S, Baylin A, Villamor E: **Food insecurity is highly prevalent and predicts underweight but not overweight in adults and school children from Bogota.** *Columbia. J Nutr* 2007, **137** (12):2747–2755.
26. Whitaker RC, Sarin A: **Change in food security status and change in weight are not associated in urban women with preschool children.** *J Nutr* 2007, **137**:2134–2139.
27. Jones SJ, Frongillo EA: **Food insecurity and subsequent weight gain in women.** *Public Health Nutr* 2007, **10**:145–151.
28. Krause RM, Winston M, Fletcher BJ, Grundy SM: **Obesity: impact on cardiovascular disease.** *Circulation* 1998, **98**:1472–1476.
29. WHO: *Obesity: Preventing and Managing the Global Epidemic in Report of WHO Consultation on Obesity 3–5 June 1997.* Geneva: World Health Organization; 1997.
30. Coates J, Swindale A, Bilinsky P: *Household Food Insecurity Access Scale (HFIAS) for Measurement of Food Access: Indicator Guide v. 3.* Washington DC: Food and Nutrition Technical Assistance Project; 2007.
31. Smith LC, Ruel MT, Ndiaye A: **Why is child malnutrition lower in urban than in rural areas? Evidence from 36 developing countries.** *World Dev* 2005, **33**(8):1285–1305.
32. Ruel MT, Garrett JL, Morris SS, Maxwell DG, Oshaug A, Engle PL, Menon P, Slack AT, Haddad LJ: **Urban challenges to food and nutrition security: a review of food security, health and Caregiving in the cities.** In *FCND Discussion Paper.* Washington DC: International Food Policy Research Institute; 1998.
33. Marie T, Garrett JL: *Features of Urban Food and Nutrition Security and Considerations for Successful Urban Programming in Paper Prepared for FAO Technical Workshop.* Rome: Food Agricultural Organization; 2003.
34. Dake FA, Tawiah EO, Badasu DM: **Sociodemographic correlates of obesity among Ghanaian women.** *Public Health Nutr* 2010, **14**(7):1285–1291.
35. Shayo GA, Mugusi FM: **Prevalence of obesity and associated risk factors among adults in Kinondoni municipal district, Dar es Salam, Tanzania.** *BMC Public Health* 2011, **11**:365.
36. GOK: *Narok District Development Plan.* Nairobi: Kenya Bureau of Statistics & National Development Office (KNBS); 2008.
37. Girma W, Genebo T: *Determinants of Nutritional Status of Women and Children in Ethiopia.* Calverton, Maryland, USA: Measure DHS; 2002.
38. UNICEF: **Strategies of improving nutritional status of children and women in developing countries.** *Indian J Pediatr* 1990, **58**(1):13–24
39. Engle PL: *Care and Child Nutrition: Theme Paper for the International Conference on Nutrition (ICN).* New York: UNICEF; 1992.
40. Griffiths PL, Bentley ME: **The nutrition transition is underway in India.** *J Nutr* 2001, **131**(10):2692–2700.
41. Cohen SS, Larson CO, Matthews CE, Buchowski MS, Signorello LB, Hargreaves MK, Blot WJ: **Parity and breastfeeding in relation to obesity among black and white women in the southern community cohort study.** *J Womens Health* 2009, **18**(9):1323–1332.
42. Harris HE, Elisson GT, Holliday M: **Is there an independent association between parity and maternal weight gain?** *Ann Hum Biol* 1997, **24**(6):507–519.
43. Cohen A, Pieper CF, Brown AJ, Bastian LA: **Number of children and risk of metabolic syndrome in women.** *J Womens Health* 2006, **15**:763–773.
44. Buschur E, Kim C: **Guidelines & interventions for obesity during pregnancy.** *Int J Gynaecol Obstet* 2012, **119**(1):6–10.
45. Kimokoti RW, Newby PK, Gona P, Zhu L, McKeon-O'Malley C, Pablo Guzman J, D'Agostino RB, Millen BE: **Patterns of weight change and progression to overweight and obesity differ in men and women: implications for research and interventions.** *Public Health Nutr* 2013, **16**(8):1463–1475.
46. Rooney BL, Schauberger CW: **Excessive pregnancy weight gain and long-term obesity: one decade later.** *Obstet Gynecol* 2002, **100**(2):245–252.
47. Kelkar G: *Rural Women's Work and Capabilities: Information for Gender-Responsibilities Development Policies, in High-Level Consultation on Rural Women & Information.* Rome: FAO; 1999.
48. Bennett L: **The role of women in income, production and intra-household allocation of resources as a determinant of child nutrition and health.** *Food Nutrition Bull* 1988, **10**(3). http://archive.unu.edu/unupress/food/8F103e/8F103E03.htm
49. GOK: *Arid Lands Resource Management Project in Drought Monthly Bulletin.* Nairobi: Ministry of State for Development of Northern Kenya and other Arid Lands; 2011.
50. Vaitla B, Devereux S, Swan HS: **Seasonal hunger: a neglected problem with proven solutions.** *PLoS Med* 2009, **6**(6):e1000101.
51. Keino S, Plasqui G, van den Borne B: **Is overweight found among the food insecure? - A study on prevalence and risk factors of malnutrition among children in Narok, Kenya.** 2013, Unpublished Manuscript.

3

Factors affecting sustainability of community food security projects in Kiambu County, Kenya

Mary Stella N Wabwoba* and Jacob W Wakhungu

Abstract

Background: Kenya is one of the countries in sub-Saharan Africa that is not able to feed its population sufficiently and it, therefore, relies on outside assistance. Many food security projects have been funded by both the Kenyan government and other development partners in an effort to mitigate against food insecurity. Unfortunately, as revealed by assessment reports, such projects leave little impact after the end of funding.

Context and purpose of the study: This study evaluated factors affecting the sustainability of community food security projects funded by various organizations between 2005 and 2009 in the Karai and Ndeiya divisions of Kiambu County, Kenya. This study was necessary because among the literature reviewed, no other study had been done on the sustainability of community food security projects. An evaluation research design was adopted and a purposive sampling method was used to select key informants from stakeholder organizations and project groups. Data were collected using face-to-face interviews with ten key informants (community-based organizations, faith-based organizations, financial institutions and the government of Kenya) and focus group discussions with twenty groups (ten women's groups, four men's groups and six mixed groups) that had benefitted from the funded projects. The data collected were analyzed using the chi-square test at the 95% confidence interval level.

Results and main findings: The findings revealed that the sustainability of community food security projects is affected by group members' participation, rainfall patterns, leadership, management and funding levels.

Conclusions and recommendations: Based on the outcome of the study, we conclude that food security projects are not sustainable. The recommendation is that group members (the beneficiaries) must participate in project planning and implementation for purposes of ownership and sustainability. Farmers need to be empowered with knowledge on irrigation and off-season intensive farming of high-value crops. We are very optimistic that this recommendation will be useful for program planners, donors, policymakers, implementers and stakeholders in project design and for funding sustainable community food projects. A further study should be done on the role of stakeholders in project sustainability.

Keywords: Sustainability, Community food security projects, Kiambu County, Kenya

Background

Over the years, the government of Kenya has invested in community food security projects as a way of helping local people improve their own lives and livelihoods. A number of communities in Kenya have been given grants and technical support by both local and international donors, with the intention of helping them combat food insecurity and reduce poverty [1]. A study carried out in 2007 revealed that all the 15 development partners who operated in Kiambu County targeted rural community food security projects [2]. Such donors included community-based organizations, faith-based organizations, financial institutions, the government of Kenya and other private organizations [3]. A total of 536 groups were funded between 2005 and 2009 by various organizations [4]. The report further revealed that some community projects were funded by as many as five donors during the same period.

Projects are intended to produce benefits that continue after some specified period of time. The goal of development assistance is to improve the quality of life and

* Correspondence: wabstella@yahoo.com
Department of Disaster Management and Sustainable Development, MasindeMuliro University of Science and Technology, Kakamega, Kenya

increase incomes. However, many projects, including those undertaken by international development organizations, fail to fulfill these objectives [5]. An impact assessment on community-funded projects showed that only 5 out of 36 groups funded in 2007 by *NjaaMarufuku Kenya* (NMK) were partially active, while the rest had become defunct and could not be traced after cessation of funding [6]. Despite the many funded food security projects in Kiambu County, there is persistent food insecurity among the rural communities [2]. The residents of Kiambu West (Kikuyuand Limuru) have high rates of malnutrition due to food insecurity [7] and yet Kiambu County has the largest number of community food security funded projects in Kenya [8]. The purpose of this research was to examine the factors that could be responsible for community projects not continuing to achieve their objectives and, therefore, forthe persistent food insecurity even among the funded groups.

Methodology

This study was conducted in Kiambu West (Kikuyu and Limuru) of Kiambu County, Kenya, which was purposively chosen because it has many community food security projects (Figure 1). Kiambu County has a total population of 834,378 of which 411,074 are males and 423,304 are females. The county covers an area of 1,207.4 km^2, it has 235,417 households and it has a population density of 417.5 persons per km^2 [9]. The county is in agro-ecological zone 2–4 in the Central Kenyan highlands with an altitude ranging from 1,400 m to 1,800 m above sea level. It experiences bimodal rainfall between March and May and in October and November with annual rainfall above 1,500 mm. The county is situated between latitude 0°75' and 1°20' south of the equator and longitudes 36°54' and 36°85' east.

Food security projects that operate in the county funded through government departments include:NMK [6], the Arid Lands Resource Management Project [10],

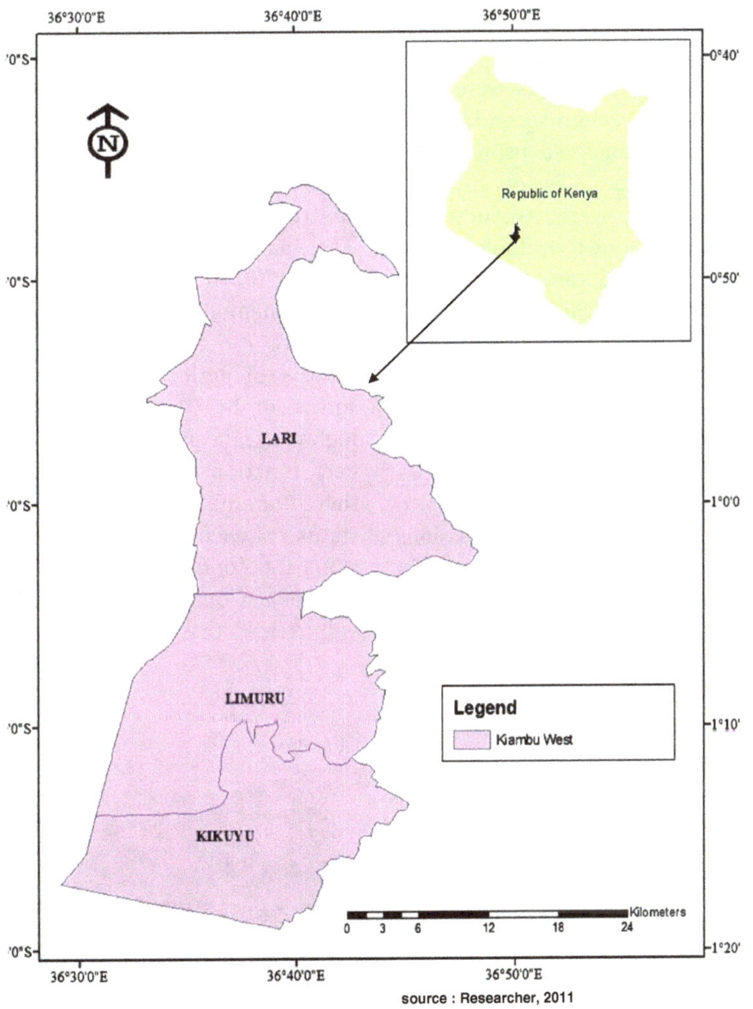

Figure 1 Kiambu west subcounties, Kenya Source: (Researcher, 2011).

Table 1 Community groups from Karai and Ndeiya divisions, Kiambu County, Kenya

Group type	Sampling method	Sample size	Names of groups
Women's group	Purposive and simple random	10	Thigio, Tiekumu, Nderu, NdiuniKarai, Mwireri, Nyathuna,, Ruku, Bathoma, Karii
Men's group	Purposive and simple random	4	Ikaitano, Gatune, Nachu, Rusingiti
Self-helpgroup (mixed gender)	Purposive and simple random	6	Renguti, Mwirimiri, jashoMukawa, Karisi, Makerecha
Total		20	

the Total War AgainstAids Project [11] and the Community Based Nutrition Program [12]. Most of these projects are concentrated in the Karai and Ndeiya divisions, which have a high incidence of malnutrition and food insecurity [13]. The study population comprised community groups, opinion leaders, non-governmental organizations, faith-based organizations and Kenyan governmentheads of departments that were involved in food security projects in Kiambu West (Karai and Ndeiya).

An evaluation research design was adopted and a purposive sampling method was employed to select key informants from the above population. Key informants were interviewed and observation checklists were used as proxy indicators of food security. Both secondary and primary data were collected and analyzed using a chi-square test at the 95% confidence level. Twenty different types of focus group discussions were conducted in theKarai and Ndeiya divisions as shown in Table 1. The variables analyzed were: group activities, functions of group members and the factors and challenges hindering goal achievement by groups.

Results and discussions

Group activities implemented by members

The study sought to discover the group activities implemented by members, which could be contributing to success or failure of projects. The focus group discussions revealed that 60% of the people had crops as their main activity, 30% practiced off-farm activities and 10% kept livestock (Table 2). The calculated chi-square of 135.45 at two degrees of freedom showed a highly significant ($P < 0.01$) variation among group activities. The researcher also observed most of the farms were occupied by crops (Figure 2). These findings are in agreement with the food security profile report for Kiambu County [14], which showed that 70% of the population is engaged in crop cultivation. Most farms (90%) are smallholdings although a few larger holdings exist in the county [15]. This was confirmed by 93% of the key informants, who said they supported farming activity. Further group discussions revealed that off-farm activities, like public transport and construction of business premises, had taken up most of the land previously used for farming. It is unlikely that the money earned is used to purchase food, hence household food security is not guaranteed.

Table 2 Main activity of members of groups in Kiambu County, Kenya

Activity	Frequency	%
Crop farming	212	60
Livestock farming	35	10
Off-farm activity	105	30
Total	352	100

Functions of group members in sustaining group projects

Focus group discussions were held to evaluate the functions of group members and the variables were: resource contribution, labor provision and attending meetings. The majority of the group members contributed resources (62%) towards running the project's activities. Other members (40%) attended meetings during which they are also trained by extension providers on various aspects of agriculture (Figure 3). The calculated chi-square of 147.01 at two degrees of freedom showed a highly significant ($P < 0.01$) variation among the members' function in sustaining group projects. Everyone in the community, regardless of gender or socio-economic status, is encouraged to provide opinions, make decisions and contribute resources towards the implementation of a Community Action Plan [16]. Oniang'o [17] emphasized that communities are extremely important

Figure 2 Mixed crop farm in Ndeiya, Limuru.

Figure 3 Focused Group discussion for Makerecha group in karai, Kikuyu.

Figure 4 Renguti self help group with pawpaw project in Kikuyu, Kiambu.

in determining the sustainability of projects, since if they contribute resources there is an element of ownership.

Factors affecting sustainability of food security projects in Kiambu County

The respondents were asked to rank how the listed factors had affected their group using ratings such as no effect, low effect, high effect, higher or highest effect in achieving their group's objectives, and, hence, project sustainability. The ranking was determined by the percentage of the respondents voting for the factor that has the highest effect in hindering the group's objective. The examined variables are listed in Table 3.

The chi-square tests for group members' participation, land tenure and gender, and rainfall showed highly significant ($P < 0.05$) variation while level of funding, politics and management and leadership were significant ($P < 0.0$). The results imply that these factors contribute towards the food security of households and, therefore, will affect the sustainability of community food security projects. However, lack of knowledge, skills and communication networks were statistically not significant ($P > 0.05$). This means that these factors do not directly contribute towards food security and are therefore not major factors in community project sustainability because the majority of the groups are empowered with projects (Figure 4). Household interviews also revealed that middle-age people were actively involved in food security and therefore community food security projects should target this age group to ensure sustainability. This results show that food security is affected by the size of the farm owned by households and hence households with more land under food crop are more food secure. The presence of food is an indicator of project sustainability [18].

Conclusions and recommendations

Based on the findings from the study, the most significant factors affecting the sustainability of community food security projects in Kiambu County are: group members' participation, rainfall patterns, land tenure and gender, level of funds allocated to the project and the individual activities undertaken by members.

The study recommends that for Kenyan communities to enjoy food security through community projects, the following measures should be adopted: involve group members in project design, implementation,resource contribution, monitoring and evaluation, to ensure ownership and hence sustainability.

Table 3 Factors affecting the sustainability of food security projects in Kiambu County, Kenya

Number	Factors/challenges	Level of effect				
		Highest effect %	Higher effect %	High effect %	Low effect %	No effect %
1.	Level of funding	33	33	32	l	0.3
2.	Lack of knowledge and skills	34	30	3	33	0.6
3.	Communication networks	2.0	65	33	0.3	**0**
4.	Land tenure and gender	40	0	34	33	30
5.	Rainfall	63	3.4	33	0	0.6
6.	Politics	34	32	33	2	0.3
7.	Collaborators and stakeholders	33	32	2	33	0.3
8.	Management and leadership	0.3	33	33	32	1
9.	Group members' participation	66	31	1	2	1

Competing interests

The authors declare that they have no competing interests.

Authors' contributions

JW conceived the study and participated in its design and coordination, helped to draft the manuscript and critically revised the text for important intellectual content. SO participated in the sequence alignment. Both authors read and approved the final manuscript.

Acknowledgement

We thank the Ministry of Agriculture staff in Kikuyu and Limuru, especially the District Agricultural Officer, Lucy Waweru, for allowing use of the office computers and printers, andthe divisional staff, H Muchui and J Njoroge, for assisting in collecting data in the Karai and Ndeiya divisions, respectively. Special thanks to our supervisors, Professor Jacob W Wakhungu and Dr Stanley Omuterema, who greatly sacrificed their time in providing guidance while developing this paper. Finally, thanks to Martin Kasina for analyzing the data.

References

1. United Nations Development Programme: *Kenya Briefing Report on Food Situation.* Nairobi: Government printers; 2010.
2. International Federation of Red Cross and Red Crescent Societies: *World Disasters Report 2009; Focus on Early Warning Early Action.* Geneva: Red Cross publishers; 2009:23.
3. IFPRI: *Reaching sustainable food security for all by 2020. Getting the priorities and responsibilities right.* Washington, D.C: IFPRI Projects; 2002:36–38.
4. DDO (2009): District Development Office: *Annual Report. Kiambu District.* Nairobi: Government printers Un published, government report; 2009:17.
5. Hodgkin: *Sustainability in Water Supply and Sanitation*, Draft wash Report February. London press; 1994:15.
6. Ministry of Agriculture: *Food Security Kenya Briefing, Update September 2010.* Nairobi: Government Printers; 2010:3.
7. UNICEF: *Determinants of Malnutrition in Children: A Conceptual Framework.* Nairobi: Government printers; 2009.
8. World Bank, Independent Evaluation Group: *Hazards of Nature, Risks to Development: An IEG Evaluation of World Bank Assistance for Natural Disasters.* Washington, DC: World Bank; 2006.
9. KNBS: *Kenya National Census Report, 2009.* Nairobi: Government Printers; 2010:319–325.
10. Ministry of State for Special Programmes: *Implementation Report Summary 2003–2010.* Nairobi government printers; 2010:12.
11. Ministry of Health: *Kiambu District Annual Report 2010*, Unpublished Government report. Nairobi: Government Printers; 2010:25.
12. Ministry of Culture and Social Services: *Project Progress Report 2007*, Unpublished Government report. Nairobi: Government Printers; 2007.
13. Ministry of Agriculture, District Agriculture Office: *2011 Annual Report. Kiambu district*, Unpublished Government report. Nairobi: Government Printers; 2011:12.
14. Central Bureau of Statistics: *Kenya Demographic and Health Survey.* Nairobi: Government Printers; 2008:15–17.
15. Ministry of Agriculture, District Agriculture Office: *Food Production Based on Cereals Production in the District for 2009.* Kiambu district: Unpublished Government report; 2009:14.
16. Mathu: *Towards a praxis model of social work: a reflexive account of practical intervention.* Nairobi University; 2005. Unpublished PhD thesis.
17. Oniang'o R: *Interventions to Address Household Food security in Kenya, ACC/SCN Symposium report.* Accra: Ghana in February 1999; 1999:6.
18. Wabwoba: *Factors affecting sustainability of food security projects in Kiambu, Kenya, Masters thesis 2012, MMUST.* Kakamega: Lambert Academic publishers; 2012.

4

What economic theory tells us about the impacts of reducing food losses and/or waste: implications for research, policy and practice

Martine M Rutten

Abstract

Background: Whereas the prevalence of hunger and food insecurity is often cited as a motivation for reducing losses and waste in agriculture and food systems, the impacts of such reduction on food security and the wider economy have not yet been investigated. This paper gives insights into these effects, the factors of influence, and derives implications for applied research, policy and practice.

Methods: We used economic theory to analyse the impacts of food loss reductions on the supply side and food waste reductions on the demand side. The analysis is graphical and uses intuitive low-dimension diagrams.

Results: The impacts of tackling food losses and waste differ from the size of food losses and waste and depend, in addition, on the extent to which they are avoidable, factors that cause them to arise (notably food prices) and the costs associated with measures to reduce them. Interactions within the food supply chain and the broader economy also affect the impacts. Trade-offs occur on the demand side where a reallocation of spending on previously wasted foods causes some producers to be worse off and some to be better off. Over time, producers tackling losses may have to incur welfare losses in the short run with gains in terms of increased revenues, if any, occurring later. Similarly, consumers may delay spending savings on previously wasted foods. As a consequence, the impacts, notably on food security and welfare, are ambiguous.

Conclusions: Further research should quantify the factors that play a role and carry out economy-wide impact analyses, employing a combination of macro, meso and micro-level tools, and presenting a comprehensive set of indicators that adequately capture broader societal impacts of tackling food losses and waste. This allows policy makers to better target policy and resources, identify complementary policies, and move beyond target-setting to addressing the underlying causes, whereby it is important to consider the whole food supply chain. Supply chain actors could contribute in terms of practical and innovative solutions where they matter most, and feed research and policy makers on the bottlenecks that explain why food losses and waste occur, and their relative importance.

Keywords: Economic theory, impacts, food losses, food waste, agriculture and food systems, food security

Background

Food waste is a contentious issue at times where hunger amongst the poor is still prevalent, and also vulnerable persons increasingly have to fear for food insecurity due to frequent food price peaks (for example, in 2007 to 2008 and 2011 to 2012) and adverse economic tidings. Given a growing population, pressures from competing claims of feed, food and fuel demands on scarce natural resources of water and land, are only likely to increase.

The numbers suggest that globally, compared to 2009, 70% more food would have to be produced to satisfy the needs of a population growing by a third in 2050 [1]. Evidence from the Food and Agriculture Organization (FAO) also suggests that close to one third of the edible parts of food produced for human consumption is lost or wasted globally, equivalent to around 1.3 billion tons per year, and that the size of food losses and waste varies greatly by type of food, country or region and the stage

Correspondence: martine.rutten@wur.nl
International Policy Department, LEI Wageningen UR, Alexanderveld 5, 2502, The Hague, LS, Netherlands

in the food supply chain [2]. Specifically, worldwide food losses and waste per year are estimated at around 30% for cereals, 40 to 50% for root crops, fruit and vegetables, 20% for oil seeds, meat and dairy and 30% for fish. Compared to other types of food, fruit, vegetables, roots and tubers suffer from relatively high losses and waste throughout the supply chain. Whereas losses and waste in industrialised and developing countries are roughly the same in terms of quantity (670 and 630 million tons respectively), in terms of value they are very different (US$ 680 and US$ 310 billion respectively), and also relative to the level produced in the market: for example, food losses and waste in Europe and North-America are around a third of production, whereas in sub-Saharan Africa and South and Southeast Asia this share is 25 to almost 40%. Moreover, in low-income countries, these so-called losses in edible food mass destined for human consumption predominantly occur in agricultural production, post-harvest and processing stages (over 40% in the latter two stages). In medium- and high-income countries these losses (over 40%) occur mostly at the end of the food chain (retail and final consumption), and are termed 'waste' [3]. The reference definition being developed by the FAO as part of the Save Food initiative [2] also includes qualitative losses such as losses in nutritional value. Reporting food losses and waste in terms of nutritional value leads to different numbers and percentages, though the patterns remain roughly the same [4].

These observations lead to the suggestion that reducing losses and waste in agricultural and food systems could relieve part of the pressures on scarce resources and enhance food security [5-8].[a] However, the economic, social and environmental impacts of reducing food losses and food waste have, so far, not been investigated in detail empirically so it is unclear whether this is the case.[b] A reason for the lack of applied studies is the absence of reliable and consistent international data on food losses and food waste, the 2011 FAO study on Global Food Losses and Food Waste [2] being the first attempt to consistently present evidence on the extent of food losses and food waste throughout the supply chain from farm to fork, by broad commodity grouping, and for all regions in the world.

The lack of reliable and consistent macro data is fed by differences in definitions and methods of data collection between different countries. First, the distinction between food losses (operating on the supply side) and food waste (operating on the demand side) is not always that clear-cut. The definition of what is edible also varies across time and countries, and it depends on the perspective taken. For example, from a food security perspective, biofuels, feed or other non-food secondary uses of what was intended for human consumption may be considered a loss, whereas from the economic perspective of the generation of value added, they are not. Alternatively, from a health perspective, food intake may be more appropriate than food consumption and anything in excess of reference levels may be interpreted as waste. The latter is a rather extreme viewpoint (everyone who is overweight may then be seen as wasting food), but it illustrates that the definition used is crucial for the determination of the size of the problem. The project, Food Use for Social Innovation by Optimising Waste Prevention Strategies (FUSIONS) [9], which runs from 2012 to 2016, aims to address the harmonisation of food waste monitoring in the EU, which will help in clarifying the definitions and making available more consistent data on food waste for EU countries.

Given the lack of applied studies, the literature on impacts of (reducing) food losses and waste has been focusing on expressing food losses and waste in terms of the value of the resources (land, water) embodied in the output that could have been realised (production, consumption, nutrition), or the negative external effects that could have been avoided (emissions), by avoiding food losses and waste (see [10] for an overview). However, this only says something about the scale of the problem, not the actual impacts, as it ignores interactions between demand and supply, the role of the price mechanism therein, and more generally, interactions between actors and sectors in the agricultural sector, the food system and the wider (national and global) economy. Given the scale of the problem, these second-order effects could be quite considerable. For example, reducing food losses on the supply side (food waste on the demand side) could lower food prices quite considerably, leading to a change in demand (supply) of food so that the resulting impact is likely to differ from the value of resources or output embodied in the losses (waste) itself.

The predominantly qualitative literature [2,4,11] reasons that food producers will profit from reducing food losses as this allows them to sell more at a lower cost so that their incomes increase. Consumers will benefit from reducing food waste as it saves them money which they can use elsewhere (on food or non-food products) and it lowers the price of the remaining food that is consumed in the market.

This paper aims to provide insights into the effects that are likely to occur if food losses and/or waste were to be reduced, to assert whether everyone is indeed better off than before, and what factors are crucial for determining the outcomes. After a brief discussion of the methodology, this paper will continue with a separate analysis of reducing food losses on the supply side and food waste on the demand side, which - as will become apparent - have distinctly different impacts, followed by a joint analysis of the influence of the types of supply and demand curves (elasticities) and other complicating factors that play a role in the outcomes. The paper then continues with a discussion in which a synthesis of the results is provided and implications for further applied research, policy and practice are derived. The final section concludes.

Methods

It is useful to start an analysis of the economy-wide impacts of reducing food losses and food waste with a theoretical framework, which guides the interpretation of the outcomes of a more complex, empirical model with added real-life complexities. This paper creates such a theoretical framework, which, to our knowledge, is missing from the literature so far. The paper provides a simple, intuitive, low-dimension diagrammatic analysis of expected impacts using standard economic theory of supply and demand and discusses what factors play a role.[c] It provides the groundwork for further applied analyses that are currently being undertaken in separate studies [12,13] and may be undertaken in future.

Results

Reducing food losses in supply

In this section we examine the economic impacts of (reducing) food losses in a low-dimension partial equilibrium analysis. Figure 1 depicts the market for a food commodity,[d] with a standard upward sloping supply curve and a standard downward sloping demand curve. The price mechanism ensures that demand equals supply. The equilibrium is reached at point A, where the price is P^0 and the quantity traded is Q^0. The depicted situation is best interpreted as capturing the full supply chain from farm to fork, but concealing the various intermediate stages in supply (for example, storage, transport, processing).

Let us assume that there are losses in the production and supply of this food commodity. In such a situation, the socially optimal supply curve, or the supply curve of this food commodity that would not have these losses, lies below the original supply curve, as depicted by Supply' in Figure 1; given the original price, P^0, more can actually be produced and supplied to the market (Q^2 at point *B*), or the original quantity, Q^0, can actually be produced at a much lower cost (P^3 at point *C*) if losses were to be absent. Note that the 'optimal' supply curve does not necessarily have to be parallel to the original supply curve, as the extent of losses may vary with the scale of production (and price). We abstract from this for ease of exposition.

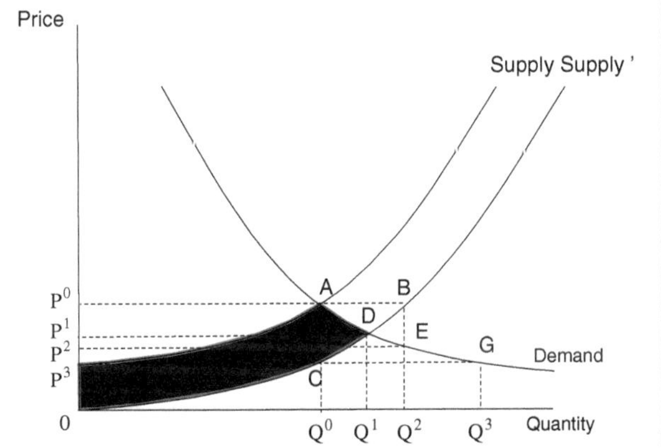

Figure 1 Impacts of reducing food losses in supply. Blue shaded area: overall welfare gain.

Impacts on price, quantity and welfare

What happens if food losses in supply for the food commodity in question are tackled? Suppliers may, for example, be induced to tackle the losses because of the emergence of a new technology, which makes this possible and worthwhile, or new policies (regulations, taxes and subsidies) that penalise and stimulate reductions in food losses. The action of avoiding the losses, given the original demand curve and given the underlying motivation of doing so, would result in a lower price, P^1, and a higher equilibrium quantity, Q^1, in the market, as given by point *D*. At this new equilibrium consumers can buy more food at a lower price, resulting in a welfare gain to consumers as measured by the change in the consumer's surplus of P^0ADP^1. Similarly, producers can sell more, but at a lower price, resulting in a change in the producer surplus of P^1D0 - P^0AP^3, which is also positive. The overall welfare gain equals the sum of the change in the producer and the consumer surplus, which amounts to the area P^3AD0, the blue shaded area between the new and old supply curve and under the demand curve.

These impacts - lower equilibrium price, higher quantity of food produced and consumed, and welfare gains for both producers and consumers - seem to be in line with the qualitative literature on impacts [2,4,11] and are encouraging from the perspective of low-income countries, where food losses on the supply side dominate. From Figure 1, it can also be seen that the size of the impacts will depend, amongst others, on how big the losses are relative to the size of the market, which as shown, varies by type of food and country or region. Whatever the extent of the losses, in terms of quantity the size of the impact, Q^0Q^1, however, is much smaller than the original size of the problem, Q^0Q^2, which is due to the change in the price. This is in sharp contrast with how the current literature approaches impacts [10].

Reducing food waste in demand

In this section we examine the economic impacts of (reducing) food waste. Figure 2 depicts the market for a food commodity, again with the supply chain from farm to fork collapsed into a standard upward sloping supply curve and a standard downward sloping demand curve and the equilibrium at point A, where the price is P^0 and the quantity traded is Q^0.

Let us assume that there are losses in the consumption of this food commodity, in that consumers waste part of what they demand. In such a situation, the socially optimal demand curve, or the demand curve that would not have these losses, lies to the left of the original demand curve, as depicted by Demand' in Figure 2; given the original price, P^0, less needs to be consumed (Q^2 at point B) so as to reach a certain level of utility if waste was to be absent, or the original quantity, Q^0, represents a much lower value to the consumer (P^3 at point C). Equivalent to the analysis on the supply side, the 'optimal' demand curve does not need to lie parallel to the original demand curve as the extent of waste in demand may vary with scale and price. We abstract from this for ease of exposition.

Impacts on price, quantity and welfare

What happens if food waste in demand for the food commodity in question is tackled? Consumers may, for example, be induced to tackle food waste because of a rising morale against waste, or new policies (regulations, taxes and policies) that penalise and stimulate reductions in food waste. Avoiding food waste in consumption, given the original supply curve and given the underlying motivation of doing so, would result in a lower price, P^1, and a lower equilibrium quantity, Q^1, in the market, as given by point D. Since producers are able to sell less and at a lower price, their welfare is negatively affected as shown by a change in the producer surplus of P^1DE - $P^0AE = - DAP^0$. Taking the difference between the area under the new and old demand curve and above the new and old price respectively, P^1DF - P^0AG, would result in a change in the consumer surplus of P^1DBP^0 - $BAGF$, which here is negative. This would ignore, however, the fact the old demand curve encompasses waste, so that consumers only realise P^0BF in value when consuming Q^0 of the food commodity at a price P^0; the remainder, $BAGF$, is lost due to wastage. The change in the consumer surplus if waste is avoided thus amounts to P^1DF - $P^0BF = P^1DPB^0$, which is now positive. The overall change in welfare that results equals P^1DBP^0 - $P^1DAP^0 = - BDA$, the red shaded area in Figure 2, which is negative.[e]

Whereas the equilibrium quantity goes down from Q^0 to Q^1, the distance Q^2Q^0, or BA, represented waste and was not eaten by consumers in the first place, so that actual food intake is going up from Q^2 to Q^1. Combined with the lower equilibrium price for food this is a

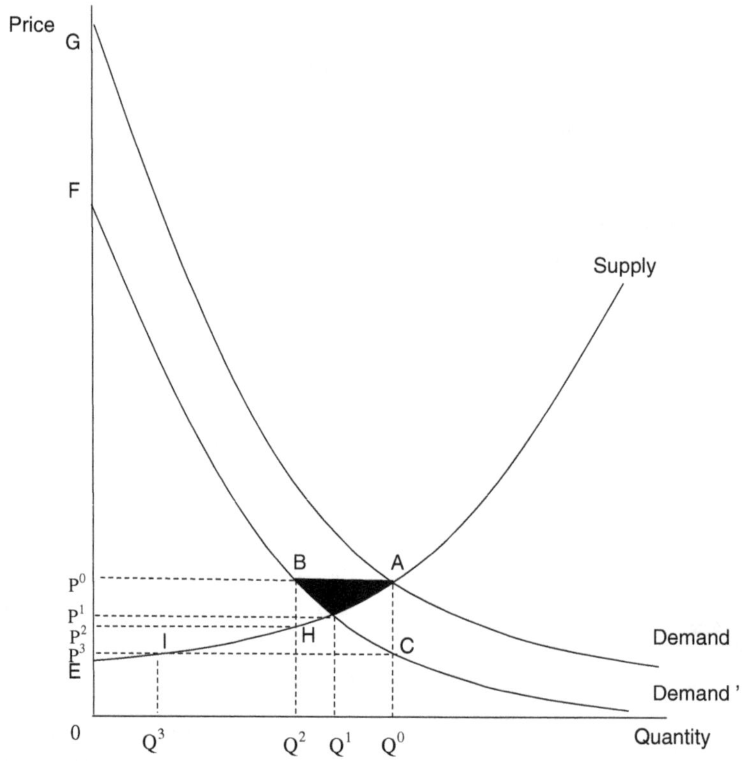

Figure 2 Impacts of reducing food waste in demand. Red shaded area: overall welfare loss.

positive outcome for consumers in this market, whose food security is going up. This outcome is in line with statements made by the qualitative literature on impacts [2,4,11] and is encouraging from the perspective of medium- and high-income countries, where food waste on the demand side is high on the policy agenda.

As before, the size of the impacts of reducing food waste in demand, in quantity terms Q^0Q^1, is influenced by the original size of the problem of food waste (Q^0Q^2) (though much smaller compared to the original size). More importantly, what is concealed in the literature, is that trade-offs occur between producers and consumers in this market, with the former being worse off, and potentially leading to negative employment effects. Reducing food waste is simply not in the interest of producers who aim to maximise profits, even if it is on commodities that are wasted.

To conclude that the overall welfare impacts of reducing food waste in demand would be negative is wrong because the analysis is still not complete. The question that remains is what consumers would do with the saved expenses on this particular food commodity, P^0Q^0 - P^1Q^1. Consumers may want to spend it now on the consumption of other commodities, and perhaps food.[f] In this case it would lead to a shift in the demand curve(s) of the respective commodity(ies) in the opposite direction of that depicted in Figure 2, that is, a shift to the right, leading to a higher price and quantity in the accompanying market(s) and a welfare gain in this (these) market(s) for producers and consumers. The overall welfare change(s) in this (these) respective market(s) would be positive and equivalent to the difference between the new and old demand curves and above the supply curve (if one were to take the example of Figure 2, *DAGF*). If consumers add the saved expenditures on previously wasted food to savings, the money could be used for consumption in future with associated utility gains then being realised.

The overall welfare impacts in the market of the food commodity in which waste is reduced and other markets combined thus depends on consumer preferences. Nonetheless, the welfare loss for producers, which occurs due to waste reduction by consumers of the food commodity in question and results in a fall in sales and price, is highly likely to be compensated for by welfare gains for producers and consumers in other markets benefiting from increased spending,[g] whether or not over time, because waste is avoided and resources that were previously spent on it are now put to a productive and welfare-generating use.

Varying the slope of demand and supply curves: inter-temporal effects

The outcomes, and thus, the size of the welfare effects, depend on the slope of the demand and supply curves. Assuming that the extent of food losses, respectively waste, is the same as before (that is, the shift in the supply or demand curve is of the same distance as before), and independent of scale and/or price, we can distinguish the following cases, discussed separately for the case of reducing food losses in supply and reducing food waste in demand.

Impacts on the outcomes of reducing food losses in supply

In the presence of a perfectly inelastic (that is, vertical) demand curve, the new equilibrium is at point *C* (same quantity, lower price), with consumers receiving all the gains from reducing food losses in the form of a lower price and a welfare gain of P^0ACP^3. In the presence of a perfectly elastic (that is, horizontal) demand curve, avoiding food losses in supply results in a new equilibrium at point *B*, where all the gains translate into an increase in the equilibrium quantity supplied and demanded (and no change in price). This results in a welfare gain to producers of P^3AB0.

Varying the slope of the demand curve thus influences the distribution of welfare gains over producers and consumers, but as shown, does in itself not lead to sign changes. With the demand for food generally being fairly inelastic, though not perfectly, the truth is likely to lie in between the two extreme cases presented.[h]

Similarly, if supply is perfectly inelastic (vertical supply curve), the equilibrium is at point *E*, resulting in a lower equilibrium price and higher equilibrium quantity compared to the analysis before. Consumers gain by P^0AEP^2, but producers here lose out by FEQ^2Q^0 - P^0AFP^2. The overall welfare result, however, is positive (area AEQ^2Q^0). Finally, a completely elastic (horizontal) supply curve results in an equilibrium at point *G*, whereby demand increases the most (to Q^3) as the price falls the most (to P^3) and all welfare gains end up with the consumers who benefit to the maximum extent possible, by area P^0AGP^3.

A vertical supply curve is representative of the short run, where it is generally difficult for producers of food to respond to price changes, whereas a horizontal supply curve corresponds to a long-run situation, where producers of agri-food commodities can respond and generally are price takers in a highly competitive market. The actual representation is likely to lie somewhere in between, but in the short run will be closer to being inelastic and in the long run will be more elastic.[i] These findings thus suggest the importance of inter-temporal effects, which are typically ignored in the literature. Specifically, overall welfare and the welfare of consumers generally goes up, whereas that of producers could go down, namely in the case of supply being relatively inelastic, that is, in the short run: the increase in sales from selling previously lost produce could be insufficient to compensate for the price decrease on existing sales.

In the long run, supply of agricultural commodities is more elastic, so then welfare gains are likely to occur and most of these end up with the consumer.

Impacts on the outcomes of reducing food waste in demand

Here we ignore what happens in other markets and focus on the market of the food commodity in which waste in demand is reduced (see Figure 2). Specifically, in the presence of a perfectly inelastic (that is, vertical) demand curve, the new equilibrium is at point *H*, resulting in a lower equilibrium quantity, Q^2, and a lower price, P^2. The change in the consumer surplus would again be positive and equal to the area, P^2HBP^0, whereas the change in the producer surplus would be negative and equal to - P^2HAP^0. The resulting overall welfare loss in this market is - *HAB*, slightly more negative than before. In the presence of a perfectly elastic (that is, horizontal) demand curve, avoiding food waste in demand results in a new equilibrium at point *I*, with an even lower quantity, Q^3, and price, P^3. This would result in a higher welfare loss in this market for producers, equal to - P^3IAP^0, but no impact on consumers. Similarly, if supply is perfectly inelastic (vertical supply curve), the equilibrium is at point *C*, where the reduction in wastage in demand fully translates into a lower equilibrium price, P^3, but has no impact on quantity. The lower equilibrium price benefits consumers by P^3CBP^0, but hurts producers, who suffer a maximum loss of - P^3CAP^0, resulting in an overall welfare loss in this market of - *CAB*. Finally, a completely elastic (horizontal) supply curve results in an equilibrium at point *B*, whereby there is only a negative impact on the equilibrium quantity, to Q^2, but no impact on price, which would result in zero impact on producers and consumers in this market.

In this simple, low-dimension diagrammatic analysis of reducing waste in demand, consumer welfare generally goes up or at best remains the same, whereas producer welfare falls or at best remains the same, resulting in an overall welfare impact ranging from negative to, at best, zero in the long term if the supply of agri-food commodities would be almost perfectly elastic. As indicated before, welfare gains to producers and consumers in other markets will result if consumers decide to spend the saved expenditures on other commodities. In sum, with elasticities differing for different commodities and in the short and long run, the truth about what will happen in reality when reducing food waste will lie somewhere in between, with trade-offs occurring between consumers who benefit, and producers of the food commodity for which waste is reduced, who lose out to the benefit of other producers who gain from increased spending now or in future.

Complicating factors

We have made various simplifying assumptions to come to our findings. Below we discuss how these complicating factors may alter the results.

The extent to which food losses and/or waste are avoidable

The impacts may be much smaller if only a part of the food losses and/or waste is avoidable. Unavoidable food losses, respectively waste, are generally considered by the literature as being inedible, such as bones, skins and peelings and represent about 19% of the food lost and wasted [14]. The remainder could possibly be avoided. The FAO uses conversion factors of the same order of magnitude to determine which part of the agricultural product is edible, specifically in the range of 0.7 to 1, with the exception of fish, which has a conversion factor of 0.5 [2].

Costs of reducing food losses and/or waste

What can feasibly or cost-effectively be avoided depends not only on the benefits but also on the costs involved in reducing food losses and/or waste compared to the status quo. If there are costs involved in reducing food losses, these will have a price-increasing and quantity-reducing effect in the market for the food commodity in question, counteracting the original shift down (or to the right) that occurs when reducing food losses in supply and counteracting the observed welfare gains. The literature remains surprisingly silent on the issue of costs of reducing food losses and waste (see [4] for some anecdotal evidence), which leads to incomplete and unfounded statements on impacts. Whereas some measures to reduce food losses on the supply side, such as improved harvesting techniques by farmers or other behavioural changes (also by other actors in the supply chain), may cost relatively little, other measures, such as investments in storage facilities and improved transport (including cold chain), may cost much more. Consistent data on costs of measures to tackle food losses are hard to find, not the least because they are likely to vary with the food commodity in question, the segment of the food supply chain, and by country due to differences in, for example, the level of development, location and climatic conditions.

Similarly, the net welfare gains may be lower if there are costs involved with reducing food waste by the consumer, which counteract the original shift down (or left) that occurs when reducing food waste in demand. All in all, although behavioural change on the part of consumers to waste less food may directly cost relatively little, it may require consumers, for example, to go more often to the supermarket to buy fresh food that is better adjusted to their needs and wants, which involves time and effort. In other words the perceived or indirect cost by the consumer may still be high.

Factors causing food losses and/or waste

Food losses may themselves be explained by underlying factors, including scale of production and price. Relatively low food prices are said to be an important cause of food losses in supply and food waste in demand [2,4,10]. If we assume that food losses decrease with greater scale (and price), the observed impacts of reducing food losses will be greater if the market is small and the price is low. This can be depicted by supply curves with and without losses, which are increasingly converging to one another as the quantity and price increases.

If we were to assume that food waste is increasing with the amount consumed and decreasing with the price paid for food, reducing food waste results in greater impacts if the market is of reasonable size (that is, the quantity traded is high and the price is low). This can be depicted by increasingly diverging demand curves with and without waste as the quantity increases and the price decreases.

Combining aforementioned observations, thus, it may well be that from the perspective of costs and benefits it is better for agri-food producers and suppliers to allow for some food losses (at a relatively low cost) rather than to take measures (at a relatively high cost and low returns) to combat them (also suggested by [4]). Similarly, for consumers the relatively low cost (price) of food may prevent her from taking action.

Interactions within the supply chain

Another simplification is that the framework presented cannot address where the losses occur in the supply chain (primary agricultural production stage, processing stage, storage and transport stages), which is shown to vary quite a bit, as do causes and measures to tackle the losses [2,4]. As a consequence, costs to tackle food losses may be borne in, say the beginning, of the supply chain, whereas the benefits may occur later in the chain. This and the time dimension associated with costs occurring upfront, and benefits - if any- occurring later, often impedes relevant actors to take action. Anecdotal evidence comes from farmers improving the quality of their crop or improving storage that reduces food losses, which subsequently cannot be sold since they do not have market access or as the market cannot absorb the additional quantity, or, if it can be sold it is sold to traders at a price that does not recuperate the original investment made. Another example is that of food processors making technical improvements that are negated by retailers when they withdraw food orders that were originally agreed upon. Finally, farmers are often not able to sell crops that do not meet marketing standards (for example, on shape) but are otherwise perfectly fine for human consumption, as they are not accepted by retailers and consumers [2,4,15].

Interactions with other markets and actors

Our analysis makes the usual *ceteris paribus* assumption, that is, that all else remains the same, which is highly unlikely. For example, reducing losses generally results in a lower price, which could increase demand elsewhere in the system, potentially leading to second-order effects. An example is food as feed (for example, from grains or vegetables and fruits) becoming cheaper if losses in production and supply fall, as a result of which meat demand may go up (as meat will become cheaper to produce due to lower intermediate input costs of using feed), which is less efficient in the use of resources compared to crops in that it uses relatively more water and land [16,17]. Similarly, biofuel use may go up. Another example is that households may waste more if food becomes cheaper, undoing the positive impact of reducing food losses on the supply side. Of course, consumers under pressure from prevailing morale may also display the opposite behaviour and reduce food waste. Opening up the framework for trade, it is likely that the increased net food supply at a lower price will find its way to export markets and is able to better compete with imports, thus improving the trade balance for the food commodity in question. This does not alter the main outcomes.

Discussion

A synthesis of the results

Table 1 integrally presents the results of the supply and the demand side analyses of tackling food losses and waste respectively. It summarises what happens to the market equilibrium (price, quantity; second and third column) and consumer and producer welfare (fourth and fifth column, and sixth column for the overall effect) in the market for the food commodity in question and for varying assumptions regarding demand and supply curves, providing the boundaries for what may happen (by row). The last column considers the impact of other factors that are of influence, where possible, relaxing some of the simplifying assumptions made throughout the analyses.

What has become evident from economic theory is that the impacts of tackling food losses in supply and food waste in demand in terms of quantity are different from the original size of the food losses and waste, and depend crucially on the extent of the food losses and/or waste relative to the size of the market, the extent to which they are avoidable, factors that cause them to arise in the first place (notably the level of food prices), the costs associated with measures to reduce them and interactions within the food supply chain and with other

Table 1 Summary of impacts of reducing losses and waste in the market for a food commodity

	Impact on market equilibrium		Welfare impacts in the commodity market			Factors of influence and impacts
Reducing losses in supply	**Price**	**Quantity**	**Consumers**	**Producers**	**Total**	Extent of food losses relative to the size of the market
Perfectly inelastic demand curve	-	Constant	+	Constant	+	If losses increase (decrease) with scale and price, then impacts are larger at higher (lower) scale and price
Perfectly elastic demand curve	Constant	+	Constant	+	+	Impacts may be smaller if not all losses are avoidable
Perfectly inelastic supply curve	-	+	+	? (– in example)	+	If loss reductions involve costs then welfare impacts will be lower. This has a price increasing and quantity reducing effect.
Perfectly elastic supply curve	-	+	+	Constant	+	Interactions within the supply chain and with other actors and/or markets ?
Reducing waste in demand	**Price**	**Quantity**	**Consumers**	**Producers**	**Total**	Extent of food waste relative to the size of the market
Perfectly inelastic demand curve	-	-	+	-	-	If waste increases (decreases) with scale and decreases (increases) with price, then impacts are larger at higher (lower) scale and lower (higher) price
Perfectly elastic demand curve	-	-	Constant	-	-	Impacts may be smaller if not all waste is avoidable
Perfectly inelastic supply curve	-	Constant	+	-	-	If waste reductions involve costs then welfare impacts will be lower. This has a price increasing and quantity reducing effect.
Perfectly elastic supply curve	Constant	-	Constant	Constant	Constant	Consumer preferences: if consumers spend savings from reducing waste: positive impact on price, quantity and welfare of producers and consumers in other commodity markets
						Interactions within the supply chain and with other actors and/or markets ?

+, positive impact; -, negative impact; Constant, implies that there has been no impact. Whereas the direction of effects is shown, their magnitudes are not, as they depend on the underlying shocks and how they impact upon each of the variables.

actors and markets. Impacts also vary with the slope of the demand and supply curves and consumer preferences play an important role on the demand side. If one takes all these factors into account, one cannot be certain a priori what the impacts will be, notably when it comes to food security and welfare. Trade-offs are shown to occur with winners and losers, most visibly on the demand side where a reallocation of spending away from previously wasted food towards other products or commodities (and perhaps food) causes some producers to be worse off and some to be better off. Trade-offs also arise over time, as in the short run, producers may have to incur costs and/or welfare losses when food losses are tackled with gains in terms of increased revenues, if any, occurring later, and consumers may delay spending savings on previously wasted foods due to market uncertainty and increased costs of living.

The analysis becomes much more complicated if food waste in demand and food losses in supply would change simultaneously, possibly in reaction to one another, and on a global scale. What exactly will happen remains an empirical question and is best investigated in further applied research. The following subsections discuss the implications for further research on the economy-wide impacts of reducing food losses and/or waste, the implications for policy and in practice.

Implications for research

The theoretical analysis provides a useful construct to shape the research agenda on the economy-wide impacts of reducing food losses and/or waste. This section aims to draw the contours of such a research agenda and what this would add to current literature, looking at it from the perspective of economic science. It provides a starting point for further applied and policy-relevant analyses, and is unlikely to be complete, because as analyses get underway, new insights will emerge that will and should influence the research agenda further.

The first and foremost implication for research is to quantify the factors that play a role. Most importantly, this includes improving data on food losses and waste and the extent to which this is avoidable throughout the supply chain for detailed agri-food commodities and by country, using commonly agreed definitions. At the country level and at the international level (for example by FAO, Organization for Economic Cooperation and Development (OECD), Eurostat and FUSIONS) a lot of effort is currently going into improving the data [9,18]. It also includes quantifying the factors causing food losses and food waste. This may, for example, result in food loss and food waste being analysed as a decreasing function of food prices. This has, to our knowledge, not been done yet. Finally, it includes estimating the cost of tackling food losses and food waste (depends on measures taken). Preliminary work is being undertaken to come to cost ranges for policy measures to address food losses and waste [19], but more work is needed to come to more precise estimates of the cost to the various actors in the supply chain, beyond that presented on an anecdotal basis [4].

Second, with this information one could investigate the impacts of reducing food losses and/or waste using applied empirical models. Although the main interest here is in the economy-wide or macro-level impacts, it is useful to combine or complement this with meso- and micro-level analyses to get more detailed impacts for the various actors in the food supply chain from farm to fork. Methodologies that are suitable for each of these levels of analysis are discussed below.

At a macro level, it seems appropriate to use computable general equilibrium (CGE) models that are firmly grounded in microeconomic theory, capturing demand and supply responses to changes in relative prices, and are able to address interactions within the food supply chain and the wider economy that are shown to be important. Applications of impact analyses are underway regarding the issue of reducing food losses in supply in the Middle East and North Africa [12] and regarding the issue of reducing food waste in demand compared with a movement towards healthier diets in the EU [13] using a global CGE model. Due to its global reach, the latter study can also address the question of how reducing food waste in industrialised countries (the EU) affects food security in poor countries (sub-Saharan Africa). Specifically, if the former reduce food waste this may increase food availability on the world market, which lowers world food prices and could benefit net food consumers (or importers) but harm net food producers (or exporters) in the developing world. These impacts will depend on trade patterns, which are influenced by whether or not tariff or non-tariff barriers to trade exist, which may prevent the free movement of food commodities altogether. In order to obtain more disaggregated impacts, a partial equilibrium (PE) model that has more detail in agri-food production and trade may be used. To our knowledge, there are no PE studies on the impacts of (reducing) food losses and waste, although the OECD has an interest in going in this direction using the AGLINK- COSIMO model framework [20]. As a PE model cannot capture economy-wide linkages, a combination of CGE and PE modelling may be most appropriate, but will be computationally more demanding. Macroeconomic CGE- or PE-type models are being increasingly coupled with models that can address non-economic impacts, such as environmental impacts (biodiversity and emissions, for example [21]) and health impacts (number of deaths avoided, health cost savings, for example [22]). These are useful directions for further work as food waste and more broadly food consumption patterns and the motivations to adjust these are strongly related to environmental and health concerns, which a pure economic focus would fail to address.

At a meso level, cost-benefit analyses along the value chain could add insights into where food losses and waste occur in the chain, why they occur, what the consequences are, and what the costs and benefits are of tackling losses, taking into account that the various segments of the food supply chain interact. Such analyses, to our knowledge, have not been carried out as yet.

At a micro level, further analyses could shed light on consequences of tackling food losses and waste for one segment of the food supply chain, for example households. Poverty impacts for different types of household can be obtained using a coupling of macro models with micro-simulation models (for example [23]), by imposing a top-down household decomposition (for example [24]) or using a fully integrated micro–macro model (for example [25]). At the household level it is important that consumer behaviour regarding food waste is appropriately captured. This is strongly related to consumer preferences regarding food and more broadly, dietary change. Although there is much literature on how consumers may reduce food waste [2,4,10], little is known about how consumers can be made to act on it and what consumers will do with savings on previously wasted food. A forthcoming study by Waste & Resources Action Programme (WRAP), looking into UK household food consumption behaviour in relation to waste [26], finds that an increase in food waste awareness (through increased media coverage) lowers the amount of food purchased whereas food intake and food sale revenues remain the same. The latter suggests that trading up takes place in that UK households tend to buy smaller quantities of higher priced foods. It contradicts economic theory somewhat in that, *ceteris paribus*, we would expect food intake to increase and some of the savings on previously wasted food (that was spent on food before) to also be spent on non-food items. Integration of such research into a general equilibrium framework would be useful so as to take into account broader food supply chain and economy-wide interactions.

Third, the results of these impact analyses need to be presented in a clear and comprehensive way, stating the assumptions made (regarding the factors incorporated and methods used), and doing justice to the research questions at hand. Two points merit further attention in this respect. These may appear trivial at first sight, but examples are included that aim to illustrate each point.

First, what are the indicators used to measure impacts? Common impact indicators include gross domestic product (GDP), sector production and prices, employment and wages, land use and prices, household income, consumption of foods (diets) and other commodities, and prices paid by households, and trade (exports, imports, trade balance). As shown from the theory, outcomes are likely to vary with the type of indicator used. Hence what may be a desirable outcome from the perspective of one indicator, may not be so from the perspective of another indicator. A clear example is that whilst consumers in industrialised countries may be better off when avoiding waste, some producers may suffer from reduced revenues and incomes, with associated negative employment impacts, to the benefit of producers and employment in other sectors. Trade-offs may extend beyond economic indicators and may also involve environmental and health indicators. For example, whilst reduced food waste may save on land use in food production, trading up on the side of consumers in terms of increased spending on higher priced and more luxurious foods may imply a more unhealthy diet and increased health care costs. With the interest in current literature being focused on broader social impacts (including economic, health and environmental impacts) it is crucial to report the impacts in terms of indicators covering these adequately. It is thus insufficient, , to merely state, as the current literature does, what the impact is likely to be only on consumers if they reduce waste (or equivalently, only on producers if they were reduce losses), without any consideration of broader impacts on society, or for that matter, without any consideration of costs. An interesting research question to look into is to find out whether there is an optimal level of food losses and/or waste taking into account both costs and benefits and to see whether this optimal level differs between private agents and society as a whole. The literature and policy makers currently suggest that food losses and waste are a public bad and should be reduced to zero. Our analyses point out that there may well be an optimum, given that there are costs and trade-offs involved in reducing food losses and/or waste.

Second, what is the benchmark or reference scenario for analysis? If we are interested in the impacts of tackling food losses and waste over time, we need to specify a so-called baseline or business-as-usual scenario to which we can contrast alternative scenarios regarding reductions in food losses and waste. The baseline traditionally assumes a continuation of current trends and no change in policies. Usually this involves capturing current population and economic growth trends, changes in dietary patterns and making assumptions on technological change processes over time. Moreover, almost certain changes in the area of trade, such as the conclusion of multilateral or bilateral free-trade agreements, or domestic policies in the near future may also be incorporated. These trends and policies are likely to affect the outcomes as the impacts on a certain indicator should be presented in difference (absolute, percentage change or change in percentage points) from the outcomes under the baseline so as to isolate the impacts of a particular analysed scenario. An example is that free-trade agreements are very likely to alter the pattern of trade between countries and therefore, the sources of foods supplied and consumed in a country; the EU-US and trans-Pacific trade agreements, which are currently being negotiated, seem particularly relevant here. Baseline construction is by no means straightforward but necessary to come to accurate projections of impacts in the future [27]. It is, therefore, not uncommon to see more than one baseline (for example, baselines with low, medium and high growth; or baselines with or without trade agreements). Developments in the baseline also qualify the results of projections. Taking the example of reduced food waste by households in industrialised countries, the losses to some food producers may appear minor in the light of expected future increases in food sale revenues from increasing demands for food.

Implications for policy

Policy makers ideally should use the outcomes of applied studies on the economy-wide impacts of reducing food losses and food waste as an input to determine what commodity and what element(s) of the food supply chain to focus on. In an international context, policy makers from the various countries in the world can discuss their respective roles in and contributions to reducing food losses and waste. This is in contrast with what is happening now; following the current literature, the focus of policies is being driven by mere assertions of the size of food waste and losses, not societal impacts (benefits and costs) and ignoring the broader policy context. The consequence of this approach is that policy efforts and resources are currently not being put where they will be most cost-effective from a societal perspective.

Making policy using the outcomes of applied studies is by no means easy. It requires taking into account the trade-offs most likely to come out of any research, weighing and ranking the outcomes of the various indicators according to the priorities of policy makers.

Naturally, other considerations such as health and environmental aspects play a role, but political and cultural aspects that cannot be modelled or quantified also play a role. Moreover, other avenues of reaching the aforementioned goals should be considered as well. For example, if the prime concern of policy makers in industrialised countries is to improve resource efficiency and reduce environmental impacts, it may well be crucial to encourage consumers to follow a more healthy diet rather than focusing only on food waste reductions. Similarly, if the interest is also to improve food security in developing countries, trade liberalisation and investment policies may be more effective and should then at least also be considered.

Second, policy makers should move beyond formulating targets for reducing food losses and food waste, such as the European Commission's target of reducing food waste by 50% in 2020 [28], and tackle the underlying causes. Food losses and food waste are nothing new and happen for a reason. Not tackling the root causes is unlikely to yield long-term sustained reductions in food losses and/or waste. With low food prices being cited as a major cause relative to high costs of tackling food losses and waste, this may require increasing the value of and appreciation for food through information campaigns, tax and subsidy policies that reduce incentives to engage in wasteful behaviour, but also financial contributions to improve infrastructure for roads, energy and markets. It may also require setting up institutions and carrying out regulatory reform to facilitate private sector investment in the agricultural sector and food industry.

Finally, policies should take the whole supply chain into consideration. If the focus is only on reducing waste by consumers, which seems to be the approach taken by the European Commission and generally most industrialised countries formulating targets on food waste in demand, the problem may move further up the supply chain to food distributors, processors and farmers, and vice versa, if the focus is on reducing food losses in the early stages of the supply chain, they may move towards the end, such as in retail [4,19].

Implications for practice

The various supply chain actors in the field could benefit from and use the outcomes of applied research by working on innovative solutions and investing in those areas (that is, commodities, stage of the food supply chain and countries) where they will have the greatest beneficial impact. In turn, they could also feed research and policy makers with information on why food losses and waste are occurring, that is, what the bottlenecks are in tackling food losses and waste, and their relative importance.

Figure 3 summarises the implications for research, policy and practice, naturally with feedback loops as research, policy and practice do not stand on their own, but inform and benefit one another. This facilitates a virtuous circle, of increasing our knowledge on the causes and impacts of food losses and food waste, enabling the fine tuning of policies, and leading to better outcomes for the various actors in the food supply chain.

Conclusions

This paper, if anything, has shown that the link from reducing food losses and/or waste to food security and welfare for producers and consumers is not as straightforward as the literature seems to suggest using standard economic theory. Many factors are shown to play a role, which should be taken on board by further applied research to investigate broader societal impacts. This would greatly improve the information base for policy making, which is currently being driven by mere considerations of the size of food losses and waste, not their impacts on society, and in turn focuses too much on addressing the size of the problem, thereby often ignoring the underlying causes. It would allow for better targeted policies and resources being devoted to areas where impacts are shown to be most beneficial.

Endnotes

[a]Food security is most commonly defined as '... when all people, at all times, have physical, social and economic access to sufficient, safe and nutritious food to meet their dietary needs and food preferences for an active and healthy life' [29].

[b]The exception is a study on healthy and sustainable diets, which also considers the issue of food waste [21]. Rather than using real data, this study assumes that 15% less food production is required to meet the same level of nutrition. This is implemented as a 15% supply chain efficiency increase on a global scale, and so cannot distinguish differential impacts of food loss reductions on the supply side and food waste reductions on the demand side. It finds that agricultural prices fall by about 4%, which increases global food consumption.

[c]As a basis, any standard microeconomic theory book can be used, such as [30].

[d]While this paper considers the economic impacts of food losses, the issue of losses, or more broadly, resource inefficiencies, is not only confined to foods but is also applicable elsewhere. This analysis can therefore be extended to other non-food commodities.

[e]Note, that if one were to mistakenly include the area *BAGF*, the overall welfare loss would amount to *DAGF*, the difference between the two demand curves and above the supply curve, which is analogous to the result of the analysis of loss reductions on the supply side. The analysis on the demand side differs from that on the supply side in that the original demand curve includes

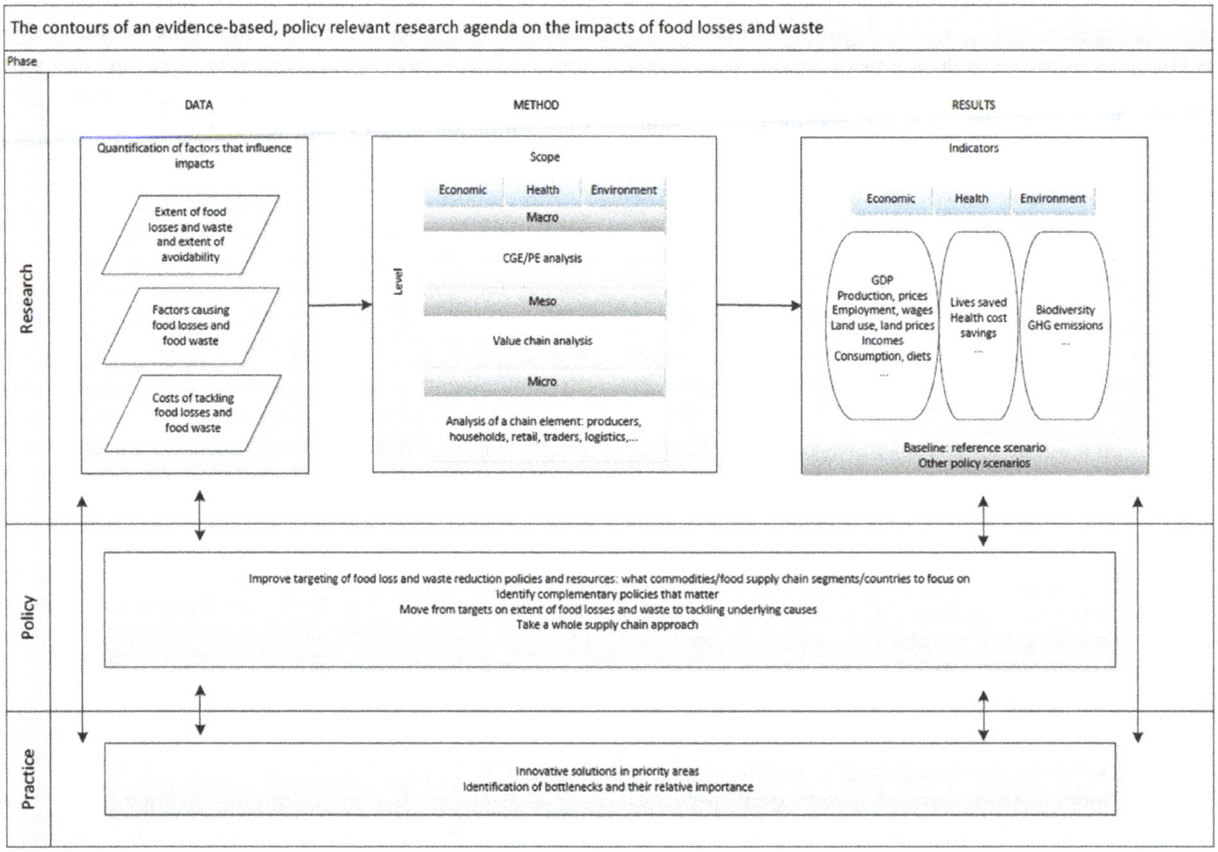

Figure 3 Summary of implications for research, policy and practice. CGE, computable general equilibrium; PE, partial equilibrium; GDP, gross domestic product.

wastage, that is, represents gross demand, whereas the original supply curve is the supply that would result after losses, that is, it represents net supply.

[f]There is some evidence that consumers may 'trade up' towards more 'luxurious', higher priced foods, such as meats [31]. This further strengthens the outcome of improved food intake of consumers reducing waste in one or more commodity markets. Note, however, that this also potentially brings with it negative health consequences associated with less healthy diets. This is further addressed in the discussion session.

[g]This is the second notable difference with the analysis of loss reductions on the supply side. Specifically, the loss reduction on the supply side is assumed to benefit the commodity in question in terms of lower costs of producing the same amount of good and/or increased outputs given costs. On the demand side however, it is almost more relevant to know what happens in other markets depending on consumer preferences and following consumer decisions on what to do with the saved expenses on the commodity that previously had wastage.

[h]Data on food-price elasticities suggest that the demand response for foods can generally be categorised as being inelastic (less than one in absolute value), but estimates generally vary by type of food and may also vary by income level [32]. Specifically for basic food commodities (that is, that form the main staple in a diet) demand will be more inelastic, whereas for more luxurious types of food (for example, soft drinks, meats and foods away from home) demand will be more elastic. Moreover, low-income populations may be more sensitive to price changes. To find price elasticities of demand for specific food commodities and specific countries one can consult an online database provided by the United States Department of Agriculture (USDA) Economic Research Service [33].

[i]Evidence on supply elasticities for agricultural commodities is mixed but the literature [34,35] suggests that the supply response in the short run is relatively inelastic (elasticity of less than one and more often than not close to zero) and in the long run more elastic (elasticity range from below one to over one), though not perfectly so.

Abbreviations

CGE: Computable general equilibrium; FAO: Food and Agriculture Organization; FUSIONS: Food Use for Social Innovation by Optimising Waste Prevention Strategies; GDP: Gross domestic product; OECD: Organization for

Economic Cooperation and Development; PE: partial equilibrium; USDA: United States Department of Agriculture; WRAP: Waste & Resources Action Programme.

Competing interests
The author declares that she has no competing interests.

Authors' contributions
MR is solely responsible for the manuscript and underlying research.

Authors' information
MR is researcher at the international policy department of LEI Wageningen UR, a leading economic research institute in the fields of food, agriculture and the natural environment in the Netherlands. MR has extensive research experience in the areas of food security, agricultural markets and international trade. She held previous positions as policy advisor in the area of international (financial) economics and as health economics and finance advisor in the context of developing countries.

Acknowledgements
The author is very grateful to Jeff Powell, Geert Woltjer, Lusiné Amaryan, Peter Nowicki, Yuca Waarts, Gerdien Meijerink, Siemen van Berkum, Adrian Tan, Francois Cohen and Clementine O'Connor for their useful comments and advice. The paper also greatly benefited from comments and suggestions by various participants of the 16th Annual GTAP conference in Shanghai (12 to 14 June, 2013), a discussion session with policy makers at the Dutch Ministry of Economic Affairs (20 June, 2013) and the 4th OECD Food Chain Analysis Network Meeting (20 to 21 June, 2013).

References

1. FAO: *How to Feed the World in 2050.* Rome: FAO; 2009.
2. FAO: *Global Food Losses and Food Waste: Extent, Causes and Prevention.* Rome: FAO; 2011.
3. Parfitt J, Barthel M, MacNaughton S: **Food waste within food supply chains: quantification and potential for change to 2050.** *Phil. Trans R Soc B* 2011, **365:**3065–3081.
4. Lipinski B, Hanson C, Lomax J, Waite R, Searchinger T: *Reducing Food Losses and Waste. Working Paper, Installment 2 of Creating a Sustainable Future.* Washington DC: World Resources Institute; 2013.
5. FAO: *Towards the Future We Want: End Hunger and Make the Transition to Sustainable Agricultural and Food systems.* Rome: FAO; 2012.
6. UNEP: *Avoiding Future Famines: Strengthening the Ecological Foundation of Food Security through Sustainable Food Systems.* Nairobi: UNEP; 2012.
7. UNEP: *The Critical Role of Global Food Consumption Patterns in Achieving Sustainable Food Systems and Food for All.* Paris: UNEP; 2012.
8. Guyomard H, Darcy-Vrillon B, Esnouf C, Marin M, Russel M, Guillou M: **Eating patterns and food systems: critical knowledge requirements for policy design and implementation.** *Agric Food Secur* 2012, **1:**13.
9. **EU FUSIONS.** http://www.eu-fusions.org/what-is-fusions.
10. Barilla Center for Food & Nutrition (BCFN): *Food Waste: Causes, Impacts and Proposals.* Parma: BCFN; 2012.
11. Lundqvist J, de Fraiture C, Molden D: *Saving Water: From Field to Fork – Curbing Losses and Wastage in the Food Chain.* Stockholm: Stockholm International Water Institute (SIWI); 2008.
12. Rutten M, Kavallari A: **Can Reductions in Agricultural Food Losses Avoid Some of the Trade-Offs Involved when Safeguarding Domestic Food Security? A Case Study of the Middle East and North Africa. GTAP Resource 4075.** https://www.gtap.agecon.purdue.edu/resources/res_display.asp?RecordID=4075.
13. Rutten M, Nowicki P, Bogaardt M-J, Aramyan L: *Reducing Food Waste by Households and in Retail in the EU: A Prioritisation Using Economic, Land Use and Food Security Impacts.* The Hague: LEI Wageningen UR; 2013. unpublished.
14. Waste & Resources Action Programme (WRAP: *The Food We Waste.* Banbury: WRAP; 2008.
15. *Personal communications with Yuca Waarts.*
16. Pimentel D, Pimentel M: **Sustainability of meat-based and plant-based diets and the environment.** *Am J Clin Nutr* 2003, **78:**660S–663S.
17. Hoekstra A: **The hidden water resource use behind meat and dairy.** *Animal Frontiers* 2012, **2:**3–8.
18. **OECD Food Chain Network: 4th network meeting 20–21 June 2013.** http://www.oecd.org/site/agrfcn/.
19. BIO Intelligence Service: *Modelling of Milestones for Achieving Resource Efficiency, Turning Milestones into Quantified Objectives: Food Waste.* Prepared for the European Commission, DG Environment; unpublished.
20. Okawa K: **Agricultural Trade and Markets, OECD: Modelling Work on Food Waste. Presentation at the 4th network meeting of the OECD Food Chain Network: 20–21 June 2013.** http://www.oecd.org/site/agrfcn/.
21. Westhoek H, Rood T, van den Berg M, Janse J, Nijdam D, Reudink M, Stehfest E: *The Protein Puzzle: The Consumption and Production of Meat, Dairy and Fish in the European Union.* The Hague: Netherlands Environmental Assessment Agency (PBL); 2011.
22. Lock K, Smith R, Dangour A, Keogh-Brown M, Pigatto C, Hawkes R, Fisberg C, Chalabi Z: **Health, agricultural, and economic effects of adoption of healthy diet recommendations.** *Lancet* 2010, **376:**1699–1709.
23. Bussolo M, de Hoyos R, Medvedev D: **Economic growth and income distribution: linking macro economic models with household survey data at the global level.** *Int J Microsimulation* 2010, **3:**92–102.
24. Hertel T, Verma M, Ivanic M, Rios A: **GTAP-POV: A Framework for Assessing the National Poverty Impacts of Global Economic and Environmental Policies. GTAP Technical Paper 31.** 2011. https://www.gtap.agecon.purdue.edu/resources/res_display.asp?RecordID=3731.
25. Bouet A, Estrades C, Laborde D: **Households Heterogeneity in a Global CGE Model: An Illustration with the MIRAGE Poverty Module. GTAP Resource 3631.** 2011. https://www.gtap.agecon.purdue.edu/resources/res_display.asp?RecordID=3631.
26. WRAP: *The Economics of Food Waste.* Prepared by Fathom Financial Consulting; in press.
27. Dixon P, Rimmer M: *Forecasting with a CGE Model: Does it Work. Centre of Policy Studies and the Impact Project, General Paper No. G-197.* 2009.
28. Commission E: *Roadmap to a Resource Efficient Europe. COM(2011) 571 final.* Brussels: European Commission; 2011.
29. FAO: *Rome Declaration on World Food Security and World Food Summit Plan of Action.* Rome: World Food Summit; 1996.
30. Jehle A, Reny P: *Advanced Microeconomic Theory.* New York: Addison-Wesley; 1998.
31. Government Office for Science: *Foresight Project on Global Food and Farming Futures. Synthesis Report C7: Reducing waste.* London: Government Office for Science; 2011.
32. Andreyeva T, Long M, Brownell K: **The impact of food prices on consumption: a systematic review of research on the price elasticity of demand for food.** *Am J Public Health* 2010, **100:**216–219.
33. *USDA Economic Research Service Commodity and Food Elasticities.* http://www.ers.usda.gov/data-products/commodity-and-food-elasticities.aspx.
34. Nerlove M, Addison W: **Statistical estimation of long-run elasticities of supply and demand.** *J Farm Econ* 1958, **40:**861–880.
35. Rao J: **Agricultural supply response: a survey.** *Agric Econ* 1989, **3:**1–22.

5

Eating patterns and food systems: critical knowledge requirements for policy design and implementation

Hervé Guyomard[1*], Béatrice Darcy-Vrillon[1], Catherine Esnouf[1], Michèle Marin[2], Marie Russel[1] and Marion Guillou[1]

Abstract

Eating patterns are important for building sustainable food and agricultural systems. This paper begins by presenting the main features of eating patterns worldwide. These eating patterns include the relative convergence of diets, more rapid food transition in emerging and developing countries, development of a more complex food chain, and substantial food losses and waste at distribution and final consumption stages. These patterns have negative consequences on health and the environment. The drivers of these patterns are examined to identify knowledge gaps, the filling of which should facilitate the design and implementation of actions and policies aimed at making food systems more sustainable.

Keywords: Eating patterns, Diets, Food losses and waste, Food systems, Food security

Introduction

According to the 1996 World Food Summit, 'food security at the individual, household, national, regional and global levels is achieved when all people, at all times, have physical and economic access to sufficient, safe and nutritious food to meet their dietary needs and food preferences for an active and healthy life' [1]. The definition broadens the initial focus of the 1974 World Food Summit on the volume and stability of food supplies [2]. It includes secured access to food for all people, especially the most vulnerable, and incorporates food safety and nutritional balance. It also reflects concern about food composition, quality and choice, as well as care related aspects of 'good' nutrition.

Global food is clearly insecure. According to the Food and Agriculture Organization of the United Nations (FAO) [3], 925 million people were undernourished in 2010, and it is now apparent that the target of the Millennium Development Goals of reducing the proportion of people who suffer from hunger by half by 2015 will not be achieved.[a] Furthermore, many people are affected by vitamin and mineral deficiencies- one out of three in developing countries according to the World Health Organization (WHO) [4]. At the same time, 1.5 billion adults were overweight in 2008, including over 200 million obese men and nearly 300 million obese women [5]. In addition, a growing number of low- and middle-income countries are facing a double burden of malnutrition. That is, these countries exhibit a persistence of under-nutrition, notably among children, along with a rapid increase in numbers of those who are either overweight or obese, as well as those who have diet-related, non-communicable chronic diseases such as cardiovascular diseases and diabetes.

Food security is not just about supply matching demand. The development of agricultural and food systems must take into account the progressive depletion of fossil energy, the protection of soil and water resources, the preservation of biodiversity and the issue of climate change. If they are to feed more than nine billion people by 2050, farmers around the world will have to produce crops using less fossil fuel in an environmentally friendly way. They will also have to provide energy and industrial commodities in place of petrochemical products, as well as providing environmental and rural services like water management, biodiversity protection, carbon sequestration, or diversified and open landscapes. In addition, food consumption habits will need to adapt. The foresight study Agrimonde [6] clearly shows that by 2050

* Correspondence: Herve.Guyomard@paris.inra.fr
[1]INRA Paris, 147 rue de l'université, 75 338, Paris Cedex 07 France

eating patterns will be a major issue for world food security. It shows in particular that 'westernized' diets cannot be generalized throughout the planet.

In this paper we will highlight the extent to which eating patterns are important for building sustainable agricultural and food systems, a research and policy area that is too often underestimated in the literature. The focus on the sustainability of eating patterns does not mean that reducing hunger is not a priority. Poverty is the principal cause of hunger, and hunger is a major cause of poverty and under-nutrition [3]. The effect of reducing poverty and hunger, however, is already well-documented in the literature (see, for example, [7,8]).

In the first section following the introduction we will present the principal features of eating patterns worldwide. These patterns include the relative convergence of world diets, more rapid food transitions in emerging and in some developing countries, and substantial food waste and losses at distribution and final consumption stages. These common evolutions have negative consequences that will be summarized in a subsequent section. Next, the main drivers of these common eating patterns will be examined in order to make recommendations about knowledge gaps, the filling of which should facilitate the design and implementation of actions and policies aimed at making food systems more sustainable.

Main characteristics of world eating patterns

Eating patterns worldwide present some common characteristics and evolutions that can be summarized as follows: world diets are converging and the food transition process that began in developed countries is now extending to a growing number of emerging and developing countries at an accelerated rate; the food chain is becoming increasingly complex with food products that are more processed, sophisticated and ready-to-eat; food products are increasingly sold in supermarkets and eaten away from home; food losses and waste are substantial; and a widening gap exists between agricultural production and producers on the one hand and food consumption and consumers on the other hand. This, in turn, is inducing a return by a growing body of consumers to local, natural and traditional agricultural products. The consequences of this "westernization' of world diets, as well as its determinants, will be described in subsequent sections.

World diets are converging

The food transition process characterizing the convergence of eating patterns includes two main steps. The first one is quantitative. It corresponds to a rise in the calorie intake with proportionally equal increases in all food products. The nutritional structure of this intake is stable. The second step, called 'diet transition', is qualitative. Once caloric saturation is achieved, diet structure changes; consumption of cereals and vegetables decreases while that of sugar, fats and animal products increases. This two-step process is illustrated in Figure 1 for the French case. As in many western European countries, the quantitative phase started with the agricultural and industrial revolutions and developed through the 19th century until sometime around 1910 to 20. From that date, consumption of cereals and potatoes decreased while consumption of fats, oils, sugar, fruit, vegetables and animal products, which was already slightly increasing during the 19th century, grew faster (Figure 1a). Changes in food diet composition translate into changes in macronutrient weight. From 1880 to 1980 the proportion of carbohydrates decreased from 70 to 45% of the total energy intake, whereas the share of lipids jumped from 16 to 42%; proteins remained constant (Figure 1b).

Developed countries achieved the second step of the food transition process over a century. Many emerging and developing countries in Asia, Central and Latin America, and to a lesser extent in Africa, are now following a similar pattern but at a considerably accelerated rate, as the diet transition is reduced to 20 years in emerging countries and 40 years in developing countries [10]. This is illustrated in Figure 2, which shows that an increasing number of countries are following a similar pattern when one compares the composition of dietary macronutrients among the years 1961 to 1963 (Figure 2a), 2001 to 2003 (Figure 2b) and 2005 to –2007 (Figure 2cc).

In Figure 3 we summarize the food transition process on a two-axis graph illustrating the common trend towards both an increasing calorie intake and a growing share of animal products. As shown in this figure, convergence is not perfect. Diet differences remain between countries for an identical level of economic development, as well as between households within the same country (notably between urban and rural households). These differences are also illustrated in Figure 4 which depicts beef, pork and poultry consumption in several European countries from 1960 to 2003. Although consumption gaps (measured by the ratio of maximal consumption per capita on minimal consumption per capita) are reduced over time, individual meat consumption levels are still significantly different. In the case of beef, consumption levels range from 13 kg per person per year in Germany to more than 26 kg in France; in the case of pork, from 28 kg per person per year in the United Kingdom to 56 kg in Germany and 62 kg in Spain; in the case of poultry, from 15 to 16 kg per person per year in Germany and Italy to 28 to 29 kg in Spain and the United Kingdom. The gaps in consumption levels of different meats are important not only because of the direct relation between total meat consumption and the environment, but also because the animal species have different impacts on the environment, notably in terms of greenhouse gas emissions [11]. A complete analysis of this issue would also require considering the supply and transformation conditions of the different animal productions.

Food products are increasingly processed, sophisticated and ready-to-eat

The ' 'westernization' of food consumption patterns cannot be characterized simply by the rise in calorie intake; by the increase in sugar, fats and animal products; and by the simultaneous decline in cereals, potatoes and pulses. Food products are also increasingly transformed, sophisticated and ready-to-eat. In 1960, 80% of French food expenditure corresponded to transformed agricultural products [12]. Since that date, the share is practically stable (83% in 2006) but the nature of the products has changed considerably. Consumption of fresh refrigerated products is 25 times higher today than in the 1960s. Ready-to-eat products began to develop in the 1990s and today represent a significant share of food expenditure. For example, consumption of ready-to-eat products from fish was five times higher in 2000 than in 1960.

In a growing number of countries, unique traditional food habits and ethnic cuisine are increasingly replaced by prepared and ready-to-eat products, soft drinks, supermarket foods and 'westernized' fast foods. In 1945, milk consumption in the United States was four times as high as that of carbonated soft drinks; 50 years later, Americans were drinking 2.5 times as much carbonated beverage as milk [13]. The frequency of eating food prepared outside the home is also well-documented for the United States, where the share grew from 26% in 1950 to 39% in 1995 [14]. Supermarkets, along with large-scale food manufacturers, have profoundly transformed agri-food markets in developed countries over the 50 years from 1950 to the beginning of the 2000s [15]. Similar changes have taken place in Central and Latin America in only one decade, from 1990 to 2000; in 2000, the share of supermarkets in the retail business in Mexico and South America (60%) was four times higher than in 1990 [16]. Since 2000, the movement has been extending to Eastern and Central Europe and to East and South-East Asia, as well as to many urban areas in Africa.

Food losses and waste at distribution and final consumption stages are substantial in countries where the diet transition has been achieved or is now occurring

At the world level less than half of the calories produced by farmers ever make it onto the dinner table (Figure 5). In the early 2000s, farmers globally produced an average of 4,600 kcal per capita per day, including nearly 600 kcal lost at the time of harvesting or lost immediately after harvest. At this stage, the percentages in losses were strikingly

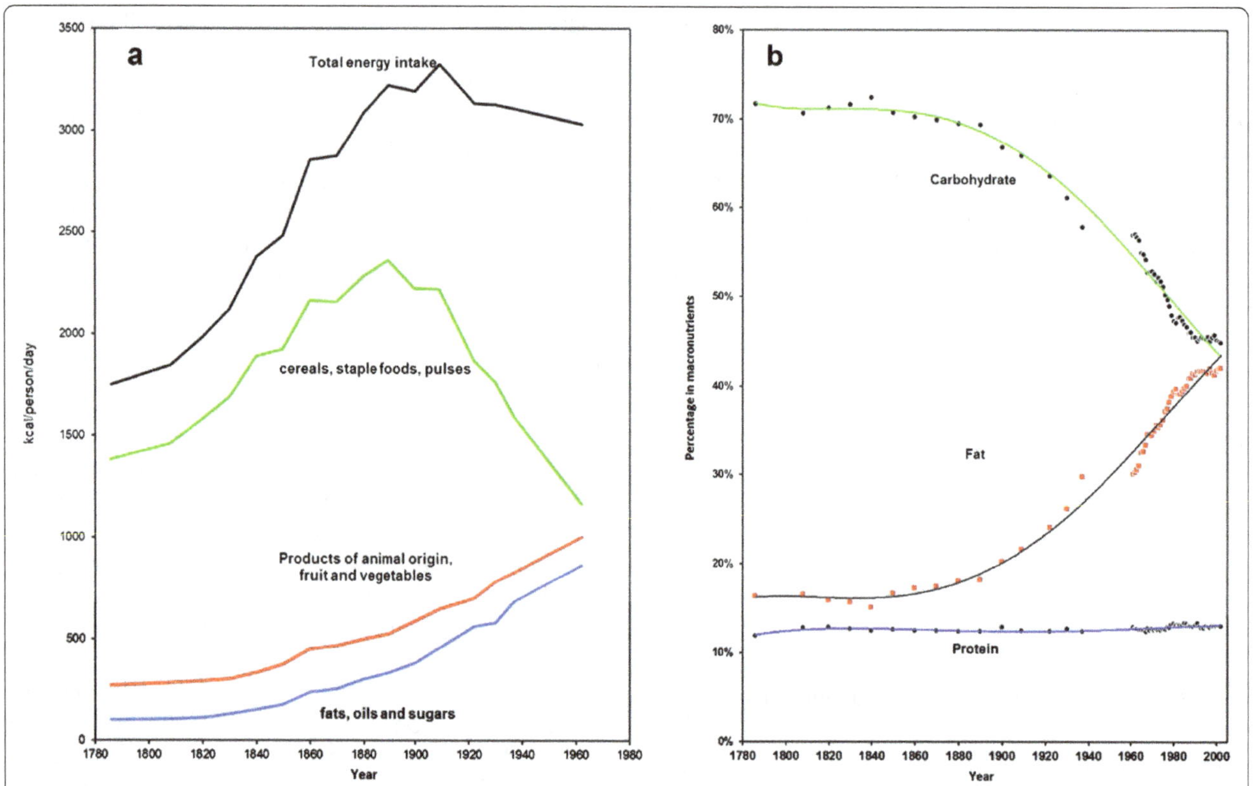

Figure 1 Secular trends in France in (a) the daily calorie intake and in (b) the structure of the diet [9].

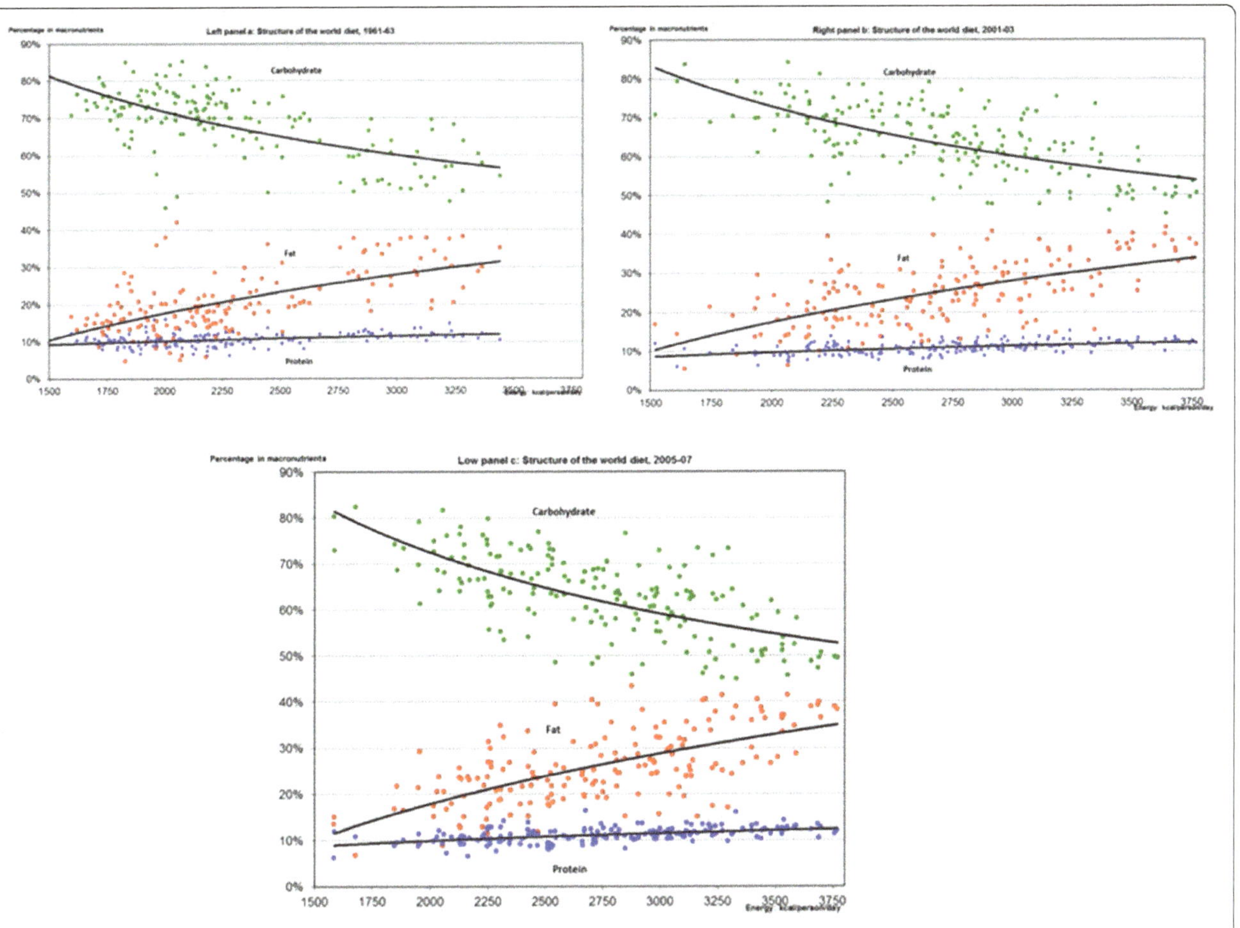

Figure 2 Structure of the world diet: (a) 1961 to 1963 , (b) 2001 to 2003 — and (c) 2005 to 2007. [9] from the FAO Stat database for a and b; elaborated by Combris from the FAO Stat database for c.

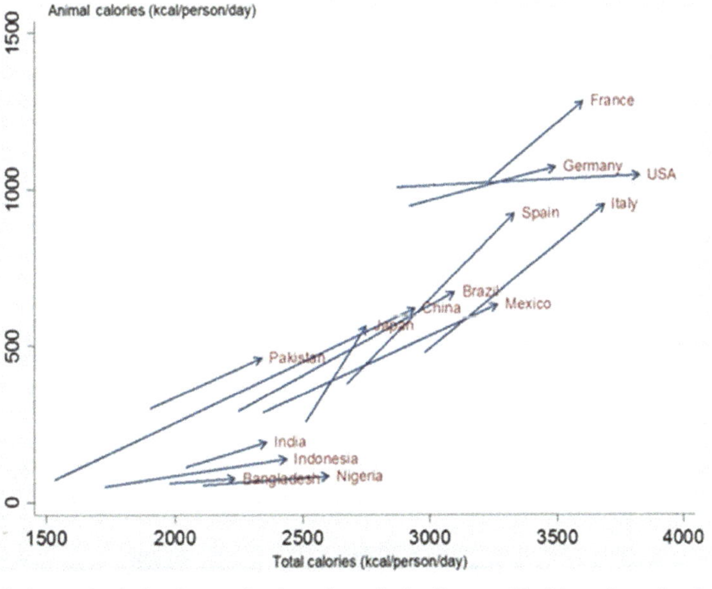

Figure 3 Evolution of total calories and calories from animal products (in kcal/person/day) in various developed, emerging and developing countries, from '1961-63' to '2003-05'. [9] **from the FAO Stat database.**

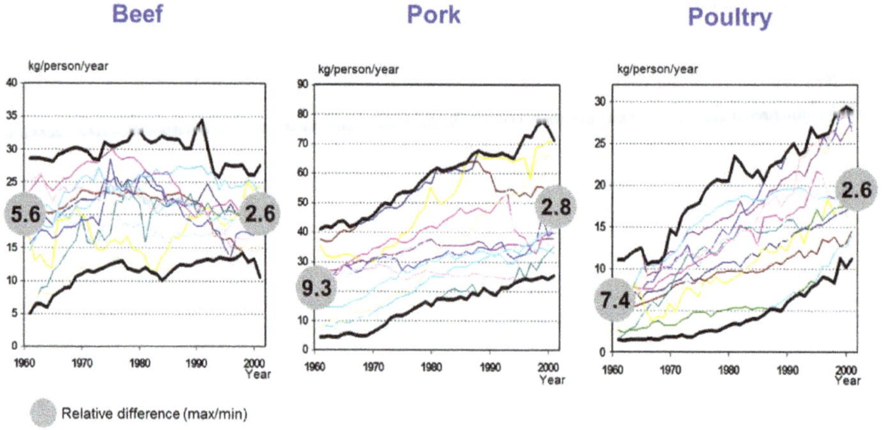

Figure 4 Beef, pork and poultry consumption in the countries of the European Union (UE-15). [9] from the FAO Stat database.

higher in developing countries in comparison to developed nations. The remaining 4,000 kcal were divided between animal feed (1,700 or 43%) and human food (2,300 or 57%). The 1,700 kcal used for animal feed produced in return 500 kcal in the form of eggs, dairy products or meat. Of the 2,800 kcal (2,300 from plant products and 500 from animal products) available for human consumption, another 800 kcal were lost during distribution and final consumption. At that stage, the percentages in losses were much higher in countries where the diet transition has occurred or is now occurring than in the developing world. Finally, on average, of the 4,600 kcal produced from plant products for each inhabitant of the world, only 2,000 ended up for actual human consumption [17,18].

These figures show that reducing losses and waste, from field to plate, is a potentially powerful lever that can be used to increase world food availability and define more sustainable food systems and diets. These losses and waste correspond to both market (pecuniary) and non-market (environmental) costs. It is relatively easy to reduce post-harvesting losses through efficient disease control and the development of infrastructure for storage, transport and marketing. At this level, it is essentially a matter of investment and budget. It is much more difficult to reduce losses and waste at distribution and final consumption levels since this requires profound changes in several strongly interconnected aspects such as individual and collective food consumption patterns, food industry behavior and food public regulations. There is a notable shortage of data available on the amount, quality and causes of food waste and losses at the household level. This point-how to reduce food

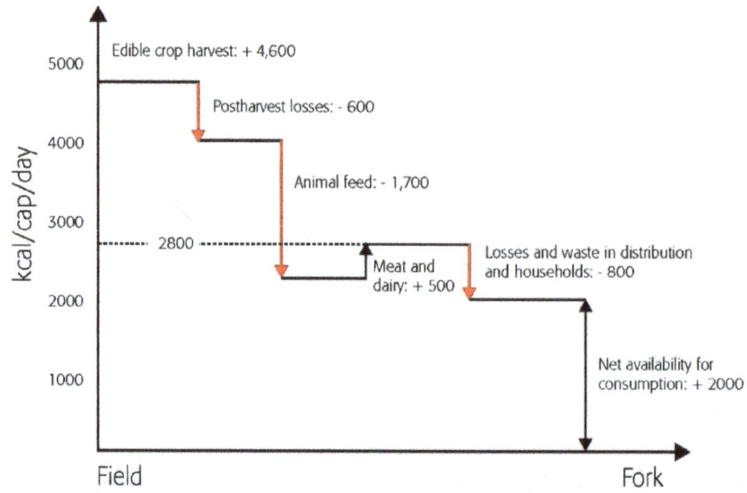

Figure 5 From field to fork: an estimation of food losses, conversion and waste in the world food chain [17,18].

losses and waste along the whole food chain-will be addressed further in a later section.

The gap between agricultural production and producers on the one hand and food consumption and consumers on the other hand, is widening

The 'westernized' food system is also based on reducing the range of primary agricultural commodities and achieving specialization for food manufacturers at two stages of the food chain, namely fractioning and formulating, in order to provide consumers with a growing variety of transformed products of regular quality. Traditional consumer skills required for selecting, preparing and storing food products are transferred to the transformation and retail stages of the food chain [19]. Consumers also have to accept that processors and retailers control food safety by conforming to public regulations and standards [20]. They need to quickly acquire new skills to face a rising range of transformed food products. In particular, consumers need to acquire a better understanding of labeling and marketing information.

This evolution toward transformed or processed food generates an increasing 'distance' between consumers and agricultural products. More generally, the gap between agricultural production, products and producers on the one hand and food consumption, products and consumers on the other, covers the three dimensions of technology, space and time [21]. The length of the food chain is growing and food products are more and more sophisticated (technological gap); there is a growing disconnect between production and consumption zones (space gap); and the delay between agricultural harvest and food consumption is extending (time gap). As this three-dimensional gap widens, it makes the final consumer more demanding and anxious about the content and safety of her food [22]. The most concerned households, often well-educated and from a higher economic class, are increasingly seeking answers to this anxiety through new ways of consumption (organic foods, local food systems and so on) that are aimed at re-establishing a link between what they eat and where and how food is produced. This interest by consumers in returning to basic food items that are local, traditional and/or natural could be exploited to improve sustainability, notably through a reduction in energy and transport costs.

Consequences: increasing risks linked to the "westernization' of food patterns

The 'westernization' of food consumption patterns, as presented earlier in this article, has impacts on the environment and on health. These impacts can be analyzed in terms of risks. Risk management requires action, notably public action. The latter will be efficient if it modifies drivers of undesirable evolutions and changes. These drivers are discussed in the following section.

Impacts on natural resources and the environment

The 'westernization' of agricultural and food systems has negative impacts on natural resources and the environment. These impacts are well-documented in the literature. Simply put, impacts are related to the so-called intensification of agricultural practices and systems (effects at the intensive margin of production) and the increasing use of land for agriculture (effects at the extensive margin of production). Since the 'westernization' of diets largely determines the evolution of agricultural practices and systems, as well as agricultural land uses, eating patterns clearly contribute to the over-use and degradation of agricultural ecosystems.

Much has been written about the negative impacts of the intensification of agriculture (for food and feed crops, as well as for grass) on natural resources and the environment (soil, water, biodiversity and so on). Modern agriculture has been successful in increasing food production over recent decades but this has led to an environmental cost due to the use of high levels of water, fertilizers and pesticides, along with a reduced number of crop species and varieties being selected. This, in turn, has led to the depletion of aquifers, increased emissions of nitrates and pesticides into the atmosphere and to declining biodiversity (see, for example, [23]).

Biodiversity

Modern agriculture has a negative impact on biodiversity through three channels: 1) the management of crops to increase their productivity through breeding, fertilizer use and the control of competitors, predators and parasites with pesticides; 2) the transformation of agricultural landscapes into new combinations and arrangements of crops and semi-natural elements; and 3) the transformation of non-agricultural and agricultural habitats [24]. One of the main pressures on biodiversity is the transformation of natural habitats to agriculture, notably through forest clearance [25]. The negative impact of intensification on biodiversity is very well-documented for birds. For example, the large shifts in agricultural management that occurred in England and Wales between 1962 and 1995 provide a plausible explanation for the decline in the farmland bird population observed over that same period [26]. This decline involves many mechanisms, including a reduced food supply for birds, fewer suitable nesting habitats and direct mortality of birds through farming operations.

Greenhouse gas emissions and climate change

Globally, agriculture, including direct and indirect land-use changes, accounts for one-third of world greenhouse gas emissions (Figure 6). Agricultural greenhouse gas emissions (14%) come from three main sources: 1) nitrous oxide emissions, which are mainly linked to the use of nitrogen fertilizers; 2) methane emissions produced by ruminant animals through the enteric fermentation process or through the anaerobic decomposition of manure; and 3) carbon dioxide emissions from the combustion of fossil fuels. The large rise in the use of nitrogen fertilizers, the number of beef and dairy animals, the number of livestock confinement systems and the agricultural use of fossil fuels have substantially increased agricultural greenhouse gas emissions over the past decades. Land-use change emissions (18%) result from the conversion of carbon-rich habitats such as forests or permanent grasslands to crop plantations and temporary grasslands for food, feed or fuel. These land-use change emissions, more specifically the indirect ones corresponding to the conversion of crops for producing biofuels that would otherwise be used for food or feed, are at the heart of the very complex debate on the sustainability of biofuels. But the question is not limited to biofuels. It also includes livestock and the rising consumption of animal products as a result of the 'westernization' of diets. Expansion of grazing land for livestock is a key factor in deforestation, especially in the Amazon, where it is estimated that 70% of previously forested land is now used as pasture, and feed crops represent the main part of the remainder [27]. Globally, FAO estimates livestock production is responsible for 18% of world greenhouse gas emissions.[b]

The growing demand for animal products

In addition, as discussed previously, animals are less efficient than crops in transforming solar energy into calories. At a global level, of the 1,700 kcal per person per day used for animal feed, animals return only 500, thus having an animal-product-to-feed conversion factor of 0.29. Conversion factors vary widely depending on the category of animals[c] as well as the production practices and systems used to produce the meat. Globally, poultry and pork production appears more efficient than ruminant animal production.

One essential characteristic of the food transition process is the increasing consumption of animal products, notably meat. Since very few countries can secure the corresponding additional protein demand by grazing, the solution lies in intensifying the cultivation of feed crops (mainly corn and soy), and expanding their cultivated area and/or their importation. As a result, decreasing the global amount of animal production and consumption appears to be a potential lever for increasing global food security and the sustainability of the world agricultural and food system. However caveats exist.

First of all, reducing the consumption of meat products is not advisable on health grounds for many households who lack protein in diets. Furthermore, aging people need a specific protein consumption to combat the age-related decrease in protein metabolism efficiency; such proteins are present in animal products. Secondly, animals, especially ruminants, present several

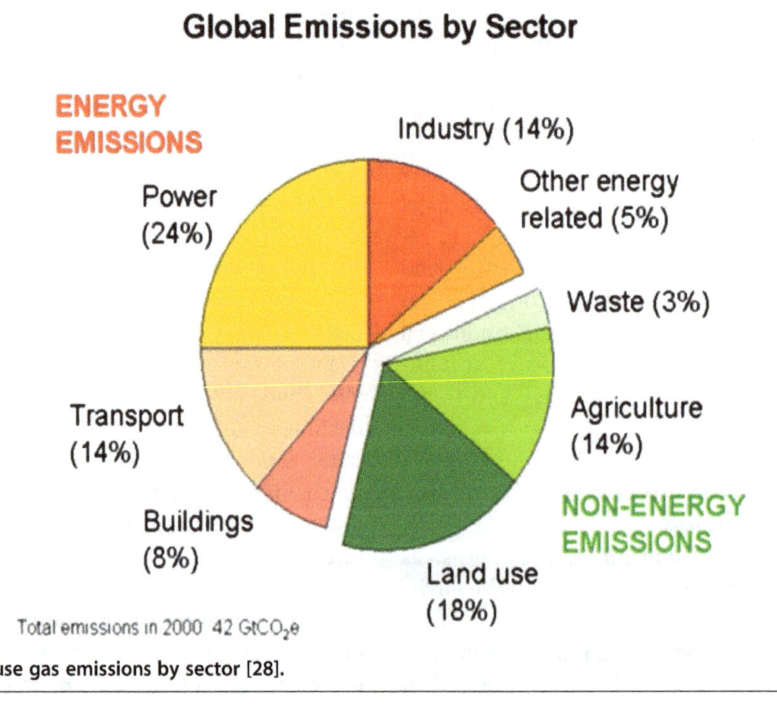

Figure 6 World greenhouse gas emissions by sector [28].

advantages from a production point of view. They can exploit large herbaceous areas that would otherwise be unproductive; they provide organic fertilizers and work as draught animals; and they are an important source of income and assets. Finally, the global food equation must be kept in mind, as well as the roles vegetal and animal production play alongside other uses of land.

Impacts on nutrition, health and well-being

The diet transition summarized earlier in this article can equally be called the nutrition transition, as the changes observed in the nature of the food items consumed result in substantial changes in the percentages of the various macronutrients that compose the energy supply of the diet. The consumption of basic food products (cereals, pulses, starchy foods) declines while that of other food products (sugar, fats, animal products and, to some extent, fruit and vegetables) increases. This translates into a quick increase in the percentage of lipids in the diet-sometimes exceeding 40% in many developed and emerging countries and households-and a sharp decrease in the percentage of complex carbohydrates, starch and fibers. The total percentage of proteins is more stable, masking a switch from plant to animal sources.

This nutrition transition is associated with changes in lifestyle, such as growing urbanization[d], increasing sedentary activities and changing modes of transport, resulting in a switch from high to low physical activity. The conjunction of inadequate situations in terms of nutrition, lifestyle and physical activity has been demonstrated to be associated with increasing risks of weight gain, obesity and diet-related chronic diseases [29,30][e].

An overweight condition or obesity basically results from an imbalance between (increasing) calorie intake and (decreasing) physical expenditure. This is illustrated in Table 1, which shows the simultaneous rise in the calorie intake and the number of overweight and/or obese people in various developed countries between 1980 and 2006. The problem is not specific to high-income countries. Numbers of overweight and obese individuals are dramatically rising in many southern countries in connection to the nutrition transition occurring, especially in urban settings. Emerging countries are the most affected (for example, Brazil, China, and India) but several of the poorest countries of the world are implicated as well, with the exception, at least for the moment, of sub-Saharan African countries. In China, obesity levels among adults are today in excess of 20%; 90 million individuals are obese and 200 million suffer from being overweight [31]. There are now more overweight and obese people in the world than underweight people [10].

Although there are methodological difficulties and constraints when assessing the long-term consequences of eating patterns on health and population levels, it is now well known and proven that these changes in the diet composition have undesirable health effects, by increasing the risk and prevalence of type 2 diabetes, coronary heart disease, cancer, osteoarthritis, work disability and sleep apnea (see, for example, [33]).

Many low- and middle-income countries are increasingly facing a double burden of under-nutrition (deficiency in energy, micronutrients or both) and infectious diseases on the one hand, and over-nutrition, obesity and degenerative diseases on the other [5]. This double burden is observed not only at the national, but also at the community or household levels. Children in low- and middle-income countries are more vulnerable to inadequate prenatal, infant and young-child nutrition. At the same time, they are exposed to high-fat, high-sugar, high-salt, energy-dense and micronutrient-poor foods, which are often less expensive than healthier foods. These dietary patterns, in conjunction with low levels of physical activity, result in sharp increases in childhood obesity while under-nutrition remains unsolved.

Obesity has a cost for society, both directly and indirectly (because of absenteeism and reduced productivity at work). This point is illustrated for full-time employees in the United States (Figure 7). Medical expenditure and indirect costs due to absenteeism and lower productivity at work rise as the numbers of overweight and obese employees increase.

The relationship between the nutritional value of diets and their environmental impacts can be ambiguous

As we have previously explained, nutrition transition has negative impacts on both the environment and on health. According to Stehfest *et al.* [35], a healthier diet would allow world crop areas to be reduced by 10%, and world grassland to be reduced by as much as 10%, with a decrease in mitigation costs of carbon dioxide emissions by 50% (in 2050 relative to a reference scenario following FAO trend assumptions).

But are things really so simple or, in other words, is it so easy to define eating patterns that are preferable from both an environmental and nutritional point of view? A recent study suggests this might not be the case [36,37]. More specifically, the study concludes that nutritionally adequate self-selected diets may engender higher greenhouse gas emissions than nutritionally inadequate diets. This can occur because nutritionally adequate diets with reduced quantities of animal products are replaced by higher quantities of plant-based products to provide the necessary nutrient intake. This shift can generate high greenhouse gas emissions, perhaps even higher than the nutritional equivalent of animal products. For the same reason, replacing (red) meat with fruit and vegetables is not always the best or most efficient option for reducing the greenhouse gas emissions of diets [38]. In these studies, the most effective

Table 1 Obese individuals, overweight individuals, and calorie intake in various developed countries, 1980 and 2005/2006 [32]

Country	Obesity among adults (ratio in percent)		Non obese but overweight adults (ratio in percent)		Obese or overweight children 7–11year old (ratio in percent)	Calorie intake in kcal/day/person (in brackets, percent of fat)	
Year	1980	2006	1980	2006	2000	1980	2005
Australia	8.3	18.7	28.0	34.4	26.2	3,051 (33%)	3,084 (39.1%)
Canada	13.8	23.1	35.4	36.1	25.1	2,946 (37%)	3,552 (37.1%)
France	6.5	11.5	26.9	31.5	19.0	3,376 (39%)	3,603 (40.7%)
Germany	na	13.6	na	36.0	16.0	3,338 (37%)	3,510 (35.9%)
Greece	na	16.4	na	41.3	31.0	3,216 (35%)	3,700 (35.7%)
Italy	7.1	10.2	27.4	35.0	36.0	3,589 (32%)	3,685 (38.4%)
Japan	2.0	3.4	15.6	21.8	17.8	2,720 (23%)	2,743 (28.0%)
Netherlands	5.1	11.3	28.2	35.2	12.0	3,071 (38%)	3,240 (38.2%)
Norway	na	9.0	na	34.0	18.5	3,350 (40%)	3,478 (37.4%)
Great Britain	7,0	24.0	29.0	38.0	20.0	3,159 (39%)	3,421 (36.5%)
USA	15.0	34.3	32.4	33.0	15.2	3,155 (36%)	3,855 (39.4%)

way of associating health and environment benefits appears to be a reduction in food quantities consumed.

Drivers of eating patterns

Food consumption depends on a large and complex set of factors related to food availability, accessibility and choice [13]. In this section, we review the various determinants that shape the nutrition transition process. Some of these determinants are specific to individuals and others depend on their environment. Some are physiological while others are linked to food characteristics, or are rooted in social interactions and cultural backgrounds. Furthermore, food

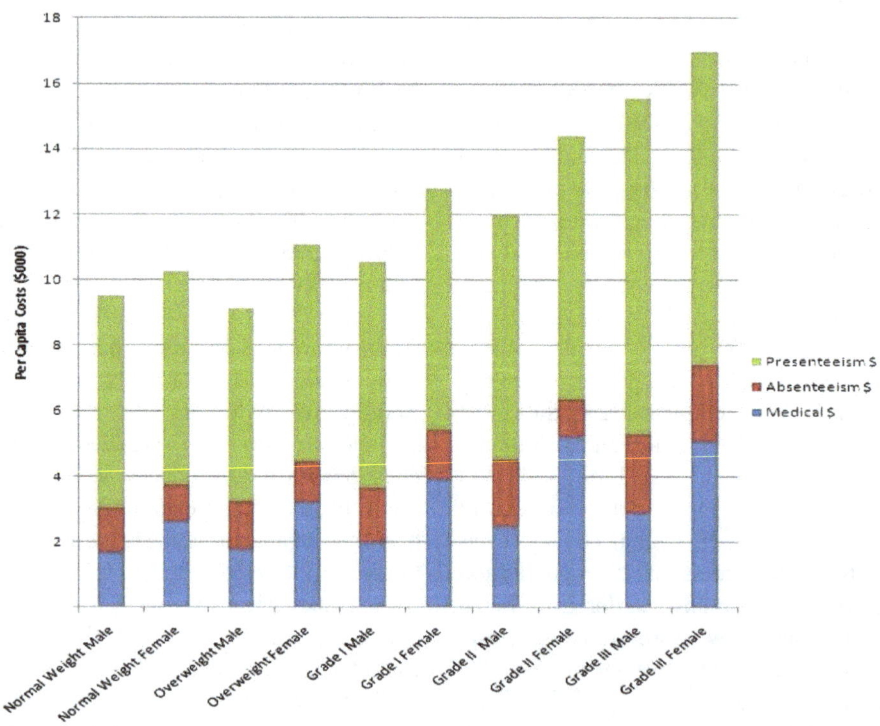

Figure 7 Direct (medical) and indirect costs due to overweight and obesity in the United States [34].

behavior is under the complex influence of a large range of short-, medium- and long-term regulation policies that involve these different drivers. Attention is focused on the underlying mechanisms of these drivers and their relative importance in order to pinpoint levers that could be used to limit deregulation sources and favor sustainable regulation pathways. An overall approach to food and diet patterns is required.

Physiological factors

Nutritional requirements

The primary reason for eating is the necessity for each individual to meet his or her nutritional needs. Energy and specific nutrients are needed for body maintenance, as well as specific functions such as growth, pregnancy or lactation. Except during infancy where maternal milk can cover all nutritional requirements, eating a combination of different food products is necessary to ensure an adequate supply of all needed nutrients. There is an almost infinite variety of food combinations that can define an adequate diet from a nutritional standpoint. This is an important consideration both from a cultural perspective and when considering food-based dietary guideline development.

Whereas biological events accounting for nutritional needs, such as cell and tissue metabolism, take place continuously over time, food intake is discontinuous in time and is organized around a few eating occasions on a daily basis.

Short- and long-term regulation of food consumption and metabolism

Food intake has a circadian variation, with food being ingested during activity periods (daytime for human) and fasting during the remaining time (rest and sleep periods). During the feeding period, food intake is discontinuous for most species. In the case of animals, the lag-time between two intake periods is a major factor of energy intake control. In the specific case of humans, the pace of feeding episodes is influenced by a complex range of social references that define the number and content of food intakes (that is, meals).

Each eating occasion is composed of three components: the pre-ingestive phase where hunger can be experienced is followed by a prandial phase during which food is ingested, leading progressively to a feeling of satiation, and then to a post-prandial phase of variable duration corresponding to satiety.

Overall food intake is controlled by both the quantity of food ingested during eating episodes (involving the satiation process) and the time elapsed since the last eating episode (involving satiety controlled on a short-term basis). It also depends on food availability, which represents an environmental (external) control mechanism.

The main mechanisms controlling food intake are increasingly well-known and understood. These mechanisms include brain structures, mainly located in the hypothalamus [39], and regulation through signals, both short- and long-term [40]. Short-term signals are sensory and digestive and are thus directly linked to food intake. The nutritional composition and consistency of food determine the satiation capacity. The long-term regulation process involves hormonal signals, primarily from leptin and ghrelin [41]. The intensity of these signals is linked to adipose tissue mass with a delayed action as compared to the time of eating.

Food intake control by physiological mechanisms is disrupted by immediate environmental factors

The whole set of regulation mechanisms described above allows self-regulation of energy intake, being particularly effective in young children in 'normal' situations [42]. This is not the case in obese people where this regulatory system seems to be altered [43].

Energy compensation can take place between two successive meals in case of temporary deficiency or excess. But dietary deficiencies are compensated far more easily than excess is managed. In a society of food abundance and choice, temporary overeating is thus more likely to be poorly managed during subsequent meals, leading to excessive weight gain and obesity.

Food intake is adjusted more effectively by eaters who are 'listening' to the physiological signals of hunger or repletion and pay more attention to what they eat. Distractions (such as eating in front of the television or in a noisy place) increase the quantity ingested during the meal and upset the energy compensation process from one meal to the next [44].

Eating triggers a sensation of enjoyment by activating a physiological system in the brain called the reward circuit. Eating enjoyment is accentuated by palatable foods, which are often fatty or sweet, high energy-dense food items. A propensity for sweet foods has been observed from birth. In obese animals and humans, recent findings have shown that addictive-type mechanisms can develop for sweet foods.

Social norms and attitudes, which vary according to age group, personal experience, social and cultural background, can shape and set dietary behavior, particularly those behaviors associated with time schedules, family meals and table manners. These social conventions also affect the physiological regulation mechanisms [45].

Economic factors

Food prices, more or less influenced by policies, notably agricultural policies, play an essential role

The industrial revolution (18th and 19th centuries) and then the agricultural green revolution have allowed world agricultural and food prices to decrease in real (deflated) terms over several decades. Over the past 50 years or more, the supply of most agricultural and food commodities has grown faster than the demand in spite of an increasing population. As a result, the real prices of agricultural and food commodities have exhibited a downward trend. This is illustrated in Figure 8, which depicts United States price indexes for maize, wheat and soybeans from 1924 to 2008, expressed in real terms by deflating with the index of prices paid by farmers [46].

This declining price trend has contributed to the alleviation of hunger in the world in relative terms (from more than 30% in the 1970s to slightly more than 15% today) and also in absolute numbers over the three decades from 1970 to 2000. This does not apply to the years since 2000, especially during the 2008 food crisis and the 2009 financial crisis. This declining price trend has also contributed to easing the nutrition transition and accelerating it in non-developed countries.

Impact of food prices on eating patterns Relative prices of agricultural and food prices can also influence the diet composition by favoring food items that are unhealthy from a nutritional point of view or, on the contrary, disadvantaging them. According to Beydoun *et al.* [47], 'food [relative] prices are associated with dietary quality, fast-food consumption and body mass index among US children and adolescents.'[f] More specifically, this study shows that lower prices of fruit and vegetables are associated with greater fiber consumption and reduced body mass index and higher fast-food prices are associated with lower consumption of fast foods and higher consumption of dietary fibers, dairy products, calcium and fruit and vegetables. As a result, modifying relative prices of unhealthy versus healthy foods appears to be a lever than can be used to improve dietary quality. From that perspective, more and more voices are raising the issue of a potential connection between agricultural policy and overweight and obese populations. Agricultural policy can keep artificially low the relative prices of cereals, oilseeds and sugar crops, as well as the prices of processed food ingredients and animal products that use these crops as feed inputs, and artificially high the relative prices of less dense foods such as fruit and vegetables [48]. For example, Schnoover and Muller [49] argue that in the United States, there is a direct relation between declining real prices for corn and soy, increasing use of fats and oils in processed food products, and the rise in obesity. Similarly, Lloyd-Williams *et al.* [50] conclude that the Common Agricultural Policy of the European Union has led to a significant increase in mortalities attributable to cardiovascular events by favoring the consumption of saturated fats.

Assessing the impact of agricultural policy on eating patterns There is however a strong debate about the health effects of agricultural policy and whether new pricing and/or subsidizing rules in the agricultural sector could help individuals adopt better eating habits. Furthermore, as noted by Mazzochi *et al.* [51], new pricing and/or subsidizing or taxing policies, while discussed regularly, are rarely enforced.

Price support policies and import trade barriers used, for example, in the European Union and the United States have a positive impact on the domestic prices of commodities supported or protected in this way and thus reduce their consumption, *ceteris paribus*. In contrast, coupled production subsidies tend to reduce domestic prices (because they incite farmers to produce more but have no direct impact on agricultural prices) and thus increase the consumption of products

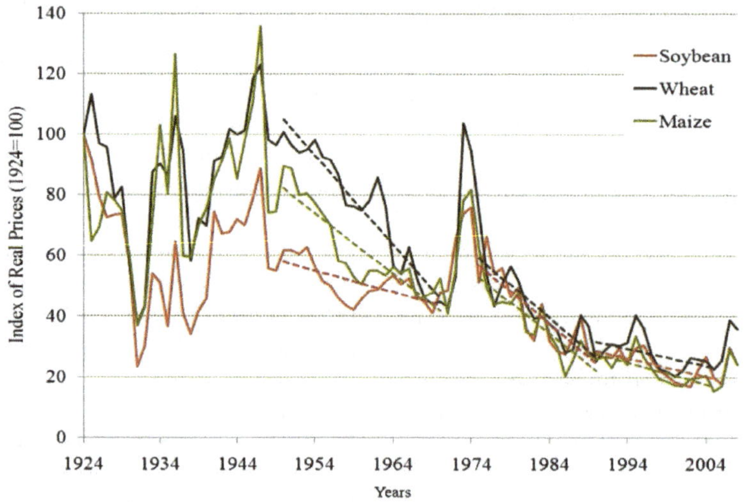

Figure 8 United States: real prices of maize, wheat and soybeans [46].

supported in this way, *ceteris paribus* again. Suppressing the whole set of agricultural policy instruments used in a given country, therefore, may have an ambiguous effect on the caloric content of diets and diet composition, depending on the relative importance of each instrument and how each agricultural product is supported or protected. This is confirmed by the work of Alston *et al.* who show that eliminating all United States agricultural policies would have only a modest impact on annual calorie consumption in that country [52]; more specifically, they claim that eliminating grain subsidies would decrease annual caloric consumption by less than 1,000 calories for each American adult whereas eliminating both grain subsidies and trade barriers would increase this annual caloric consumption by around 200 calories for each American adult.

However, this does not mean that the lever of agricultural policy is neither relevant nor efficient for improving diets. The problem is not with production subsidies, trade barriers or consumption taxes *per se*. It is rather with the disconnect between the current forms of farm programs as they are applied in a vast majority of countries and the long-term policies that should be set up to promote healthier diets and more generally, the sustainable development of agricultural and food systems, from field to fork. In other words, the problem is with the inconsistency of current (agricultural) policy, which very often neglects spillover effects. This can be illustrated by the work of Bonnet and Réquillart who analyze the impact of the European sugar reform on the consumption of sugar sweetener beverages in the European Union [53]. The reform mainly consists of a significant decrease in the domestic price of sugar. It is aimed at gaining consistency (between policy instruments used in the sugar sector and in other agricultural sectors) and compatibility with World Trade Organization rules. But in doing so, the reform is at odds with what is recommended by nutritionists and public health authorities. Specifically, the reform would raise the French consumption of regular soft drinks by more than one liter per person per year and the consumption of added sugar by 124 g per person per year. Furthermore, the substitution game between brands would favor products with the highest sugar content and the rise in consumption levels would be more significant in overweight and obese households.

Household income is a key determinant of eating patterns

It is now well established that rising household incomes lead to more unhealthy diets by favoring an increased consumption of fats, oils, sugar, animal-based products and processed foods [54].

Of course, this direct causality should not invite us to blame an increase of individual incomes. As noted by Kearney [13] quoting Marmot [55], 'in most industrial countries (for example, the United States or the United Kingdom), the effects of increased income have generally been considered as beneficial, resulting in better quality diets, better healthcare, lower morbidity and mortality from infectious diseases and lower risk of obesity.' In developed countries where the nutritional transition is achieved and in emerging or developing countries where it is occurring, the conventional wisdom is that the poorest families have the worst diets from a nutritional standpoint. This is mainly because lower-income people have less access in their immediate neighborhood to affordable healthy foods like fruit and vegetables that are relatively more expensive than high-calorie foods. Relative to richer and more-educated households that can largely choose their lifestyle, poorer and less-educated families have fewer food choices. This shows that even if income is a key parameter, it cannot and should not be analyzed independently of the whole set of food consumption drivers. These include food prices but also include many individual variables such as socio-economic status, culture or religion, as well as environmental or global determinants such as urbanization, globalization and marketing.

This means that a systemic approach to food consumption behaviors is required; otherwise one risks coming to erroneous conclusions and making inefficient policy recommendations. This point can be illustrated by noting that children from British families with an annual income of between £22,000 and £33,000 are 10% more likely to be overweight or obese than those from families with an income of less than £11,000 [56]. This is mainly because of the rise of highly paid working mothers in the first group of families who are then forced to leave a nanny, child minder or someone else in charge of their children's diet and physical exercise. The authors of this study conclude that 'long hours of maternal employment, rather than lack of money, may impede young children's access to healthy foods and physical activity. Policies supporting work-life balance may help parents to reduce potential barriers.'

Other drivers: urbanization, food retailing and industry, marketing and advertising, food environment, sociology and culture

Urbanization Historically, urbanization is strongly related to industrial and agricultural revolutions. A decline in physical activity results in lower energy needs, while at the same time working ability, life span and life quality increase. Today the challenge consists of feeding an increasingly urbanized world population and a growing number of very large megacities (especially in emerging and developing countries) in a sustainable way, while improving the situation of farming and rural households that still represent the bulk of poor and hungry people. The more a society is urbanized, the more it challenges the sustainability of its agricultural and food system, mainly in terms of transport, supply chain and modified diets.

The influence of urbanization on eating patterns is partly linked to the distance between consumers and the various supply sources (shops, restaurants, dispensers, *et cetera*) [57]. In developed countries, the numbers of supply sources, their relative importance, and their repartition have undergone important changes in the last 50 years. In the United States, for example, the increasing concentration of fast-food restaurants accounts for the frequency of overweight and obese individuals [58]. Evolutions are today similar and faster in a growing number of emerging and developing countries.

Food retailing and industry

In most developed countries, the food supply has gone from specialized shops to larger and non-specialized supermarkets and shopping malls. In a country like France it took 40 years for this change to occur. Although supermarkets and shopping malls were only 5% of the total food expenditure in the 1970s, they now represent 70%. A similar evolution is occurring in many emerging and developing countries where uniform and centralized supermarkets and shopping malls are increasingly and quickly replacing small shops. The movement starts from basic foods like corn, oil or sugar, and extends to fresh ones. In emerging and developing countries, supermarkets and shopping malls are no longer the domain of the richer households of capital cities. They are now consumption places for urban crowds with increasingly 'westernized' lifestyles [16].

The worldwide rise in supermarkets and shopping malls comes with the development of transnational food corporations (franchises and manufacturers) like Danone, Nestlé or McDonalds. These transnational food corporations are significant drivers of the nutritional transition and the 'westernization' of food consumption behaviors [59]. The 'fast-food cultural infiltration' means that fast-food restaurants can now very easily be found in shopping malls, railway stations, airports, and schools. In highly populated communities, you can drive just a few kilometers and pass several fast-food restaurants and convenience stores. The increase in numbers of these stores comes with the development and concentration of brand names that have changed the relationship between consumers and producers and led to changes in profit sharing to the detriment of agricultural producers.

Declining physical activity and growing food marketing

Many activities now widely undertaken by adults and children involve very little physical activity. In a growing number of cultures, the most popular leisure activities, especially for children and teenagers, are watching television or videos, playing electronic or computer games, and surfing the internet. These passive forms of entertainment are replacing more traditional and more physical recreational activities. This ever-increasing sedentary lifestyle is accompanied by growing marketing and advertising expenditures from both transnational food corporations and supermarkets. These are essential factors for the 'westernization' of diets. According to Willett [60], exposure to TV advertising might even be the single largest determinant of child obesity in the United States.

Food environment, sociology and culture

People make decisions based on their food environment, community, family, culture, and history. All these miscellaneous factors influence food consumption behaviors and habits in a way that may or may not be favorable to more healthy diets. It is, for example, particularly important to create food environments that make it easier for individuals and families to choose healthy diets and engage in physical activity. This is more likely to happen when one is provided with food options that are low in calories, fats and added sugar, and that are rich in fruit and vegetables. In France, the convergence of eating patterns meets resistance from familial and cultural habits. In 1997, four out of five French people were having lunch at home on weekdays, a percentage increasing to 90% for dinner [61]. Recent observations suggest that the first percentage is slightly decreasing while the second remains constant. Today 75% of French people have lunch at home but with considerable variation according to age, location and socio-economic status.

A common wisdom is that traditional meals are losing ground in France. Such an assertion is quite difficult to prove and feeds a controversy around 'the breakdown of the French meal pattern'. Sociology research shows that many factors interact around the French meal pattern, such as deregulation of family life, more individualized lifestyles and more independent teenagers [62]. Such evolutions could foster more individualized food intakes as illustrated by meal-trays used in front of the television, in sitting rooms or even with guests.

Despite the global convergence of eating patterns, several types of behaviors co-exist inside one pattern. Nutritional implications of such disparities are expressed through social differences in terms of health. Many diet-related diseases affect lower social categories more frequently and social inequalities in health are growing [63-66].

To reduce those inequalities, the study of main consumption trends should take account of various balances depending on life standards. Thus, in France, the relative shares of animal and vegetal foodstuffs are known to vary according to income, education level and socio-economic status. For example, the consumption of animal foodstuffs is lower in upper social categories [67,68]. Similarly, nutritional messages are known to have different impacts depending on education level and social background of the person receiving the message [69].

Achieving greater sustainability of food systems through healthier eating patterns: knowledge gaps and action recommendations

As highlighted by the foresight exercise Agrimonde [6], defining healthier and environmentally friendly diets would require changing behaviors along the whole supply chain, from consumption habits to agricultural practices and systems. Apart from the adjustment of quantities consumed in relation to a better nutritional balance of diets, reducing losses and waste along the whole food chain and saving fossil and natural resources are major targets.

Reducing food losses and waste along the whole food chain

The FAO estimates that achieving global quantitative food security by 2050 would require an increase in world agricultural production by as much as 70% [3]. As established in the recent United Kingdom foresight exercise on food and farming [70], halving losses and waste along the food chain represents an economy of 25% of current agricultural production. Other estimates are even more optimistic. Harvest and post-harvest losses (primary losses) are much more important in the developing world, while losses at distribution and final consumption stages (secondary losses) are greater in developed countries, especially those countries that have experienced or are experiencing the food transition.

Factors for secondary food losses are multiple and involve a large set of complex mechanisms. These mechanisms depend on exogenous factors (for example, climate and safety) as well as human factors, including management practices. These factors concern the different actors and the different stages in the food chain, that is producers and market regulations applied at this level (for example, withdrawals of fruit and vegetables because of quality standards), the food industry (in relation to issues such as supply chain management, packaging, safety rules and the recycling of wasted quantities), the retailing stage (where storage conditions and marketing are key parameters), and the final consumer (through the mechanisms which define her food consumption behavior). Secondary food losses could be reduced by increasing the use of products in human food, by using them for animal feed, and by recycling, especially for energy production.

Improving the utilization of raw materials

Priority should be given to the use of agricultural products for food. Extending and diversifying preservative processes can be used to delay the consumption of highly perishable products such as fruit and vegetables. Recycling technologies of co-products are efficient and well-developed. For example, the content of milk proteins can be fully utilized though traditional outlets like cheese and thanks to membrane technologies, the residual whey proteins can be simultaneously used as food additives in a very large range of products. It is not only a technological issue. In parallel, public policies should be developed for facilitating a reduction in losses and improving the food use of raw materials. Education of all actors in the food chain, from producers to final consumers, is a key parameter for the long term. It could be efficiently supplemented by shorter-term regulation on prices.

A large part of world crop production is used for animal feed. Farm and company animals also use an extensive variety of co-products from first- and second-transformation of agricultural products, notably oilcakes (from oilseeds), corn gluten feed and corn germ meal (from corn). Using organic waste for feeding animals is also a promising solution that would allow a significant decline in the tension on arable lands.

Developing new processes

As the carbon circular economy develops, bio-refineries play a central role since they use plants and organic waste from both industrial and domestic processes [71]. Bio-refineries go beyond processes currently used in the food industry, as they include renewable energy and green chemistry to optimize whole plant use. Based on improvements in production and recycling technologies, they are promising tools to reduce food losses and waste through a reorganization and management of resource use and recycling, and an increased utilization of co-products. Perspectives offered by white and green biotechnologies are enlarging the range of utilization and valorization possibilities thanks to the degradation of lignocellulose and the development of new sources of oils, solvents or biomaterials.

Understanding and changing domestic food practices

According to Redlingshöfer and Soyeux [72], food losses and waste at home represent 42% of total secondary losses and waste in the European Union (89 million tons per year, i.e., 179 kg per person); the food industry is second (39%, but includes by-products that are generally well-used), catering is third (14%), and retailing is fourth (5%). As a consequence, reducing food losses and waste at home is a potentially powerful lever. To achieve this objective and more generally to improve food management at home, a number of simple actions are available. For example, on her famous blog "No More Dirty Looks", Siobhan O'Connor presents ten ways to reduce the food that American households throw out, and what to do with it if it really is past its prime.[g] Solutions include 'store food properly', 'make juice', 'cook then freeze' but also 'find places that will make use of your waste' or 'shop like a Parisian'.[h] More generally, this list of intuitive recommendations shows that changing individual food practices is difficult as far as the food environment is concerned (for example, not everyone can shop like a Parisian!).

The use of food at home as well as the sociological representation of what food means may be very different from one country to another, even for those at the same level of economic development. It is also by changing the food environment, retailing and food industry practices, especially in terms of advertising, labeling, marketing or packaging, and influencing public policies and education that it will be possible to significantly reduce food losses and waste at home, and along the whole food chain. Efficiency requires combining a large variety of actions that involve all actors in the food chain.

Defining and adopting healthier and more sustainable diets

'Westernized' diets are unhealthy because of an excessive calorie intake and/or because they are unbalanced from a nutritional standpoint. Defining healthy and sustainable diets is a challenge. Defining the conditions allowing people to adopt them is even more difficult. As in the case of food losses and waste, there is no unique solution that could apply to all individuals and households in all countries. Combined actions that work simultaneously on demand and supply factors will be more efficient.

Assessing the nutritional and environmental impacts of diets simultaneously

Assessing the nutritional and environmental impacts of different diets is imperative. To that end, a research effort is required to develop methods aimed at assessing the nutritional value of food products and their integration in eating patterns. Related to that, research at a sufficiently detailed level is needed on the nutritional qualities of plant and animal products; for example, determining how to restrict the consumption of animal products without adversely affecting health. Research should also target methods for assessing the environmental impacts of diets. To a large extent, research work developed in that domain concerns carbon dioxide and greenhouse gas emissions. Such research should be extended to other environmental dimensions, notably by including the impact of diets on fossil energy, water and biodiversity. Furthermore, the definition of the carbon footprint of diets should take into account greenhouse gas emissions that are linked to direct and indirect land-use changes. Research efforts currently developed in that domain are mainly applied to biofuels; they could easily be extended to diets as well.

Such a research agenda implies a substantial effort in terms of methods and data which should be coordinated and harmonized at an international level. Existing databases should be shared and extended, notably by incorporating information from the private sector on processed food products.

Reducing the quantities

As previously noted, the challenge of reducing the intake quantity of over-nourished people is essential. It implies not only a modification of consumer habits but also a modification of the economic model of the supply chain because the main driver for supply chain development is currently an increase in quantities sold. Research is needed to develop new economic models of the supply chain, to assess the sustainability performances of these models, and to work out how they could be implemented.

Better understanding of food consumption behaviors and their determinants to favor desirable changes

There is no single diet that can be considered as nutritionally adequate. Several diets can achieve the same nutritional requirements in terms of energy and specific nutrients. Even if detailed information is largely lacking for the moment, these various diets, while adequate from a nutritional point of view, will very likely have differing impacts on the environment. But it is also likely that there is no unique diet that is preferable from both a nutritional and environmental standpoint when all the environmental dimensions are taken into account.

Combined actions on demand and supply factors

Changing food consumption behaviors is a very difficult task as eating encompasses a large range of dimensions including historical, cultural, sociological and even emotional aspects. Changing food consumption behaviors requires simultaneously mobilizing the various levers identified earlier in this article through a combination of actions affecting both the demand and supply side. An increasing number of success stories in different countries confirm that it is through a combination of actions that it will be possible to efficiently and durably change food consumption behaviors. These actions should comprise both short-term objectives and long-term targets.

Enhancing food intake control Research in progress will allow better identification of the food properties such as nutrient content, texture, and sensorial properties that have an impact on satiety/satiation, an important biological event that controls food intake. Nevertheless, it is necessary to establish to what extent these are key effects for the control of body weight and composition. Furthermore, given the importance of the food consumption environment, it is crucial to act simultaneously on environmental factors to induce sustainable food-related behavioral changes. Research and experiments should be conducted to identify the best ways to act on these proximate environmental drivers. Public policy could then have an impact on public catering environments, especially on those disturbance factors such as noise and short eating times. Consistent information also should be provided concerning disturbance factors associated with the home, such as how watching television while eating affects consumption patterns.

Acting on (relative) prices on foods Modifying the relative prices of the different food products by taxing the 'bad' foods and subsidizing the 'good' ones is part of the solution, even if the distinction between what is a 'bad' *versus* a 'good' food is not simple, from both a nutritional and environmental point of view. It is very likely however that public policies that have either a direct or indirect impact on eating patterns, and more generally on food and agricultural systems, are insufficiently consistent. Thus, improving the consistency of public policies is a priority. This could be achieved by following two basic economic policy principles: first, by seeking a consensus (or at least a politically acceptable compromise) on policy objectives, taking into account priorities and trade-offs, and second, by targeting the objective at the source using a distinct instrument for each goal (following the so-called targeting principle). Furthermore, modifying the relative prices of the various foods would also influence supply by reducing the availability of more expensive goods (their prices should go up) and increasing the availability of less expensive goods (their prices should decrease).

Taking into account the specificities of individuals, social groups and countries The efficiency of economic instruments will be enhanced if these instruments are used in conjunction with other actions aimed at defining a better food environment and/or by modifying behavioral habits. To do this, policies must target individuals, their families and their social communities. As there are critical periods for changing eating patterns during the life span, such as childhood, marriage, child birth or aging, efficiency will be gained by targeting these periods and taking into account the characteristics and needs of the corresponding populations. More generally, as there is a range of consumer types in each country and consumption behavior varies during a consumer's life cycle, the research and policy agenda should address detailed knowledge of food consumption behaviors of individuals in relation to their position in the life cycle, their socio-economic status and the food environment. This research agenda includes the importance of job-related physical activity in explaining weight gaps; gender differences in food consumption, behavior and weight; the role of the matching market, such as marriage or a birth, in determining behavior and weight; and, on a more global scale, the analysis of factors that contribute to maintaining differences in food consumption behavior between countries while other factors are maintained or controlled [73].

Acting on demand factors Action requires a better understanding of the multi-factorial relationship of consumers *vis-à-vis* their food, how this relationship evolves, and drivers of changes. Action should aim to inform consumers on the risks of following unhealthy diets, of being overweight and of being obese. Any action taken should try to influence food consumption behaviors through nutritional education, sensorial education (especially for children and teenagers), social marketing (because this technique can modify not only individual behaviors but also social norms[i]) as well as through the use of understandable food labeling for nutritional and environmental purposes. Efficiency will be enhanced if all these actions are combined and short-term measures are consistently included in a long-term action plan.

Acting on supply factors In a large majority of cases, acting solely on the demand side is not sufficient. It is equally important to intervene on the supply side, more specifically in the food industry, the retailing sector and the consumption environment. In that regard, many analysts suggest that the free market is not optimal (see, for example, [74-76]). Policy intervention is thus required and justified from a public economic point of view. Instruments that public authorities can use for that end are multiple, from simple interdictions to operations developed jointly with the food industry and the retailing actors. These should be encouraged as their efficiency is enhanced by the adhesion of actors to program objectives. The research and policy agenda should also address the issue of food marketing and its regulation (see, for example, [75]), and the question of the 'architecture' of food choices, that is the decrease in the availability and visibility of unhealthy foods and conversely, the increase in the availability and visibility of healthier options. This availability/visibility issue concerns shopping malls, fast-food restaurants, supermarkets, and all buying and consumption collective places such as schools, canteens and work places. Several success stories in that domain are well-documented (see, for example, [77-79]).

Saving fossil energy and natural resources

Reducing losses and waste along the food chain should permit the conservation of fossil energy and other natural resources. In general, increasing the sustainability of eating patterns and food systems requires a decrease in the use of fossil energy and natural resources, notably of water and land. Although the research needs and policy recommendations illustrated herein apply to energy, they extend to other natural resources such as water and land as well.

A recent report from the United States Department of Agriculture [80] summarizes the issue in the specific case of the United States, but the analysis extends to all countries with 'westernized' eating patterns and food systems. According to the authors, 'energy is used throughout the United States food supply chain, from the manufacture and application of agricultural inputs, such as fertilizers and irrigation, through crop and livestock production, processing, and packaging; distribution services, such as shipping and cold storage; the running of refrigeration, preparation, and disposal equipment in food retailing and foodservice establishments; and in home kitchens. Dependence on energy throughout the food chain raises concerns about the impact of high or volatile energy prices on the price of food,[j] as well as about domestic food security and the Nation's reliance on imported energy. Use of energy in the food chain could also have environmental impacts, such as through carbon dioxide emissions.'

In the United Kingdom, the food system represents 13% of total energy consumption. The details of the energy consumed in the food system are provided for all activities from field to fork (Figure 9). Transportation and logistics consume the highest amount of energy (30%), followed by processing (17%) and then catering (16%). Figure 9 also highlights the small contribution of agriculture to energy consumption, with only 5%, while food cooking at home represents 9%. Similar information is provided for the United States in Figure 10. In that country, food activities at home come first with 31% and agriculture second with 22%, a percentage much higher than in the United Kingdom. Differences in data, system boundaries, measurement methods and category definition (for example, one study includes all 'home activities' while the other considers only 'food cooking at home') can explain in part this difference between the situation in the United Kingdom and the situation in the United States. Furthermore, these percentages do not include the energy content of agricultural imports and exports that vary considerably between the two countries and can also explain some differences. More generally, this simple comparison suggests that the research agenda should address data and methodology needs to ensure that such breakdowns are robust and comparable. It should also address the detailed analysis of differences between countries and factors that could explain these discrepancies, especially with the aim of identifying and generalizing the best practices. Here also, a holistic approach is required.

Energy use in food transforming activities is of the same order of magnitude in both the United Kingdom (17%) and the United States (16%). Following the energy crisis of the 1970s, the food industry has developed and adopted new technologies to reduce its dependence on energy. Significant progress in the thermal processing of foods (multi-stage unit operations, energy recycling, notably by using co-products of the first-transformation process, replacement of non-thermal technologies) has been achieved. Less research effort has been devoted to home food activities and the energy yields of culinary cooking techniques are far from being optimized. As a result, a priority of the research agenda is the question of home food management, including an analysis of losses, waste and energy consumption in culinary processes. Transport and energy use for that purpose are also of importance, notably because of the increasing distance between production and consumption places, as well as the concentration of food retailing.

Conclusions

A change in eating patterns is a key condition for ensuring world food security. According to the Agrimonde foresight study [6], if daily food consumption per person keeps steadily increasing, it will reach 3,600 kcal in 2050 and the world will have to produce no less than 62 G kcal for feeding nine billion people at that date, but if daily food consumption per person remains limited to 3,000 kcal in 2050, only 42 G kcal (32% less) will need to be produced.

Acting on eating patterns, alone, however, is not sufficient because the food challenge requires actions on both the supply and demand side. On the agricultural production side, in addition to reducing post-harvest losses, it will be necessary to increase crop yields, especially in world regions where they are currently low, and it is necessary to do so in a sustainable way. In other words, agricultural practices and systems used worldwide should radically change [6,70]. On the food consumption side, a first priority is to reduce waste and losses at distribution and final consumption stages by improving the utilization of raw materials for food and feed, by developing new transformation processes based on the concept of bio-refineries, and by improving our understanding of food consumption behaviors in order to induce desirable changes and define adapted public regulations. It will be also necessary to change current diets that are excessive and/or

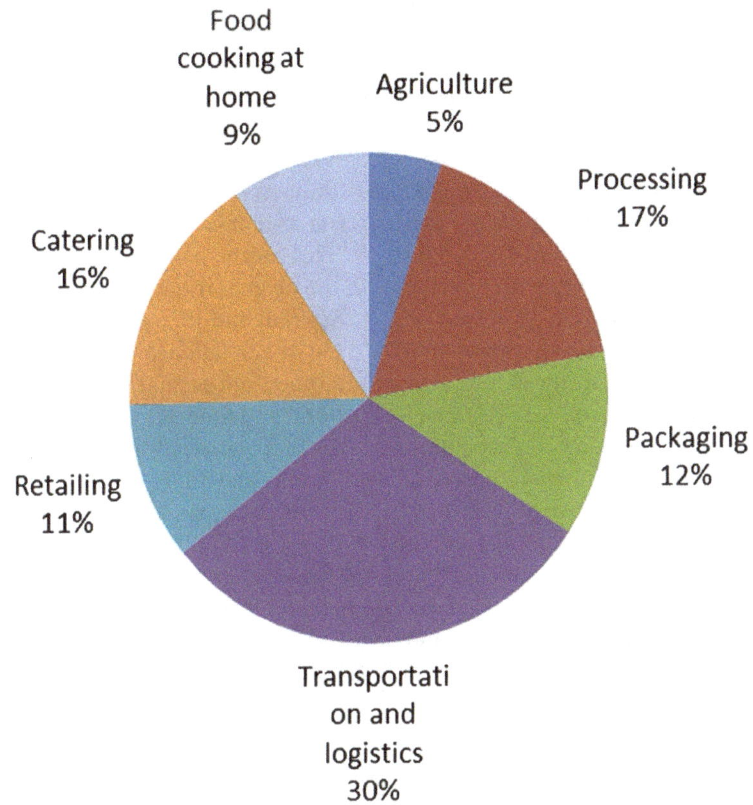

Figure 9 Breakdown of energy consumption of the United Kingdom food chain according to activities (2002) [81].

unbalanced and more generally, to reverse the global trend towards the 'westernization' of eating patterns worldwide as the latter presents numerous shortcomings and drawbacks, notably in terms of health, impacts on the environment and use of natural resources.

This does not mean setting up a common eating pattern for every part of mankind. Although eating patterns have tended to converge over the past decades, they remain diverse throughout the world and are determined by a complex set of physiological, economic, historical,

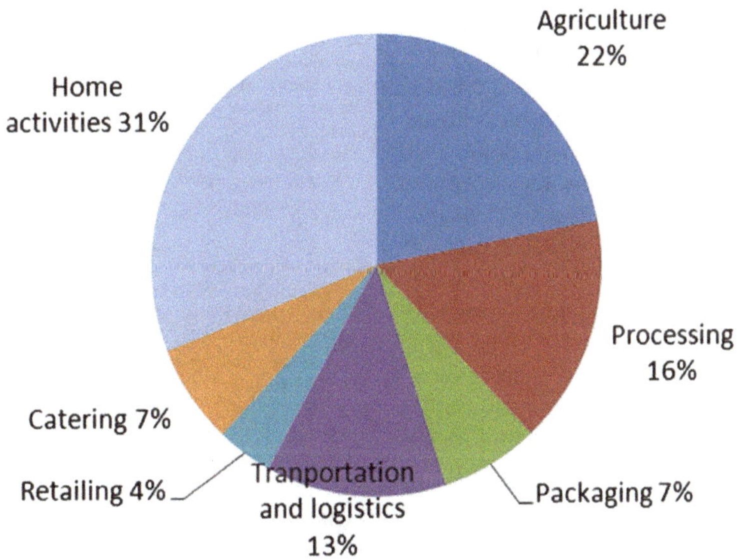

Figure 10 Breakdown of energy consumption of the United States food chain according to activities (2002) [82].

cultural and sociological factors. This diversity should be exploited further to define healthier and more sustainable diets.

To that end, a holistic approach is required. Reducing losses and waste along the whole food chain should result in significant savings of fossil energy and other natural resources, and those savings are a key condition to increasing the sustainability of both eating patterns and food systems. More generally, eating patterns and food systems could be sustainable only if they rely on a thrifty use of all non-renewable natural resources. It would be necessary to assess the nutritional and environmental impacts of alternative diets simultaneously in order to set the targets and identify the trade-offs. Research in this area should focus on data needs and the definition of robust methodologies. Far from being imposed, healthier and more sustainable diets must be based on changes accepted by everyone. This requires a better understanding of food consumption behaviors and of their various determinants in order to favor desirable changes through actions and regulations on buying and consumption behaviors, the food environments, the relative prices of foodstuffs as well as the behavior of the food and retailing industry. These actions and regulations will have to take into consideration the specificities of individuals, social groups and countries.[k]

Many actions need to be taken to ensure food security throughout the world. And many of those actions have something to do with eating patterns. In other words, future food security cannot be achieved unless food patterns have been identified and modified. Here is food... for thought!

Endnotes

[a]The Millennium Development Goals were developed out of the eight chapters of the Millennium Declaration signed in September 2000. There are eight goals with twenty-one targets and a series of indicators for each target: eradicate extreme poverty and hunger (with target 1A: halve the proportion of people who suffer from hunger); achieve universal primary education; promote gender equality and empower women; reduce child mortality rate; improve maternal health; combat HIV/AIDS, malaria and other diseases; ensure environmental sustainability; and develop a global partnership for development. For more details, see [83].

[b]This estimate sums up emissions of the livestock commodity chain at the various production stages. The production stages include deforestation for pasture and feed crops, animal production, and the processing and transportation of animal products.

[c]For example, in the United States, 2.6 kg of feed are needed to produce 1 kg of chicken, but 7 kg of feed are needed to produce 1 kg of beef. [84].

[d]See [85] on the consequences of urbanization for food and farming.

[e][29] analyzes the nutrition transition and its health consequences in the specific case of India.

[f]Title of their article published in 2010 in the *Journal of Nutrition*.

[g]http://www.good.is/post/10-ways-to-reduce-food-waste-at-home/.

[h]This means having only the basics at home and going to the neighborhood grocery store before supper.

[i]See [86] and [87].

[j]On the relationship between crude oil and agricultural/food prices, and the impact of this relation on agricultural and food price volatility, see [88].

[k]In that perspective, we recommend the special issue of Public Health Nutrition [89] on the Mediterranean diet, which is presented and analyzed as an example of a sustainable diet in which nutrition, biodiversity, local food production, culture and sustainability are strongly interconnected.

Competing interests
The authors declare that they have no competing interests.

Authors' contributions
HG led the study, reviewed background literature, structured the concepts, developed the arguments and wrote most of the manuscript. BDV, CE, MM, MR and MG reviewed background literature, developed important concepts, helped design the arguments and wrote parts of the manuscript. All authors read and approved the final manuscript.

Authors' information
HG is director of research and scientific director in charge of agriculture at INRA with expertise in economics, agricultural market modeling and agricultural policy analysis as well as in research management. BDV is director of research and deputy head of the 'nutrition, chemical food safety and consumer behavior' research division at INRA with expertise in nutrition and research management. CE is deputy scientific director in charge of food and nutrition at INRA with expertise in nutrition, sustainable food systems and research management. MM is professor and president of the INRA research centre of Toulouse, with experience in food process engineering and research management. MR is a scientific project manager at INRA with an academic background in food history and food science, a doctorate in the historical archaeology of these topics, and experience in research management. MG is CEO of INRA with expertise in physico-chemistry of bio-transformations and great experience in public policy and research management.

Acknowledgements
This review was funded by INRA and conducted in support of the Commission on Climate Change, Sustainable Agriculture and Food Security which was convened by the CGIAR Research Program on Climate Change, Agriculture and Food Security (http://ccafs.cgiar.org/commission). We thank all the participants of the Bellagio workshop (5–9 September 2011), especially Christine Negra and Sonja Vermeulen, where an earlier draft of this paper was discussed. We also thank Antoine Momot, John Kearney and two anonymous referees for very helpful comments.

Author details
[1]INRA Paris, 147 rue de l'université, 75 338, Paris Cedex 07 France. [2]INRA Toulouse, 24 Chemin de Borde Rouge, CS 52627, 31326 Castanet Tolosan, Cedex, France.

References

1. FAO: *Rome Declaration on World Food Security and World Food Summit Plan of Action*. Rome: World Food Summit; 1996.
2. Nations U: *Report of the World Food Conference: 1974 November 5–16; New York*. Rome: United Nations; 1974.
3. FAO: *The State of Food Insecurity in the World*. Rome: Addressing Food Insecurity in Protracted Crises; 2010.
4. *Childhood and Maternal Undernutrition*. http://www.who.int/publications/cra/chapters/volume1/part2/en/index.html.
5. WHO: *Obesity and Overweight. Fact Sheet N° 311*. Geneva 2011.
6. Paillard S, Treyer S, Dorin B: *Agrimonde, Scenarios and Challenges for Feeding the World in 2050*. Paris: Quae; 2010.
7. FAO, IFAD, WFP: *Reducing Poverty and Hunger: the Critical Role of Financing for Food, Agriculture and Rural Development*. Rome: Paper prepared for the International Conference on Financing for Development. Monterrey, Mexico; 2002. http://www.ifad.org/events/monterrey/e/jointe.pdf.
8. IFAD: *IFAD Strategic Framework 2011–2015*. Rome: IFAD; 2011. http://www.ifad.org/sf/strategic_e.pdf.
9. Combris P: **Le poids des contraintes économiques dans les choix alimentaires.** *Cahiers de Nut et de Diététique* 2006, **41**:279–284.
10. Popkin BM: **Global nutrition dynamics: the world is shifting rapidly toward a diet linked with non-communicable diseases.** *Am J Clin Nutr* 2006, **84**:289–298.
11. Combris P, Maire B, Réquillart V: **Consommation et consommateurs.** In *DuALIne -durabilité de l'alimentation face à de nouveaux enjeux: Questions à la recherche*. Edited by Esnouf C, Russel M, Bricas N. Paris: Quae; 2011:26–43.
12. Etiévant P, Bellisle F, Dallongeville J, Etilé F, Guichard E, Padilla M, Romon-Rousseaux M: *Les comportements alimentaires. Quels en sont les déterminants ? Quelles actions, pour quels effets ?* Paris: INRA, Expertise Scientifique Collective; 2010.
13. Kearney J: **Food consumption trends and drivers.** *Phil Trans Royal Soc B* 2010, **365**:2793–2807.
14. Swinburn BA, Caterson I, Seidell JC, James WPT: **Diet, nutrition and the prevention of excess weight gain and obesity.** *Public Health Nutr* 2004, **7**(1A):123–146.
15. Reardon T, Swinnen J: **Agrifood sector liberalisation and the rise of supermarkets in former state-controlled economies: a comparative overview.** *Dev Pol Rev* 2004, **22**:515–523.
16. Reardon T, Timmer CP, Barrett CB, Berdegué J: **The rise in supermarkets in Africa, Asia and Latin America.** *Am J Agric Econ* 2003, **85**:1140–1146.
17. Smil V: *Feeding the World. A Challenge for the Twenty-First Century*. Cambridge: MIT Press; 2000.
18. Lundqvist J, de Fraiture C, Molden D: *Saving water: from field to fork -curbing losses and wastage in the food chain. SIWI Policy Brief*. Stockholm: SIWI; 2008.
19. Dubuisson-Quellier S: **Consumer involvement in fair trade and local food systems: delegation and empowerment regimes.** *GeoJournal* 2008, **73**:55–65.
20. Dubuisson-Quellier S: **From consumerism to the empowerment of consumers: the case of consumer oriented movements in France.** *Sustainability* 2010, **2**:1849–1868.
21. Colleu S, Guyomard H, Soler LG: *Les consommateurs à la recherche d'une ré-identification et d'une réappropriation de leur alimentation*. Paris: INRA; 2010.
22. Soler LG, Réquillart V, Trystram G: **Organisation industrielle et durabilité.** In *DuALIne -durabilité de l'alimentation face à de nouveaux enjeux: Questions à la recherche*. Edited by Esnouf C, Russel M, Bricas N. Paris: Quae; 2011:85–95.
23. European Commission: *Sustainable Food Consumption and Production in a Resource-Constrained World. Third SCAR (Standing Committee on Agricultural Research)*. Brussels: European Commission; 2011.
24. Firbank L, Petit S, Smart S, Blain A, Fuller R: **Assessing the impacts of agricultural intensification on biodiversity.** *Phil Trans Royal Soc* 2008, **363**:777–787.
25. Jenkins M: **Science.** *Prospects for biodivers* 2003, **302**:1175–1177.
26. Chamberlain DE, Fuller RJ, Bunce RGH, Duckworth JC, Shrubb M: **Changes in the abundance of farmland birds in relation to the timing of agricultural intensification in England and Wales.** *J Appl Ecol* 2000, **37**:771–788.
27. FAO: *Livestock's Long Shadow: Environmental Issues and Options*. Rome: FAO; 1996.
28. IPCC: *Fourth Assessment Report: Climate Change*. 2007. http://www.ipcc.ch/publications_and_data/publications_and_data_reports.shtml.
29. Shetty P: **Nutrition transition in India.** *Public Health Nutr* 2002, **5**:175–182.
30. Technical Report Series WHO: *Diet, Nutrition and the Prevention of Chronic Diseases*. Geneva: WHO; 2003:2003.
31. Popkin BM: **Will china's nutrition transition overwhelm its health care system and slow economic growth?** *Health Aff* 2008, **27**:1064–1076.
32. Etilé F: **Food consumption and health.** In *Oxford Handbook of the Economics of Food and Agricultural Policy*. Edited by Lusk J, Roosen J, Shogren J. Oxford, UK: Oxford University Press; 2011:317–346.
33. Visscher TLS, Seidell JC: **The public health impact of obesity.** *Annu Rev Public Health* 2001, **22**:355–375.
34. Finkelstein EA, Strombotne KL, Popkin BM: **The costs of obesity and the implications for policymakers.** *Choices 3rd quarter* 2010, **25(3)**.
35. Stehfest E, Bouwman L, Vuuren DPV, Elzen MGJD, Eichout B, Kabat P: **Climate benefits of changing diet.** *Clim Chang* 2009, **95**:83–102.
36. Darmon N, Soler LG: **Impact carbone et qualité nutritionnelle des choix alimentaires: une étude de cas.** In *DuALIne -durabilité de l'alimentation face à de nouveaux enjeux: Questions à la recherche*. Edited by Esnouf C, Russel M, Bricas N. Paris: Quae; 2011:44–58.
37. Vieux F, Soler LG, Supkova M, Russel M, Redlingshöfer B, Darmon N: *Is there a relation between the nutritional adequacy of diets and their greenhouse gas emissions? A study based on self-selected French diets*. Paris: INRA; 2011.
38. Vieux F, Darmon N, Touazi D, Soler LG: *Food consumption and greenhouse gas emissions: changing food consumption patterns or consuming less?* Paris: INRA; 2011.
39. Woods SC, D'Alessio DA: **Central control of body weight and appetite.** *J Clin Endocrinol Metab* 2008, **Suppl 1**:37–50.
40. Berthoud HR: **Interactions between the « cognitive » and « metabolic » brain in the control of food intake.** *Physiol Behav* 2007, **91**:486–498.
41. Cowley MA, Smith RG, Diano S, Tschop M, Pronchuk N, Grove KL, Strasburger CJ, Bidlingmaier M, Esterman M, Heiman ML, Garcia-Segura LM, Nillni EA, Mendez P, Low MJ, Sotonyi P, Friedman JM, Liu H, Pinto S, Colmers WF, Cone RD, Horvath TL: **The distribution and mechanism of action of ghrelin in the CNS demonstrates a novel hypothalamic circuit regulating energy homeostasis.** *Neuron* 2003, **37**:649–661.
42. Birch LL, Deysher M: **Caloric compensation and sensory specific satiety: evidence for self-regulation of food intake by young children.** *Appetite* 1986, **7**:323–331.
43. Blundell JE, Lawton CL, Hill AJ: **Mechanisms of appetite control and their abnormalities in obese patients.** *Horm Res* 1993, **Suppl 3**:72–76.
44. Bellisle F, Dalix AM, Airinei G, Herberg S, Peneau S: **Influence of dietary restraint and environmental factors on meal size in normal-weight women: a laboratory study.** *Appetite* 2009, **53**:309–13.
45. Rogers PJ: **Eating habits and appetite control: a psychobiological perspective.** *Proc Nutr Soc* 1999, **58**:59–67.
46. Alston JM, Pardey PG: **Theme overview: agricultural productivity and global food Security in the long run.** *Choices 4th quarter* 2009, **24(4)**. http://www.choicesmagazine.org/magazine/article.php?article=90.
47. Beydoun MA, Powell LM, Chen X, Wang Y: **Food prices are associated with dietary quality, fast food consumption, and body mass index among U.S. children and adolescents.** *J Nutr* 2010, doi:10.3945/jn.110.132613.
48. Schaffer H, Hunt DB, Ray DE: *US agricultural commodity policy and its relationships to obesity, Background Paper developed for the Wingspread Conference on Childhood Obesity, Healthy Eating and Agriculture Policy, Racine*. Wisconsin: United States; 2007.
49. Schnoover H, Muller M: *Food without thought: how US farm policy contributes to obesity*. Minneapolis: Institute for Agriculture and Trade Policy; 2006.
50. Lloyd-Williams F, O'Flaherty M, Mwatsama M, Birt C, Ireland R, Capewell S: **Estimating the cardiovascular mortality burden attributable to the Common Agricultural Policy on dietary satured fats.** *Bull World Health Org* 2008, **86**:535–542.
51. Mazzochi M, Traill WB, Shogren JF: *Fat Economics: Nutrition*. Oxford: Oxford University Press; 2009.
52. Alston JM, Rickard BJ, Okrent AM: **Farm policy and obesity in the United States.** *Choices 3rd quarter* 2010, **25(3)**. http://www.choicesmagazine.org/magazine/pdf/block_51.pdf.
53. Bonnet C, Réquillart V: **Does the EU sugar policy reform increase added sugar consumption? An empirical evidence on the soft drink market.** *Health Econ* 2011, **20**:1012–1024.
54. Périssé J, Sizaret F, François P: **Effet du revenu sur la structure de la ration**

alimentaire. *FAO, Bull Nutr* 1969, **7(3)**:1–10.

55. Marmot M: **Income inequality, social environment, and inequalities in health.** *J Policy Anal Manage* 2001, **20**:56–159.
56. Hawkins SS, Cole T, Law C: **Maternal employment and early childhood overweight: findings from the UK millennium cohort study.** *Int J Obes* 2008, **32**:30–38.
57. Popkin BM, Duffey K, Gordon-Larsen P: **Environmental influences on food choice, physical activity and energy balance.** *Physiol Behav* 2005, **86**: 603–613.
58. Inagami S, Cohen DA, Brown A, Asch M: **Body mass index, neighbourhood fast food and restaurant concentration, and car ownership.** *J Urban Health* 2009, **86**:683–695.
59. Hawkes C: **The role of foreign direct investment in the nutrition transition.** *Public Health Nutr* 2005, **8**:357–365.
60. Willett W: *Eat, drink and be healthy: the Harvard medical school guide to healthy eating.* New York: Free Press; 2002.
61. Volatier JL: **Le repas traditionnel se porte encore bien.** *Consommation & Modes de Vie* 1999, **132**:1–4.
62. Chauffaut D: *L'évolution des relations entre générations dans un contexte de mutation du cycle de vie.* Paris: CREDOC; 2001:163.
63. Guignon N, Collet M, Gonzales L, de Saint Pol T, Guthmann JP, Fonteneau L: **La santé des enfants en grande section de maternelle en 2005–2006.** *Etudes & Résultats* 2010, **737**:1–8.
64. Mackenbach JP, Stirbu I, Roskam AJR, Schaap MM, Menvielle G, Leinsalu M, Kunst AE: **Socioeconomic inequalities in health in 22 European countries.** *N Engl J Med* 2008, **358**:2468–2481.
65. Charles MA, Eschwege E, Basdevant A: **Monitoring the obesity epidemic in France: the Obepi surveys 1997–2006.** *Obesity* 2008, **16**:2182–2186.
66. Leclerc A, Chastang JF, Menvielle G, Luce D: **Socioeconomic inequalities in premature mortality in France: have they widened in recent decades?** *Soc Sci Med* 2006, **62**:2035–2045.
67. Caillavet F, Lecogne C, Nichèle V: ***La consommation alimentaire: des inégalités persistantes mais qui se réduisent.*** *In Cinquante Ans de Consommation en France.* Paris: INSEE; 2009:49–62.
68. Recours F, Hébel P: *Les populations modestes ont-elles une alimentation déséquilibrée?* Paris: CREDOC; 2006.
69. Regnier F: **Obésité, goûts et consommation: intégration des normes d'alimentation et appartenance sociale.** *Rev Fr Soc* 2009, **50**:747–773.
70. The Government Office for Science, London: *Foresight. The Future of Food and Farming (2011).* London: Final Project Report; 2011.
71. Colonna P, Fournier S, Touzard JM: **Systèmes alimentaires.** In *DuALIne -durabilité de l'alimentation face à de nouveaux enjeux: Questions à la recherche.* Edited by Esnouf C, Russel M, Bricas N. Paris: Quae; 2011:59–84.
72. Redlingshöfer B, Soyeux A: **Pertes et gaspillages: les connaître et les reconnaître pour les réduire et les valoriser.** In *DuALIne -durabilité de l'alimentation face à de nouveaux enjeux: Questions à la recherche.* Edited by Esnouf C, Russel M, Bricas N. Paris: Quae; 2011:112–129.
73. Philipson T: **The world-wide growth in obesity: an economic research agenda.** *Health Econ* 2001, **10**:1–7.
74. Hawkes C: **Regulating and litigating in the public interest: regulating food marketing to young people worldwide -trends and policy drivers.** *Am J Public Health* 2007, **97**:1962–1973.
75. Harris JL, Pomeranz JL, Lobstein T, Brownell KD: **A crisis in the marketplace: how food marketing contributes to childhood obesity and what can be done.** *Annu Rev Public Health* 2009, **30**:211–225.
76. Stanley F, Daube M: **Should industry care for children? Public health advocacy and law in Australia.** *Public Health* 2009, **123**:283–286.
77. Seymour JD, Yaroch AL, Serdula M, Blanck HM, Khan LK: **Impact of nutrition environmental interventions on point-of-purchase behavior in adults: a review.** *Prev Med* 2004, **Suppl 2**:108–136.
78. Knai C, Pomerleau J, Lock K, McKnee M: **Getting children to eat more fruit and vegetables: a systematic review.** *Prev Med* 2006, **42**:85–95.
79. Lawrence S, Peterken R, Burns C: **Fresh kids: the efficacy of a health promoting schools approach to increasing consumption of fruit and water in Australia.** *Health Promot Int* 2007, **22**:218–226.
80. Canning P, Charles A, Huang S, Polenske KR, Waters A: *Energy use in the US food system.* Washington DC: USDA-ERS; 2010.
81. Smith A, Watkiss P, Tweddle G, McKinnon A, Browne M, Hunt A, Treleven C, Nash C, Cross S: *The validity of food miles as an indicator of sustainable development: Food Miles Final Report.* London: DEFRA; 2005.
82. Heller MC, Keoleian GA: ***Life cycle-based sustainability indicators for assessment of the US food system,*** **Report CSS00-94 of the Center for Sustainable Systems.** *Ann Arbor* 2000, .
83. We Can End Poverty: *Millennium Development Goals. A Gateway to the UN System's Work on the MDGs.* 2015. http://www.un.org/millenniumgoals/bkgd.shtml
84. Leibtag E: **Corn prices near record high, but what about food costs?** *Amber Waves* 2008, **6**:10–15.
85. Satterthwaite D, McGranahan G, Tacoli C: **Urbanization and its implications for food and farming.** *Phil Trans Royal Soc B-Biological Sciences* 2010, **365**:2809–2820.
86. Kotler P, Armstrong G: *Principles of Marketing.* New Jersey: Prentice Hall International; 2005.
87. Henley N, Raffin S: **EPODE (Ensemble, Prévenons l'Obésité des Enfants or Together, Let's Prevent Childhood Obesity') case study: preventing childhood obesity.** *J Mark Manage*, . in press.
88. Hajkowicz S, Negra C, Barnett P, Clark M, Harch B, Keating B: **Food price volatility and hunger alleviation - can Cannes work?** *Agric Food Secur* 2012, **1**:8. doi:10.1186/2048-7010-1-8.
89. Selected Conference Proceedings of the VIIIth Barcelona: **International Congress on the Mediterranean Diet, Barcelona, Spain, 24–25 March 2010.** *Public Health Nutr* 2011, **14**(Suppl 12A):2271–2349.

6

Pollination and biological control research: are we neglecting two billion smallholders

Peter R Steward[1*], Gorm Shackelford[1], Luísa G Carvalheiro[1,2], Tim G Benton[1], Lucas A Garibaldi[3] and Steven M Sait[1]

Abstract

Food insecurity is a major world problem, with ca. 870 million people in the world being chronically undernourished. Most of these people live in tropical, developing regions and rely on smallholder farming for food security. Solving the problem of food insecurity is thought to depend, in part, on managing ecosystem services, such as the pollination of crops and the biological control of crop pests, to enhance or maintain food production. Our knowledge regarding regulating ecosystem services in smallholder-farmed (or dualistic) landscapes is limited and whilst pollination has been the focus of considerable research, the provision of natural enemy services, important for every crop worldwide, has been relatively neglected. In order to assess whether ecosystem-service research adequately represents smallholder-farmed landscapes, whilst also considering climatic region and national economic status, we examined the constituent studies of recent quantitative reviews relevant to biological control and pollination. No regulating ecosystem service meta-analysis, to our knowledge, has focussed on smallholder agriculture despite its importance to billions of peoples' local food security. We found that whilst smallholdings contributed 16% of global farmland area and 83% of the global agricultural population (estimated using Food and Agriculture Organisation's (FAO's) World Census of Agriculture 2000) only 22 of 190 studies (12%) overall, came from smallholder-farmed landscapes. These smallholder studies mostly concerned coffee production (16 studies). Individual reviews of biological control were significantly and strongly biased towards data from large-scale farming in temperate regions. In contrast, pollination reviews included more smallholder studies and were more balanced for climate regions. The high diversity of smallholder-farmed landscapes implies that more research will be needed to understand them compared to large-scale landscapes, but we found far more research from the latter. We highlight that these skews in research effort have implications for sustainable intensification and the food security of billions in the developing world. In particular, we urge for balance in future ecosystem-services research and synthesis by greater consideration of a diverse range of smallholder-farmed landscapes in Africa and continental Asia.

Keywords: Biological control, Ecosystem service, Sustainable intensification, Smallholder, Pollination, Food security

Background

Global food insecurity is receiving increasing attention from researchers and policy makers [1-5]. An increasing human population and rising demand for more varied, high-quality diets is placing pressure on agro-ecosystems and biodiversity across the globe [1,2,4]. To prevent widespread food insecurity arising from the expected increase in human population size, predictions suggest that agricultural land will need to increase crop production by 60% or more by 2050 [6]. There are around 870 million hungry people today, nearly all of whom (98%) live in developing countries and half of them are from smallholder households [7,8]. As we currently produce enough calories to feed the world [7,9] yet still have hunger, producing additional food in food secure areas will not solve global food insecurity alone [5]. Furthermore, the long-term sustainability of the industrial intensification of agriculture (high inputs, low crop and landscape diversity) has been questioned [10,11]. Where industrial agriculture exists it is often associated with soil degradation and even desertification in arid regions [2], thus, in the future, we may suffer declining production in some currently productive areas. On-going industrial intensification of agriculture, typically in developing nations, is linked to negative social and economic impacts, including poverty and loss of local

* Correspondence: bsprs@leeds.ac.uk
[1]Africa College and School of Biology, University of Leeds, Leeds LS2 9JT, UK

food security [12,13]. Sustainable intensification, the pursuit of higher or more sustainable yields with fewer negative consequences for the environment [1], is the conceptual solution to reducing global food insecurity and meeting future food demands, and it includes the management of ecosystem services (ecological intensification) [1,2,5]. In this review, we examine how much ecosystem service research is derived from developing nations, tropical climates and smallholder farming landscapes, where local food security is at stake.

Smallholder-farmed landscapes

Smallholder farming is a major source of food production and income in many countries [14,15] and for the global rural population in general. Global estimates suggest there are 2.1 to 2.5 billion people involved in farming smallholdings and 500 million smallholdings, mostly in developing nations [16,17] (Additional file 1: Table S1). The majority of the population growth forecast for 2050 will occur in sub-Saharan Africa and South Asia [18] (Additional file 2: Figure S1), where food insecurity is currently rife and more than a third of agricultural land is composed of smallholdings [7] (Additional file 1: Table S1). Thus, ensuring that sustainable intensification benefits the world's undernourished population living in developing countries, particularly those that live in smallholder households, would promote global food security and poverty reduction [19].

There is no single definition of a smallholding, but the common understanding is that the unit of land management is small. Several reports arbitrarily use a definition of two hectares or less [14,15,20] but larger holdings (for example, three to five hectares) will still create very complex landscapes compared to large-scale farming. Smallholder-farmed landscapes are, therefore, greatly sub-divided and potentially have a high diversity of crops at relatively small spatial scales (Figure 1A). The higher potential for small-scale intermixing of crop and non-crop habitats (due to landscape configuration not composition) means that in a smallholder-farmed landscape the average distance of a crop-plant to a "natural" area that could enhance ecosystem service delivery can be much shorter than in a large-

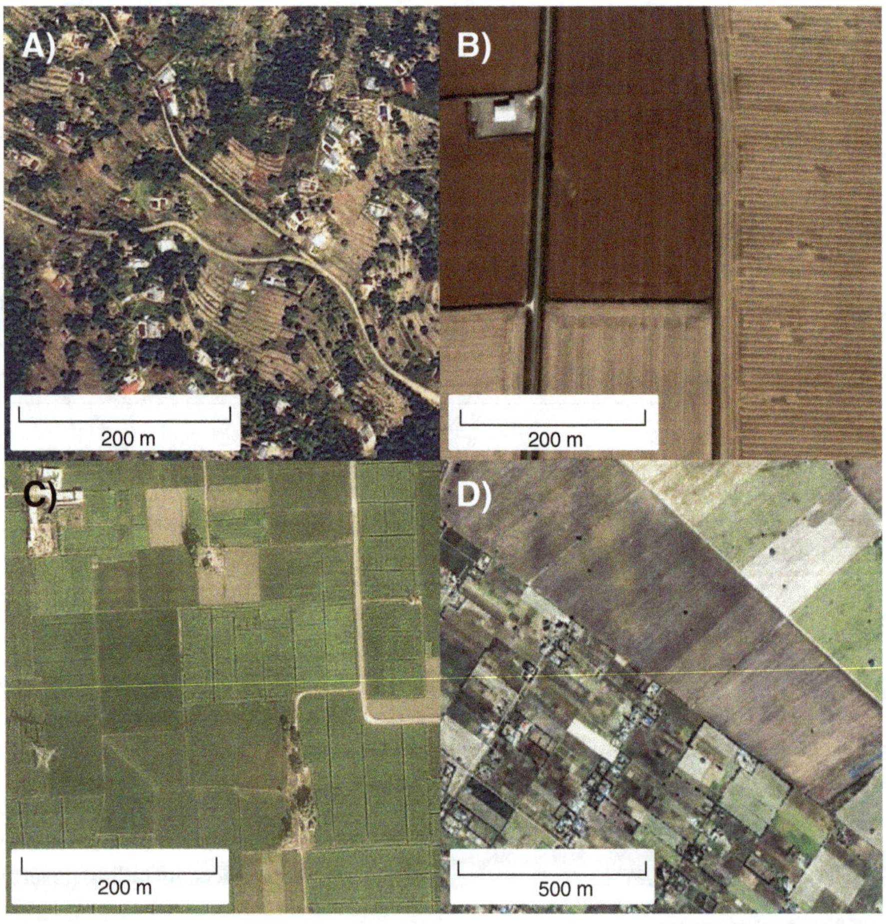

Figure 1 Examples of farming landscapes. A) Locally complex upland smallholdings, Taita, Kenya; **B)** Large-scale commercial farming, Norfolk, UK; **C)** Locally simple lowland smallholdings, Punjab, India; **D)** Dualistic farming with smallholdings and large-scale commercial farming, Nakuru, Kenya.

scale farming landscape. As smallholder-farmed landscapes can be highly heterogeneous within and between landscapes (Figure 1), there are a variety of such landscapes that could be described as smallholder. For example, they can be commercial or subsistence, polycultural or monocultural, and with high or low input (Additional file 3: Figure S2). Smallholdings can merge together to form an extensive area of contiguous agriculture or they can be isolated patches surrounded by other land-uses or natural habitats.

Ecological intensification and regulating ecosystem services of pollination and biological control

The landscape provides a range of natural resources that are valuable to people. In an analogy with economics, natural "capital" (such as soils and forests) provides ecosystem services (ES) as flows (or interest) of value. These include food, forage, fibre and fuel in the case of forests, and nutrients, water and carbon storage in the case of soils. ES are broadly separable into four categories: provisioning services (food, fuel, water), regulating services (carbon storage regulates climate, plant cover regulates flood risk), supporting services (soil microbes support nutrient cycles and aid food production) and cultural services (for example, recreational, spiritual and educational values of a landscape). Pollination of crops and pest control, by natural enemies of crop pests, are regulating ecosystem services that contribute to food production. The notion of enhancing ES to increase crop yield (rather than using pesticides, fertilisers or other intensive agricultural practices) is known as "ecological intensification" [21]. It is well-established that insufficient pollination and biological control services can limit crop production when other factors such as soil nutrients and water are sufficient [21]. Evidence is now emerging to support the theory that pollination and biological control respond to similar drivers, such as floral resources and landscape structure [22-24], and it is therefore logical to consider them together.

Recent quantitative reviews [for example, 25–28] have investigated factors that affect pollination and biological control services or providing species, such as the influence of landscape complexity and management practices, and have shown that variability in ES provision is likely to be context dependent [25-28]. For example, the diversity of both pollinators and natural enemies seems to be higher in complex agro-ecosystems, and pollination services are generally stronger and more stable on farms near natural habitats [28,29]. The results of quantitative reviews have a key role to play in synthesising the evidence and parameterising models that can predict ES provisioning (such as InVEST, see [30]), and therefore they contribute to the design of sustainable farming landscapes and the policy interventions that bring them about. The extent to which the body of the work synthesised in such reviews can suggest management interventions across a range of agricultural systems clearly depends on the range and representativeness of the studies included. Geographical biases have recently been demonstrated in reviews relevant to pollination ecology suggesting our understanding of pollination is poor in developing regions such as sub-Saharan Africa [31]. We build on these findings, with regards to food security, by focussing on regulating ES research relevant to crop productivity and by evaluation of the farming landscape where data were collected, particularly with regard to smallholdings. In this study we specifically ask if the constituent studies of recent reviews relevant to agricultural pollination and biological control adequately represent farming landscapes (smallholder-farmed *vs.* large-scale farming), global biomes, regions (as defined by the World Bank), and national income statuses.

Methods

In May 2013, we used a Web of Science topic search for "[landscape OR disturbance OR diversity OR crop yield OR fruit set OR food production] AND [pollinat* OR natural enem* OR biological control OR CBC OR pest control] AND [meta-analysis OR review]" (Year >2005) to find recent quantitative reviews relevant to crop pollination and biological control (note CBC means conservation biological control). Reviews were excluded (Additional file 4: Table S2) if they had a regional focus, were relevant only to a specific crop, did not use meta-analysis or quantify/model a trend or pattern, or used few agricultural studies (less than 50%). Any additional reviews relevant to pollination and biological control referenced in selected reviews were also included in the screening process. The agricultural studies used in each review (Additional file 5: Table S3) were selected for further analysis (n = 190). We excluded studies with no focus on crops (n = 63) (for example, those conducted in natural habitats), no field component (those conducted in labs or greenhouses), or those conducted in plantation forests. The selected reviews and the studies therein did not necessarily consider ES impacts on crop yields directly. Often it was the response of ES providers that was the focus of a study, with no or limited quantification of impacts on food production. However, some reviews did link ES providers to yield, such as Garibaldi *et al.* (2013), which linked wild pollinators to fruit-set. Thus, when we refer to "pollination" reviews or studies, these are relevant to pollination, but they did not necessarily directly quantify it, and might instead have focused on the abundance and diversity of pollinators. The same is true for "biological control" reviews or studies, which did not necessarily directly estimate pest suppression, but might instead have used proxies such as natural enemy abundance and diversity.

Studies were manually assigned to a farming landscape (smallholder or large-scale; Table 1) based on descriptions in the publication, satellite imagery from BingMaps [32] and GoogleMaps [33] and, in some cases, direct correspondence from authors. It is possible that some landscapes may have changed since the date of fieldwork in a study, but we found no discordance between author descriptions and satellite imagery. It should be noted that the relative (%) composition detailed in Table 1 classifies landscapes from land that is under temporary arable agriculture and (or) permanent cropland. These comprise crops that are sown or planted once and crops that are not replanted after each harvest, such as cocoa, coffee and rubber (but not timber).

From an ES perspective we believe that a 2-ha maximum field size (as often used in United Nations Environment Programme (UNEP) or Food and Agriculture Organisation (FAO) reports) was too small to define smallholder-farmed landscapes because areas with many fields of 3 ha still have high local and landscape complexity. Therefore, we used a 3-ha maximum field size, which gave conservative results (given our *a priori* assumption that smallholdings will be underrepresented) that slightly over-estimated the number of studies from smallholder-farmed landscapes (considering our expected values are derived from [17] using a maximum field size of 2 ha). Landscapes with ≥33% of (arable and permanent) cropland areas composed of fields <3 ha and <33% of cropland areas composed of fields >5 ha were considered to be smallholder-farmed landscapes. Landscapes with ≥33% of cropland areas composed of fields <3 ha and ≥33% of cropland areas composed of fields >10 ha were considered to be dualistic landscapes. To provide a sufficient contrast to smallholder or dualistic landscapes, large-scale landscapes were defined as having ≤33% of cropland areas composed of fields <5 ha and ≥33% of cropland areas composed of fields >10 ha. If a landscape had large areas under a single management, but small field sizes, such as a large-scale commercial orchard subdivided into small sub-units, these were classified as large-scale. Author descriptions of the farming system were particularly important in the characterisation of orchard and plantation systems which were more difficult to define from satellite imagery. If it was unclear whether a field was pasture or cropland then it was considered cropland.

Many studies defined their location as a single spatial point and in these cases the landscape was considered to be an area within a 1 km radius of the point. If the location of a study was defined as a general area (for example, "West of Göttingen, Germany"), the study landscape was estimated from the dominant farming system for that area. Where study sites gave high resolution spatial references for multiple sites, the landscape was defined from all the sites. If multiple landscape types were present, the study was included in multiple landscape categories.

To assess economic and biogeographic biases, studies were classified according to national income and global region as per World Bank Databank 2011 data [18], and climate, using ArcMap v10.0 [34] and The Nature Conservancy's terrestrial global ecoregions map [35]. To generate expected values for studies by the World Bank income group, the World Bank region (including all national incomes and not just developing nations), and biome, we assumed that the number of studies in a category would be proportional to the area of cropland contained in that category. FAOSTAT [9] was used to calculate 2011 cropland area (combined area of arable and permanent croplands) for World Bank income groups and regions. Cropland area per biome was extracted from the GlobCover2009 landcover map [36] using The Nature Conservancy's (TNC's) terrestrial ecoregions map [35] in ArcMap V10.0 (ESRI, Redlands, California, USA) [34]. Where crop cover for a pixel was defined by a range, the central value of that range was used. As figures were unavailable to describe the proportion of smallholding area that was cropland, we generated expected values for global and regional areas of smallholder farming (large-scale vs. smallholder only) by assuming that the number of studies in a category would be proportional to the area of agriculture contained in a category. Expected values for the number of studies from different farming landscapes were estimated using the World Census of Agriculture 2000 [17] in conjunction with 2011 FAO national estimates of agricultural area and agricultural population (Additional file 1: Table S1). It should be noted that FAO census data were collected over the period from 1995 to 2005 and for many countries data were deficient or the structural nature of agriculture was poorly assessed (Additional file 1: Table S1). This is an issue that will hopefully be addressed in future censuses and research. Chi-square or exact multinomial goodness-of-fit tests were used to compare observed against expected proportions. Exact multinomial tests used MonteCarlo simulations with one billion trials to generate significance values. Statistical analyses were performed in R 3.0.1 (R Foundation for Statistical Computing, Vienna, Austria) [37].

Table 1 Criteria used to classify farming landscape

	Field size		
Farming landscape	3 ha	≥5 ha	≥10 ha
Smallholder	≥33%	≤33%	-
Dualistic	≥33%	-	≥33%
Large-scale	-	≤33%	≥33%

Farming landscapes were classified by the percent of a studies' landscape composed of different sized fields (arable and permanent cropland). These criteria were suitable for all the studies used in our selected reviews, but they do not encompass all combinations of field size.

Results

We analysed seven quantitative reviews (Tables 2 and 3) relevant to crop pollination and four relevant to biological control containing a total of 190 studies (Additional file 5: Table S3).

Farming landscape

Overall, smallholder studies accounted for 12% (n = 22) of the pooled studies (7%, n = 7, of biological-control studies and 17%, n = 15, of pollination studies; Figure 2D). Both globally and in developing nations the expected proportion of smallholder studies was much higher than observed when considering the agricultural population, but not for the agricultural area (Figures 2D and 3). For both services, there were far fewer smallholder studies than expected, given the size of the agricultural population in each farming landscape. Given the size of the agricultural area, in contrast to the agricultural population, biological control was insufficiently studied but there were approximately as many studies as expected for pollination. Considering individual reviews (Additional file 6: Figure S3, studies within a review, not the overall study pool), the mean ratio of large-scale to smallholder studies for pollination was 5.4 (SD 4.5) and 18.3 (SD 8.4) for biological control. Most (10/11) reviews also differed significantly from expected values for farming landscape when considering agricultural population, but when considering global cropland area no significant differences were apparent.

When considering only smallholder studies, the regional distribution was uneven (based on agricultural area; Figure 3A). Most regions apart from 'Latin America and the Caribbean' had fewer than expected smallholder biological-control studies based on smallholder area or population. Smallholder pollination studies exceeded expectations based on area for 'East Asia and the Pacific' and 'Latin America and the Caribbean' but were lacking for 'sub-Saharan Africa'. Nearly all regions were deficient for smallholder pollination studies when considering the agricultural population, again with the exception of 'Latin America and the Caribbean'.

Regions

Pooling the constituent studies of the quantitative reviews showed that 86% (n = 88) of the biological-control studies and 55% (n = 45) of the pollination studies came from Western Europe and North America (Figures 4 and 2A). Both percentages were significantly higher than expected compared to the 34% of global cropland contained within the World Bank Regions that encompassed these areas (Figure 2A). There were no biological-control studies in Africa or continental Asia and fewer pollination studies than expected given that these regions contain approximately half of global cropland with an agricultural population of approximately two billion [18]. For information on the importance of a region to various global statistics (for example, crop production, population, biodiversity and so on) see Additional file 2: Figure S1.

Biome (climate)

With respect to the area of global cropland, the temperate region contributed double the expected number of pooled studies (77%, n = 78, for biological control and 58%, n = 45, for pollination; Figure 2C). The number of biological control studies was a quarter of that expected (11%, n = 11) and, whilst higher, the number of pollination studies was also less than expected (27%, n = 22). Mediterranean

Table 2 Quantitative reviews and meta-analyses of pollination and biological control selected for this review

	Author	Theme	Total studies*	Agricultural studies
Pollination	Ricketts *et al.* [58]	Distance to natural habitat	22	22
	Winfree *et al.* [39]	Disturbance	50	27
	Williams *et al.* [59]	Disturbance	21	11
	Garibaldi *et al.* [29]	Isolation from natural habitat	29	29
	Garibaldi *et al.* [53]	Crop pollination by insects	43	43
	Kennedy *et al.* [50]	Local and landscape effects	34	34
	Shackelford *et al.* [22]	Local and landscape complexity	19	19
	Unique studies (n)		**115**	**88**
Biological control	Letourneau *et al.* [60]	Natural enemy diversity	63	30
	Chaplin-Kramer *et al.* [61]**	Landscape complexity	45	44
	Veres *et al.* [25]	Landscape complexity	25	24
	Shackelford *et al.* [22]	Local and landscape complexity	28	28
	Unique studies (n)		**138**	**102**

*The number of studies we present may differ with those presented by a review as some combined studies from different years at the same location into single entries. These were spilt for the purposes of this review. Other quantitative reviews (QRs) split single studies into multiple entries when more than one crop was investigated. Here we consider them as a single study.
**46 studies were used in this review but one PhD thesis (O'Rourke, 2010) was omitted due to lack of access.

Table 3 Significance of developing nations (as classified by the World Bank as low to upper-middle income) to global population and agriculture statistics (for regional information see Additional file 2: Figure S1)

	World	Developing nations	All other nations
Agricultural land (million km^2)*	48	74%	26%
Cereal production (million metric tons)*	2,587	71%	29%
Land under cereal production (million km2)*	7.0	80%	20%
Population (million)*	6,974	84%	16%
Agricultural population (million)*	2,598	98%	2%
Yield gap (percentage difference between actual and potential crop yields in high input systems)**	57%	63%	38%
Agricultural population using holdings <2 ha (million)***	2,147	99%	1%
Smallholdings <2 ha as percentage of global agricultural area***	16%	15%	1%

*2011 [9,18]. **2000 [56]. ***Values estimated using data from the World Census of Agriculture 2000 [9, 17] (Additional file 1: Table S1).

studies were more numerous than expected from cropland area for both pollination and biological control. Individual reviews on pollination contained studies with the expected proportions (given cropland area) for temperate and tropical biomes, whereas the coverage within three biological-control reviews was significantly and strongly biased towards temperate biomes (Additional file 6: Figure 3). The remaining biological control review by Letourneau *et al.* (2009) was also significantly skewed to temperate studies, but much less so, and it did explicitly compare tropical and temperate studies. Overall, nine studies were from biomes that were of low productivity (cold or dry) or small in extent: seven were from deserts and xeric bushlands and two were from boreal and taiga, a

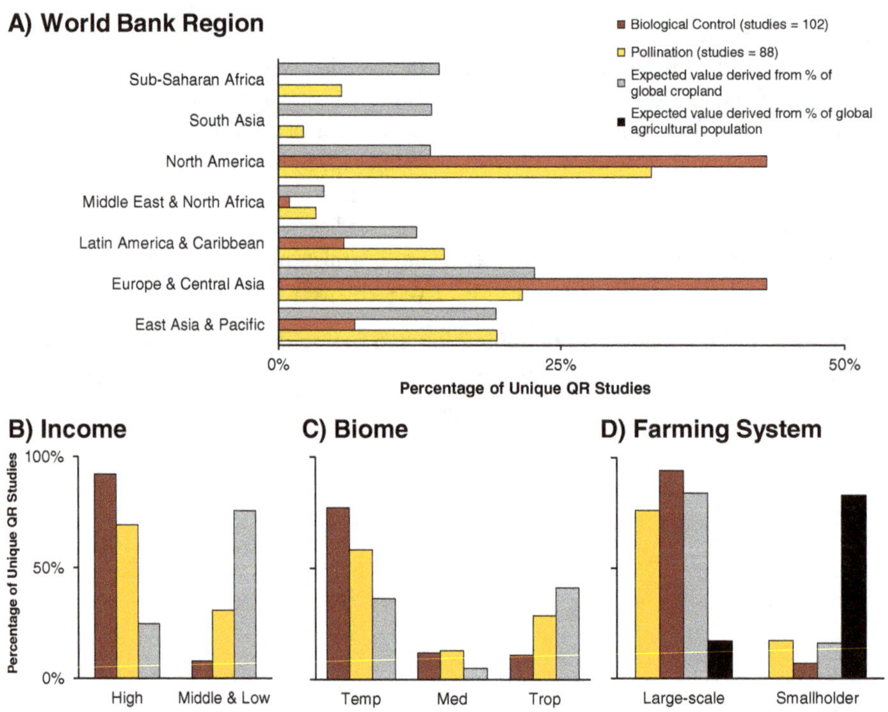

Figure 2 Analyses of pooled studies for biases in region, income group, biome and farming landscape. Unique pollination and biological control studies (relevant to agriculture), pooled from all quantitative reviews, are compared by percentage for World Bank regions **(A)**, World Bank income groups **(B)**, biomes **(C)** and farming landscapes **(D)**. In **C)** Trop = tropical and sub-tropical; Temp = temperate; Med = Mediterranean. Expected values for each factor level were derived from the percentage of the total area of global cropland **(A and B)** [9]; the percentage of the total area of global cropland **(C)** [35,36]; and the percentage of global agricultural area **(D)** [17,18] or the percent of the global agricultural population in each farming landscape. **A) Region** Pollination (Poll):$\chi^2 = 37.9$, $P < 0.001$; Biocontrol (BioC): $P < 0.001$ (Exact Multinomial Test). **B) Income** Poll: $\chi^2 = 95.8$; $P < 0.001$; BioC: $\chi^2 = 253.0$, $P < 0.001$. **C) Biome** Poll: $\chi^2 = 18.3$, $P < 0.001$; BioC: $\chi^2 = 78.1$, $P < 0.001$. **D) Farming System** percent Global Agriculture: Poll: $P = 0.541$, $\chi^2 = 0.4$; BioC: $P = 0.014$, $\chi^2 = 6.09$; percent Global Agricultural Population: Poll: $P < 0.001$, $\chi^2 = 237.5$; BioC: $P < 0.001$, $\chi^2 = 412.4$ (Note α for tests in D is 0.025 due to a Bonferoni correction for multiple testing).

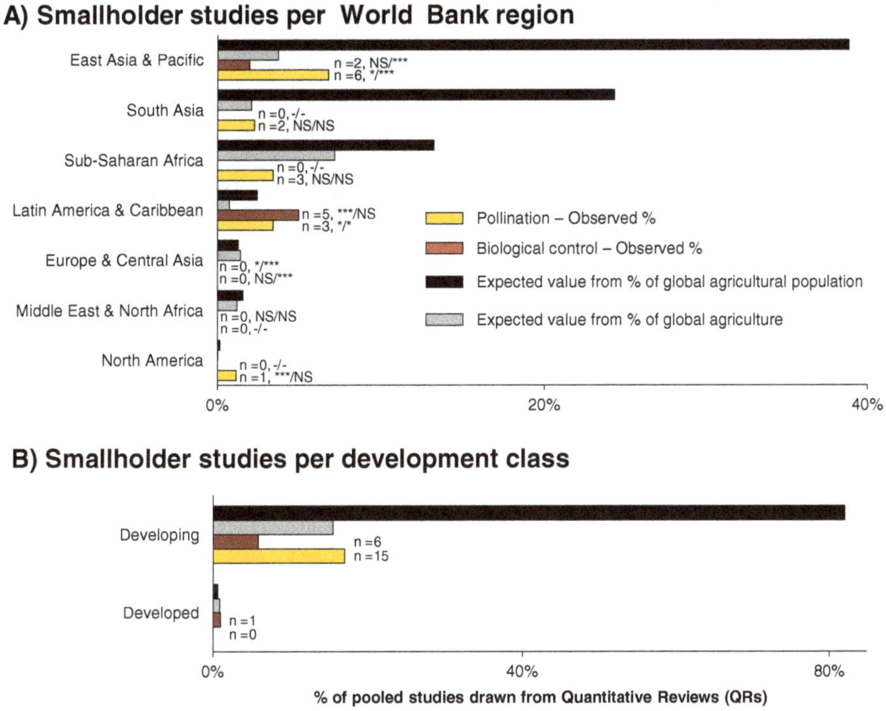

Figure 3 Analysis of only smallholder studies by region and development class. Pollination and biological control studies from smallholder farming landscapes as a percent of all unique pooled studies for each service are compared by **A)** World Bank regions and **B)** development class. Expected values for A and B were derived from [9,17] and were generated by multiplying the global percent of agricultural land/agricultural population for a region/development class by the within factor percent that was estimated to be smallholder.

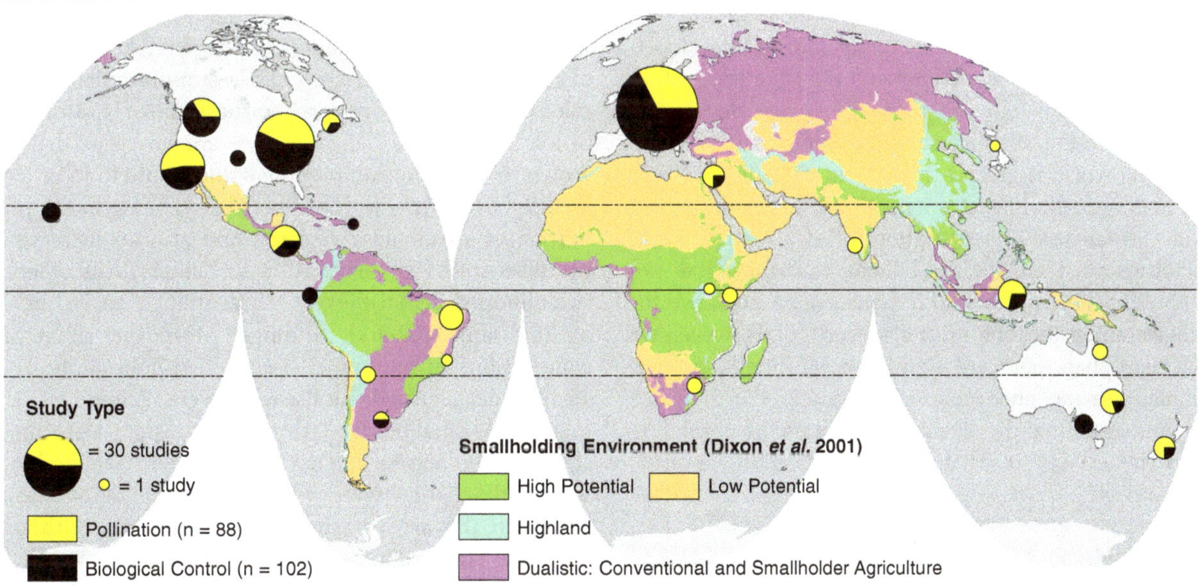

Figure 4 Distribution of pollination and biological control studies relevant to food production used in quantitative reviews. Smallholding Environment (coloured areas) is a broad classification of smallholding type (see [54] for definitions) for countries that are not defined as OECD-high income nations (greyed areas) by the World Bank. These colours state the likely type of smallholding to be found if present and do not reflect the presence/absence or importance of smallholding in an area. High potential environments (in terms of crop yield) are a combination of irrigated farming systems, wetland rice-based farming systems, and rain-fed farming systems in humid areas of high resource potential. Low potential environments are those of low current productivity or potential because of extreme aridity or cold. Highland environments are steep, highland areas.

further study was from the montane grasslands and shrublands biome.

Income (development)

Studies from developed regions accounted for 82% (n = 155) of pooled studies (92%, n = 94, of biological control and 69%, n = 61, of pollination studies; Figure 2B). The number of pooled pollination studies from developing countries was approximately equal to expectations based on cropland area (Figure 3B) although with varied geographic distribution (Figure 3A). However, biological control (pooled) was insufficiently studied in developing countries (Figure 3B). On an individual basis, most reviews (10 out of 11) also had significantly fewer studies than expected from developing regions (Additional file 6: Figure S3).

Discussion

Food security and sustainable intensification

Agricultural growth is particularly effective in improving food security, especially in low income areas [38], and sustainable intensification is a pathway for realising this [2,10,19]. Sustainable intensification includes enhancing or conserving ecosystem services for the role they can play in maintaining and increasing crop production. This aspect of sustainable intensification is known as "ecological intensification" [21]. Ecological intensification is as important in the developing world as it is in the developed world. Whilst most of our understanding of some ecosystem services - exemplified here by pollination and biological control - comes from the temperate and developed world, the bulk of the world's agricultural land, production and human population can be found in the developing world. In addition, the diversity of farming systems in the developing world is greater than the more uniform large-scale and typically intensive agriculture in the developed world. Sustainable intensification of developing world agriculture must include the billions of farmers that are smallholders [19,38] for whom management guidance for ecological intensification must also work [8]. However, we show here that there are significant farming-landscape, regional, climatic and economic biases in the evidence base underpinning the likely contribution of regulating ecosystem services to sustainable intensification via quantitative review.

Farming landscape

Only 12% of pooled studies came from smallholder-farmed landscapes as most quantitative reviews were based on data from large-scale farming and typically used three or fewer studies from smallholder-farmed landscapes. The extent to which management interventions derived from research conducted in large-scale farms in developed regions, such as the US and Western Europe, can be generalised to benefit food security in other regions is likely to be limited when there are significant contextual differences in the farming system, climate (for example, differences in extremes and modality of temperature and rainfall between temperate and tropical regions) and biogeography (for example, differences in regional species pools).

Compared to large-scale farming landscapes smallholder-farmed landscapes are much more diverse in terms of local and landscape complexity, management intensity and the interactions between them (Additional file 7: Figure S4). Smallholder landscapes, particularly in areas of subsistence farming, are likely to have a high richness and diversity of crop types, both spatially and temporally [16]. In smallholding areas typical of sub-Saharan Africa (Figure 1A,D), dwellings and associated livestock, trees, gardens, paths and boundary features generate local complexity, but this is much reduced in large-scale systems. Also, smallholder landscapes typically will not reach the levels of management intensity that occur in large-scale farms. If research efforts are concentrated in large-scale systems then we are unlikely to be fully exploring the gradients of farmland landscape complexity and management intensity and their relationship with ecosystem service providers and function.

As well as differences between smallholder and large-scale farmed landscapes, there are clearly cultural and contextual differences within the smallholder system (for example, coffee agroforestry in humid South-East Asia vs. maize farming in semi-arid sub-Saharan Africa). However, of the 22 smallholder studies in the quantitative reviews, 16 came from a single context - coffee systems. Coffee is a cash crop, a perennial crop and a stimulant that has no calorific value. Thus, research in these coffee landscapes is of low relevance to landscapes of annual crops that are grown for local consumption and contribute to local food security. Therefore, we need both an increase in the proportion of smallholder research and also an increase in the diversity of research. All two billion smallholders - two billion decision makers - are unlikely to be served by the same research findings. Moreover, since our expectations of how much research should come from smallholder farmed landscapes were based on the number of smallholder farmers, the results were more significantly biased against smallholder farmed landscapes than they were when our expectations were based on the area of smallholder farmland. Although the area of smallholder farmland is more relevant to total global food production, we emphasize that the number of smallholders, who constitute a large proportion of the undernourished, would seem to be more relevant to local food security. Thus, the combination of diversity and food insecurity in smallholder systems means that research biases against these systems are all the more acute.

Region

We found that 55% of pollination and 86% of biological control studies came from North America and Western Europe. Regional biases such as these might pose a problem for generalising to other under-represented regions. The responses of different bee species to disturbance [39,40] and the applicability of a generalised ES relationship (for example, the relationship between flower visitation and distance from natural habitat) may depend upon the species present in a region. In some cases, functional groups of service-providing species may be absent or substantially different between regions [31]. For example, bumblebees (*Bombus* spp.) are a well-studied genus of wild pollinator, for which management options have been developed in Europe and North America [41], but they are absent from sub-Saharan Africa [42]. Furthermore, the balance between ecosystem services and dis-services (services that reduce productivity or increase production costs, such as herbivory) flowing from a natural habitat to a nearby farmland may differ between regions. Quantitative reviews have often found that natural habitats benefit ecosystem services, such as pollination in nearby farmland [for example, 32]. However, the species that can move from natural habitats into croplands can fundamentally differ between regions. Proximity to a natural area in sub-Saharan African and South Asia can expose a farmer to crop-raiding elephants and primates that can reduce crop yields and create human-wildlife conflict. In contrast, in Europe and North America, regions to which research is biased, crop-raiding is a minor problem. When studying the net benefits of natural habitat in certain sub-Saharan African contexts, we might learn that crop raiding (an ecosystem disservice) tends to outweigh the benefits of enhanced pollination and biological control services on farms near natural habitats. The consequences of crop raiding are likely to impact large landholders less than smallholders, as an individual smallholder can easily lose the majority of their harvest to an elephant or troupe of baboons. Although ecosystem dis-services could strongly affect the design of sustainable farming landscapes (for example, crops that are unpalatable to primates might be used to buffer a habitat with many baboons and monkeys), they are poorly considered in the published literature (but see [43]).

Biome (climate)

Two-thirds of pooled studies came from the temperate region despite tropical croplands occupying a larger area than temperate croplands. Obviously, there are profound climatic differences between biomes that shape the assemblages of pollinators and invertebrate pests and their natural enemies. In general, the effects of climate and climate change on pollinators are much better understood than are effects on other groups of ecosystem-service providers [44], and our analyses reflect this. We show individual quantitative reviews relevant to pollination tend to balance data from tropical and temperate regions whereas biological-control reviews were all significantly and often strongly biased to the temperate zone.

There can be major differences between tropical and temperate biomes in the spatial and temporal availability of resources important for ecosystem-service providers in natural habitats (habitats that may enhance ES in nearby crops). For example, the plant community in aseasonal tropical lowland forests has a continuous pollination period throughout the year compared to just late spring and summer in northern temperate forests [45]. Temperate forests also have higher mean flower longevity and a larger proportion of wind-pollinated plants than aseasonal tropical forests [45]. As the functional significance (flowers providing nectar and pollen) of a forest to pollinators differs between and within biomes, this is likely to be the case for other habitats (including agricultural land), functions (for example, nesting sites) and ecosystem service providers too. As such, interventions for ecological intensification that require manipulating or conserving natural areas in a farming landscape should carefully consider what climate the intervention was derived from and the implications of any functional differences in habitats between derived and target climates (see [44]).

Income (development)

It is not surprising that more than 80% of studies were conducted in developed regions since funding for science is higher there. This may also reflect a publication bias in that researchers from developed nations may be more likely to publish their work in English-language peer-reviewed journals. The consequences of the over-abundance of studies from temperate regions and large-scale landscapes (particularly the U.S. and Germany) were discussed above.

Insights from large-scale studies

Studies of ecosystem services from large-scale farming landscapes (typically temperate) provide insight into the aspects of an agro-ecosystem that should be conserved when a complex and/or low-intensity system is faced with intensification. For example, local management options are likely to have more positive effects on service providing insects in agricultural landscapes of intermediate complexity [46-48], but less so in small-scale landscapes comprising many other habitats in addition to agricultural fields [39,49-51]. The interaction of local and landscape factors can be important for ES delivery as was the case for flower visitation and production in commercial South African sunflower fields where the enhancement of floral diversity within fields ameliorated the negative effects of isolation from natural habitat

[52]. Cultural species (dependent on crop habitats) of pollinators and natural enemies might be negatively affected by landscape complexity, whilst ecotone species and dispersers (dependent on non-crop habitats) might be positively affected [22].

The most recent reviews relevant to pollination should be commended for considering complexity at multiple spatial-scales. For example, Garibaldi *et al.* 2013 [53] and Kennedy *et al.* 2013 [48] stratify their study selections to incorporate a range of farming-landscapes for meta-analysis making it much easier to generalise their findings to multiple contexts. However, even in recent reviews, smallholding landscapes are not considered explicitly. As such, determining when generalisations can and cannot be applied to a type of smallholder landscapes across multiple regions (and thus climates) should be a priority for ES science.

Conclusions

Quantitative reviews are essential for modelling and predicting ES provisioning in the design of sustainable farming landscapes, for directing the policies required to adapt our current farming practices and advancing ecosystem service theory. However, the constituent studies of recent reviews relevant to agricultural biological control and, to a lesser extent, pollination, were biased towards large-scale landscapes and/or global biomes (temperate), regions (North America and Western Europe) and national economic statuses (high-income, developed nations). Differences (spatial and temporal) in management intensity and local complexity between smallholder-farmed and large-scale farming landscapes may cause ecosystem service-providing insects to respond differently to disturbance and management interventions. The high local complexity of smallholder-farmed landscapes could promote beneficial species and, consequently, the ES they provide. In this scenario the conservation of ES whilst increasing crop production will be required (we might term this "ecological conservation") compared to large-scale landscapes where ES are diminished and their restoration or replacement is required (ecological intensification). Biogeographic differences between regions in terms of climate and service- and disservice-providing species pools may also present problems for the generalisation of findings and application of ecological intensification. Thus, generalisations from the quantitative reviews included here to smallholder-farmed landscapes and, for biological-control reviews, to tropical landscapes, should be made with caution, especially in regions where little research has been conducted.

More specifically, research bias in reviews affects their general application to informing about sustainable intensification. The large number of pollination and biological control studies from temperate large-scale farming landscapes suggests we are well placed to improve ES and trial ecological intensification there. However, lower data availability from other farming landscapes and/or climates, notably tropical (for biological control reviews) and smallholder, means we may struggle to use current reviews to inform ecological intensification in such data-deficient regions. This problem is greatest in tropical regions with diverse farming landscapes and high agricultural populations, such as South Asia and sub-Saharan Africa, where smallholdings contribute more than a third of the agricultural area (Additional file 1: Table S1). Data deficiency for regions that contribute much of the world's population, crop production and hunger (for example, South Asia, China, Russia and Africa) poses further problems to improving food security with ecological intensification (and ecological conservation).

Investing globally in smallholder research for multiple crops and finding more projects and publications from regions where there is little information (such as China, Russia, South Asia and Africa) is essential. This is especially the case for research relevant to biological control, which appears more biased than pollination-relevant research. To increase their global relevance, quantitative reviews investigating landscape or local effects on ES should consider the coverage of their datasets with regards to region, climate and farming landscape. Following on from this, future empirical studies should target smallholder systems, with quantification of climate and complexity in time and at multiple spatial-scales. Classification of existing studies that do not present landscape information could also provide new data and this is becoming easier. Open-access satellite imagery (for example, Landsat8 or GoogleEarth) and software (R, QGIS, GRASS) now enable *post-hoc* classification of farming landscapes, and detailed global datasets for biodiversity, food production and development are available from the IUCN, World Bank and FAO. International collaboration is needed to search for, translate (if necessary) and disseminate the ES datasets that no doubt exist in the developing world. We suggest that non-English language publications and agricultural institutions that may often be overlooked will be productive. Data regarding the cost of ecosystem dis-services in agricultural areas near protected areas in South Asia or sub-Saharan Africa may already be available from social and developmental disciplines. If so, this should be integrated into current assessments of the net ES value of natural habitats.

Global datasets regarding the structure of farmland are incomplete and inconsistent. Therefore, our estimates of smallholder area in some regions, such as sub-Saharan Africa, were based on limited data and we hope the pending World Census of Agriculture 2010 will improve the situation. Furthermore, our classification of landscapes into broad structural categories was necessarily simplistic. From an ES perspective, smallholding landscapes could better be defined using statistical measures of configuration (see [50]) and information regarding composition and management. An appropriate classification scheme for small-scale farming (perhaps building on [54]) could be used as a guide to ecological intensification/conservation.

In the face of global climate change and challenges to food security, it is important to understand these issues for diverse environmental conditions and landscapes that fully represent the global farming constituency. Further investigation of the conditions that characterize smallholder-farmed landscapes would provide crucial information regarding the resilience of such landscapes to environmental disturbance across multiple ecosystem services.

Availability of supporting data

See Additional file 1: Table S1 and Additional file 5: Table S3.

Additional files

Additional file 1: Table S1. Word Census of Agriculture 2000 farm area and population coverage (Table 1) and estimates (Table 2). Methods for estimating the area of smallholder farmland are presented here along with tabulated results.

Additional file 2: Figure S1. Relative significance of agricultural, human population and biodiversity indicators between World Bank Regions. Each factor (represented by a coloured line) is scored by dividing the regional value by the highest value recorded for that factor (maximum score is therefore 1), with the exception of yield gap which is the actual percentage for each region. Factor values were derived as follows: *Biodiversity and threats* (green sector): **AgInc** = increase in agricultural area for 2001 to 2011 [9]; **Forest** = area of forested land [1]; **En-T** = threatened endemic species [2], En-Sp = endemic species [55]. *Food production and demand* (red sector): **YGap** = percentage gap between actual and potential production [56]; **AgPop** = agricultural population [4]; **Pop2011** = total population [4]; **Pop2050** = predicted 2050 population increase [18]; **Crop** = area of global cropland [1]. *Knowledge* (blue sector) was derived from the numbers of studies contributing to quantitative reviews (Figure 2): **Poll** = studies relevant to pollination; **BioC** = studies relevant to biological control. *Farming System* (purple sector) bars represent a qualitative estimation of importance to production and livelihoods based on reports [14,15] and the informed opinion of the authors: **SH/D** = smallholder and dualistic farms and landscapes, **C** = large-scalefarms.

Additional file 3: Figure S2. Smallholding land-use scenarios. Complexity is a combination of spatio-temporal compositional and configurational heterogeneity (see [57]). Intensity refers to management intensity including factors such as chemical and inorganic fertiliser inputs, irrigation and mechanisation, lower intensity need not lead to lower yields when sustainable management options are used. Scenarios: AD is high intensity and high local complexity; AC is high intensity and low local complexity (main text Figure 1C); BD is low intensity and high local complexity (main text Figure 1A); and BC: low intensity and low local complexity. Due to 'conventional' agricultural intensification it is probable that of global farms more will fall in the region between AC to BD than AD to BC.

Additional file 4: Figure S4. Interaction of local complexity and management intensity between farming landscapes. Here we theoretically compare large-scale farming landscapes (green) to smallholder farming landscapes (red). Points suggest the location of the farming systems shown in Figure 1 (main text) within this space. There is overlap between farming systems, but there is a greater diversity of smallholder-farmed landscapes hence they occupy a much greater area than large-scale landscapes. Configurational heterogeneity (geometry) limits both farming types for local complexity. A relatively small maximum field size for smallholder landscapes puts a lower limit on local complexity and vice-versa for large-scale farms. Limits for both axes are hypothetical and will vary with context.

Additional file 5: Table S3. Studies relative to agriculture included in our analyses. The "Income" column refers to the World Bank Income status for the country or countries a study was conducted in: H = High Income; UM = Upper Middle Income; LM = Lower Middle Income; and L = Low Income. High income nations were considered to be developed economies and all other nations developing economies. The "ES" column refers to the ecosystem service a study was relevant to.

Additional file 6: Figure S3. Study ratios for individual quantitative reviews. For each quantitative review included in our analyses we show the ratio of studies (relevant to agriculture) for: 1) **Income**, high-income nations vs. low- and middle-income nations (derived from World Bank categories); 2) **Landscape**, large-scale vs. smallholder landscapes; and 3) **Biome**, temperate regions vs. tropical regions. There were zero smallholder or tropical studies contributing to Veres *et al.* (2011). Chi-squared (or Exact Multinomial Test when values <5 present) significance values were calculated in R 3.0 and are presented by: NS = $P > 0.05$; * = $P \leq 0.05$; ** = $P < 0.01$; and *** = $P < 0.001$. Tests compared observed values with expected values which were derived from: 1) (**Income**) Estimates of the percent of the 2012 global arable and permanent cropland [9]; 2) (**Landscape**) Estimate of total farming system area [2] converted into a percent of 2012 global arable and permanent cropland area [1] (left value) and estimate of the percent of global number farms that were in each class [16] (right value); and 3) (**Biome**) percent of global cropland [36].

Additional file 7: Table S2. Excluded quantitative reviews.

Abbreviation

ES: Ecosystem service.

Competing interests

The authors declare that they have no competing interests.

Authors' contributions

PRS was responsible for the conception and design of the study. PRS, TGB, GS, LGC, LAG and SMS were involved in interpretation of results and drafting the manuscript. PRS acquired the data, carried out the analysis, and coordinated the manuscript. All authors read and approved the final manuscript.

Acknowledgements

PRS is supported by a BBSRC doctoral training grant.

Author details

[1]Africa College and School of Biology, University of Leeds, Leeds LS2 9JT, UK. [2]Naturalis Biodiversity Center, P.O. Box 9517, Leiden 2300RA, The Netherlands. [3]Sede Andina, Universidad Nacional de Río Negro (UNRN) and Consejo Nacional de Investigaciones Científicas y Técnicas (CONICET), Mitre 630, San Carlos de Bariloche, CP 8400, Río Negro, Argentina.

References

1. Foresight: *The Future of Food and Farming.* London: The Government Office for Science; 2011.
2. The Royal Society: *Reaping the Benefits: Science and the Sustainable Intensification of Global Agriculture.* London: The Royal Society; 2009.
3. Bruinsma J: **The resource outlook to 2050.** In *Expert Meeting on How to Feed the World in 2050, Proceedings of the Expert Meeting on How to Feed the World in 2050 24-26 June 2009, FAO Headquarters, Rome.* Rome: FAO; 2009.
4. Godfray HC, Beddington JR, Crute IR, Haddad L, Lawrence D, Muir JF, Pretty J, Robinson S, Thomas SM, Toulmin C: **Food security: the challenge of feeding 9 billion people.** *Science* 2010, **327:**812–818.
5. De Schutter O: **Promotion and Protection of All Human Rights, Civil, Political, Economic, Social and Cultural Rights, Including the Right to**

Development: Report of the Special Rapporteur on the Right to Food, Olivier De Schutter. In *Promotion and Protection of All Human Rights, Civil, Political, Economic, Social and Cultural Rights, Including the Right to Development: Report of the Special Rapporteur on the Right to Food, Olivier De Schutter.* Geneva, Switzerland: United Nations General Assembly; 2008.

6. Alexandratos N, Bruinsma J: **World Agriculture towards 2030/2050: the 2012 revision.** In *ESA Working Paper No. 12–03.* Rome: FAO; 2012.
7. FAO: *FAO Statistical Yearbook 2013: World Food and Agriculture.* Rome: FAO; 2013.
8. Dobie P, Yuksel N: *Halving Hunger: it Can Be Done.* London: Earthscan; 2005.
9. **FAOSTAT.** [http://faostat3.fao.org/home/index.html].
10. Pretty J: **Agricultural sustainability: concepts, principles and evidence.** *Phil Trans R Soc B* 2008, **363:**447–465.
11. Millennium Ecosystem Assessment: *Ecosystems and Human Well-Being.* Washington DC: Island Press; 2005.
12. Bacon CM, Getz C, Kraus S, Montenegro M, Holland K, Bacon CM, Getz C, Kraus S, Montenegro M, Holland K: **The social dimensions of sustainability and change in diversified farming systems. *Synthesis*, part of a Special Feature on A Social-Ecological Analysis of Diversified Farming Systems: Benefits, Costs, Obstacles, and Enabling Policy Frameworks.** *Ecol Soc* 2012, **17:**41.
13. Kremen C, Miles A: **Ecosystem services in biologically diversified versus conventional farming systems: benefits, externalities, and trade-offs.** *Ecol Soc* 2012, **17:**40.
14. Salami A, Karama A, Brixiova Z: *Working Paper 105: Smallholder Agriculture in East Africa: Trends Constraints and Opportunities.* Tunis, Tunisia: African Development Bank; 2010.
15. Singh R, Kumar P, Woodhead T: *Smallholder Farmers in India: Food Security and Agricultural Policy.* Bangkok, Thailand: FAO; 2002.
16. IFAD, UNEP: *Smallholders, Food Security, and the Environment.* Rome: IFAD; 2013.
17. FAO: *World Census of Agriculture: Main Results and Metadata by Country (1996–2005).* Rome: FAO; 2000:2010.
18. **World Bank Databank.** [http://databank.worldbank.org]
19. The Montpellier Panel: *Sustainable Intensification: a New Paradigm for African Agriculture.* London: The Montepellier Panel; 2013.
20. IFAD: *Rural Poverty Report 2011: New Realities, New Challenges, New Opportunities for Tomorrow's Generation.* Rome: IFAD; 2011.
21. Bommarco R, Kleijn D, Potts SG: **Ecological intensification: harnessing ecosystem services for food security.** *TREE* 2013, **28:**230–238.
22. Shackelford G, Steward PR, Benton TG, Kunin WE, Potts SG, Biesmeijer JC, Sait SM: **Comparison of pollinators and natural enemies: a meta-analysis of landscape and local effects on abundance and richness in crops.** *Biol Rev* 2013, **88:**1002–1021.
23. Pontin DR, Wade MR, Kehrli P, Wratten SD: **Attractiveness of single and multiple species flower patches to beneficial insects in agroecosystems.** *Ann Appl Biol* 2006, **148:**39–47.
24. Krewenka KM, Holzschuh A, Tscharntke T, Dormann CF: **Landscape elements as potential barriers and corridors for bees, wasps and parasitoids.** *Biol Conserv* 2011, **144:**1816–1825.
25. Veres A, Petit S, Conord C, Lavigne C: **Does landscape composition affect pest abundance and their control by natural enemies? A review.** *Agr Ecosyst Environ* 2013, **166:**110–117.
26. Winfree R, Kremen C: **Are ecosystem services stabilized by differences among species? A test using crop pollination.** *Proc Biol Sci* 2009, **276:**229–237.
27. Tscharntke T, Batáry P, Dormann CF: **Set-aside management: how do succession, sowing patterns and landscape context affect biodiversity?** *Agr Ecosyst Environ* 2011, **143:**37–44.
28. Diekotter T, Wamser S, Wolters V, Birkhofer K: **Landscape and management effects on structure and function of soil arthropod communities in winter wheat.** *Agr Ecosyst Environ* 2010, **137:**108–112.
29. Garibaldi LA, Steffan-Dewenter I, Kremen C, Morales JM, Bommarco R, Cunningham SA, Carvalheiro LG, Chacoff NP, Dudenhöffer JH, Greenleaf SS, Holzschuh A, Isaacs R, Krewenka K, Mandelik Y, Mayfield MM, Morandin LA, Potts SC, Ricketts TH, Szentgyörgyi H, Viana BF, Westphal C, Winfree R, Klein AM: **Stability of pollination services decreases with isolation from natural areas despite honey bee visits.** *Ecol Lett* 2011, **14:**1062–1072.
30. Kareiva P, Tallis H, Ricketts TH, Daily GC, Polasky S: *Natural Capital: Theory and Practice of Mapping Ecosystem Services.* Oxford: Oxford University Press; 2011.
31. Archer C, Pirk C, Carvalheiro L, Nicolson S: **Economic and ecological implications of geographic bias in pollinator ecology in the light of pollinator declines.** *Oikos.* In press.
32. **Bing maps.** [http://www.bing.com/maps/]
33. **Google maps.** [https://maps.google.com/]
34. ESRI: *ArcGIS Desktop. Release 10.* Redlands, CA: ESRI; 2011.
35. The Nature Conservancy: *Global Ecoregions, Major Habitat Types, Biogeographical Realms and the Nature Conservancy Terrestrial Assessment Units as of December 14, 2009.* Vienna, Austria: The Nature Conservancy; 2009.
36. Arino O, Ramos J, Kalogirou V, Defourny P, GlobCover AF: *ESA Living Planet Symposium 2009.* Bergen, Norway: European Space Agency; 2010.
37. R Development Core Team: R: *A Language and Environment for Statistical Computing.* Vienna, Austria: R Foundation for Statistical Computing; 2013 [http://web.mit.edu/r_v3.0.1/fullrefman.pdf]
38. McGuire S: **The State of Food Insecurity in the World 2012. Economic growth is necessary but not sufficient to accelerate reduction of hunger and malnutrition.** *Adv Nutr* 2013, **4:**126–127.
39. Winfree R, Aguilar R, Vazquez DP, LeBuhn G, Aizen MA: **A meta-analysis of bees' responses to anthropogenic disturbance.** *Ecology* 2009, **90:**2068–2076.
40. Cariveau DP, Williams NM, Benjamin FE, Winfree R: **Response diversity to land use occurs but does not consistently stabilise ecosystem services provided by native pollinators.** *Ecol Lett* 2013, **19:**835–950.
41. Dicks LV, Showler DA, Sutherland WJ: **Bee conservation: evidence for the effects of interventions.** In *Synopses of Conservation Evidence.* Exeter, UK: Pegasus Publishing; 2010.
42. Michener CD: *The Bees of the World 2nd Edition.* Baltimore, MD: Johns Hopkins University Press; 2000.
43. Zhang W, Ricketts TH, Kremen C, Carney K, Swinton SM: **Ecosystem services and dis-services to agriculture.** *Ecol Econ* 2007, **64:**253–260.
44. Cock MJ, Biesmeijer JC, Cannon RJ, Gerard PJ, Gillespie D, Jimenez JJ, Lavelle PM, Raina SK: **The implications of climate change for positive contributions of invertebrates to world agriculture.** *CAB Rev* 2013, **8:**1–48.
45. Bawa KS: **Plant-pollinator interactions in tropical rain forests.** *Annu Rev Ecol Syst* 1990, **21:**399–422.
46. Tscharntke T, Klein AM, Kruess A, Steffan-Dewenter I, Thies C: **Landscape perspectives on agricultural intensification and biodiversity - ecosystem service management.** *Ecol Lett* 2005, **8:**857–874.
47. Tscharntke T, Tylianakis JM, Rand TA, Didham RK, Fahrig L, Batary P, Bengtsson J, Clough Y, Crist TO, Dormann CF, Ewers RM, Fründ J, Holt RD, Holzschuh A, Klein AM, Kleijn D, Kremen C, Landis DA, Laurance W, Lindenmayer D, Scherber C, Sodhi N, Steffan-Dewenter I, Thies C, van der Putten WH, Westphal C: **Landscape moderation of biodiversity patterns and processes - eight hypotheses.** *Biol Rev Camb Philos Soc* 2012, **87:**661–685.
48. Scheper J, Holzschuh A, Kuussaari M, Potts SG, Rundlöf M, Smith HG, Kleijn D: **Environmental factors driving the effectiveness of European agri-environmental measures in mitigating pollinator loss – a meta-analysis.** *Ecol Lett* 2013, **16:**912–920.
49. Bengtsson J, Ahnstrom J, Weibull AC: **The effects of organic agriculture on biodiversity and abundance: a meta-analysis.** *Appl Ecol J* 2005, **42:**261–269.
50. Kennedy CM, Lonsdorf E, Neel MC, Williams NM, Ricketts TH, Winfree R, Bommarco R, Brittain C, Burley AL, Cariveau D, Carvalheiro LG, Chacoff NP, Cunningham SA, Danforth BN, Dudenhöffer JH, Elle E, Gaines HR, Garibaldi LA, Gratton C, Holzschuh A, Isaacs R, Javorek SK, Jha S, Klein AM, Krewenka K, Mandelik Y, Mayfield MM, Morandin L, Neame LA, Otieno M, *et al*: **A global quantitative synthesis of local and landscape effects on wild bee pollinators in agroecosystems.** *Ecol Lett* 2013, **16:**584–599.
51. Batáry P, Báldi A, Kleijn D, Tscharntke T: **Landscape-moderated biodiversity effects of agri-environmental management: a meta-analysis.** *Proc Biol Sci* 2011, **278:**1894–1902.
52. Carvalheiro LG, Veldtman R, Shenkute AG, Tesfay GB, Pirk CW, Donaldson JS, Nicolson SW: **Natural and within-farmland biodiversity enhances crop productivity.** *Ecol Lett* 2011, **14:**251–259.
53. Garibaldi LA, Steffan-Dewenter I, Winfree R, Aizen MA, Bommarco R, Cunningham SA, Kremen C, Carvalheiro LG, Harder LD, Afik O, Bartomeus I, Benjamin F, Boreux V, Cariveau D, Chacoff NP, Dudenhöffer JH, Freitas BM, Ghazoul J, Greenleaf S, Hipólito J, Holzschuh A, Howlett B, Isaacs R, Javorek SK, Kennedy CM, Krewenka KM, Krishnan S, Mandelik Y, Mayfield MM,

Motzke I, *et al*: **Wild pollinators enhance fruit set of crops regardless of honey bee abundance.** *Science* 2013, **339**:1608–1611.
54. Dixon JA, Gibbon DP, Gulliver A: *Farming Systems and Poverty: Improving farmers' Livelihoods in a Changing World.* Rome: FAO; 2001.
55. **IUCN Redlist.** [www.iucnredlist.org/]
56. FAO/IIASA: *Global Agro-Ecological Zones (GAEZ v3.0).* Laxenburg, Austria: IIASA and Rome: FAO; 2012.
57. Fahrig L, Baudry J, Brotons L, Burel FG, Crist TO, Fuller RJ, Sirami C, Siriwardena GM, Martin J-L: **Functional landscape heterogeneity and animal biodiversity in agricultural landscapes.** *Ecol Lett* 2011, **14**:101–112.
58. Ricketts TH, Regetz J, Steffan-Dewenter I, Cunningham SA, Kremen C, Bogdanski A, Gemmill-Herren B, Greenleaf SS, Klein AM, Mayfield MM, Morandin LA, Ochieng' A, Potts SG, Viana BF: **Landscape effects on crop pollination services: are there general patterns?** *Ecol Lett* 2008, **11**:499–515. Erratum in: *Ecol Lett* 2008, **11**:1121. Potts, Simon G [added].
59. Williams NM, Crone EE, Roulston TH, Minckley RL, Packer L, Potts SG: **Ecological and life-history traits predict bee species responses to environmental disturbances.** *Biol Conserv* 2010, **143**:2280–2291.
60. Letourneau DK, Jedlicka JA, Bothwell SG, Moreno CR: **Effects of natural enemy biodiversity on the suppression of arthropod herbivores in terrestrial ecosystems.** *Annu Rev Ecol Syst* 2009, **40**:573–592.
61. Chaplin-Kramer R, O'Rourke ME, Blitzer EJ, Kremen C: **A meta-analysis of crop pest and natural enemy response to landscape complexity.** *Ecol Lett* 2011, **14**:922–932.

7

Self-sufficiency in rice and food security: a South Asian perspective

Ghose Bishwajit[1*], Sajeeb Sarker[4], Marce-Amara Kpoghomou[2], Hui Gao[1], Liu Jun[3], Daogen Yin[1] and Sharmistha Ghosh[5]

Abstract

South Asia is world's most densely populated region and houses the largest population of undernourished people. Despite a period of marked economic growth averaging 6% a year over the past 20 years, it remains the world's second poorest region with more than 500 million people living on less than US$1.25 a day. Yet, there has been a considerable improvement in food security steered by the Green Revolution, use of high-yielding rice varieties, increasing investments in agriculture, improved fertilizer use and irrigation infrastructure and the potential for further increase remains high. Firstly this paper scrutinizes the role rice has been playing in the economy and food security of SouthAsia so far and that it is still the most potential means to improve the food security situation and tackle severe undernutrition as other sectors are, until now, far less furnished to address this issue. This paper probes into various economic and historical perspectives of rice economy and culture in this region, and shows that self-sufficiency in rice production is paramount to its domestic food security, and thereby proposes that emphasis should be given on increased rice production which is decelerating amid the upsurge of modern economic sectors. **Objective:** The objectives of this study are twofold. First, it attempts to show the general situation and production trend of rice. Then by relating it to the current status and future potential, it proposes that reaching self-sufficiency in rice production is the paramount item on the food security agenda in this region.

Keywords: Food security, Rice, South Asia, Self-sufficiency

Introduction

South Asia, which is also known as the Indian subcontinent, houses well over one fifth of the world's population and nearly one fifth of world's undernourished people. Despite unprecedented economic progresses in past three decades, millions of people remain food-insecure. The Green Revolution has immensely influenced various aspects of food security in India, tripling its food grain production between 1960 and 2000 [1] and thus halving the percentages of food insecurity and poverty. According to the World Food Summit (1996), food security exists when all people, at all times, have physical and economic access to sufficient, safe and nutritious food to meet their dietary needs and food preferences for an active and healthy life. In Asia, food security has traditionally been defined as maintaining stable prices for rice in the major urban markets of a country [2] where it is the staple food of more than 50% of the population. For South Asia the figure is around 70% (FAOSTAT) which is highest in the world, and hence food security is essentially a reflection of rice security in this region. And thus achieving self-sufficiency in rice production is an effective way to promote national level food security. Indeed, this analysis considers the many supporting facts. Firstly, the countries in this region share significant similarities in terms of livelihood, diet, culture, lifestyle and socio-economic status and are characterized by widespread poverty, undernutrition and low literacy rate. Thus, it assumes implementation of a somewhat identical measure to face an issue as local and as global as food insecurity. Secondly, it is a developing region and the economy is chiefly agrarian. No other sector of the economy is likely to bring a sustainable level of food security for the fast growing population as effectively as the rice industry is doing at present. Last but not least, natural disasters, political corruption, malgovernance and transnational crime are rife in

* Correspondence: brammaputram@gmail.com
[1]Department of Nutrition and Food Hygiene, School of Public Health, Tong Ji Medical College, Wuhan, China

South Asia and continue to cripple development issues and opportunities. No significant interventions are possible in such cases to ameliorate the condition of food security, irrespective of production status.

Data sources

This study focuses on five South Asian countries including India, Nepal, Bangladesh, Sri Lanka and Pakistan and main data sources are from their ministries of agriculture. Other sources include the World Bank, FAO Agricultural Statistics, WHO, Bangladesh Bureau of Statistics (BBS), International Rice Research Institute (IRRI), International Food Policy Research Institute (IFPRI), Bangladesh Rice Research Institute (BRRI), Global Agricultural Information Network (GAIN), National Center for Agricultural Economics and Policy Research (NCAP), Bangladesh Agricultural Research Council, Consultative Group on International Agricultural Research (CGIAR), United States Department of Agriculture (USDA) and various conference and research papers. The reason why India and Bangladesh appear more frequently in the analysis is mainly because they rank among top five global producers and the two largest producers and consumers of rice in South Asia, and also because their rice consumption plays a vital role in the rice economy of South Asia and the whole world.

Rice profile of South Asia

In rice production, South Asia has made a quantum leap since the spread of the Green Revolution. With 41% arable land, it is currently the second largest producing region in the world [3]. In 2010/2011 it produced 139.7 million tons of rice which is the second highest on record to date [4]. India has become the world's second largest rice producing nation with the largest rice harvesting area in the world [5]. Rice occupies about 24% of total cropped area of the country and contributes 42% of total food grain production and 45% of total cereal production. Around 65% of the total population in India depends on rice and it accounts for 40% of their food production [6]. Bangladesh, being the third largest economy in South Asia, is the world's sixth largest rice producer [7]. In 2000 to 2001, rice alone occupied 75% of the cultivable land [8] in Bangladesh which increased to 79.4% in 2008 [9]. In 2011, Bangladesh produced an all time record of 34.25 million tons of rice [10]. Rice also accounts for 94% of the cereals consumed and constitutes 92% of the annual food grain production [11]. According to data from the last 40 years, the per capita rice consumption rate in Bangladesh is 153.02 Kg per person per year. In Pakistan, wheat is the staple food yet they are the world's fourth largest producer of rice. Paddy rice production in Sri Lanka for 2011/2012 recorded an all time high of 4.869 million metric tons [6]. Sri Lanka currently produces 2.7 million tons of rough rice per year which supplies 95% of the domestic requirement. South Asia imported 1.84 million tons of rice in 2011, with Bangladesh the largest buyer [4]. Only India and Pakistan are self-sufficient in rice production; Sri Lanka, Bangladesh and Nepal are not yet self-sufficient and depend on exports from other countries (Figures 1 and 2).

Share of rice in agriculture and GDP

Agriculture has historically been a dominant sector in South Asian economy, employing about 60% of the labor force and contributing 22% of the regional GDP [13]. Figure 3 shows share of rice in total GDP for 2010/2011. In 2000/2001, the agricultural GDP share of the total GDP of Bangladesh at the current price was 30% [14] and rice alone occupied 75% of the cultivable land [8] whichbecame 79.4% in 2008 [9]. In Nepal, agriculture remains the principal economic activity, employing 66% of the population and providing 39% of GDP [15]. In 2007, rice alone accounted for 87.5% of the agriculture share of its total GDP [16]. Sri Lanka's rice sector contributes 30% to the agricultural GDP [17]. Rice is the single most important crop occupying 34% (0.77 /million hectares) of the total cultivated area in Sri Lanka. In 2003, India and Bangladesh together contributed 28% (22 and 6 respectively) of global rice production and 33.57% in 2009 (26 and 7.5 respectively). However, the share of rice in total agricultural output (Figure 4) and total GDP has been declining. In 1961, rice accounted for 8.4% of GDP in South Asia, declined to 2.7% in 2007.

Why rice?

There is no direct answer to this question. However, the response provokes a counter-question; what else is there to substitute rice? The answer is simple- nothing. Rice is the cheapest and most effective means available in this region that is likely to eradicate acute undernutrition. Many studies have revealed that there is a great potential to increase the rice production in South Asia. India, Bangladesh and Sri Lanka are among the most disaster-prone countries in the world. Recurrent floods, cyclones, earthquakes, landslides and droughts hugely affect the production in these countries. Still they have maintained a steady growth for the last three decades (Figure 5). And now with the advent of modern breeding strategies, which has proved to be much more efficient than conventional techniques, and environmental stress-tolerant varieties, the whole scenario for rice production appears better than ever.

Rice provides around 60 to 70% of calories and 50 to 55% of protein intake in the South Asian population. Many micronutrient supplementation projects have been carried out, but to no avail. Inadequate calorie intake hampers proper micronutrient metabolism. Though dependence on rice on the scale seen in South Asia is nutritionally undesirable, it is also true that rice is still the

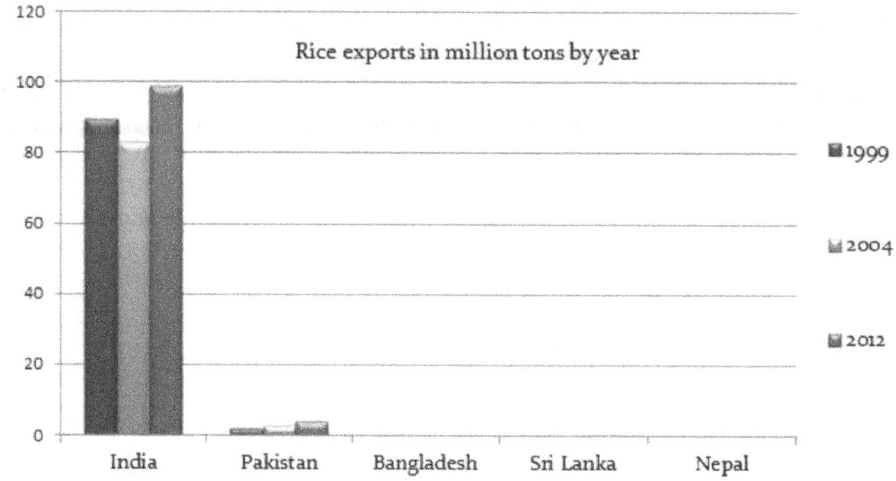

Figure 1 Shows that India is the major exporter of rice in South Asia selling about 98 million tons in 2012. Pakistan is the only other exporting country but the volume is much lower compared to India. Source: United States Department of Agriculture [12].

most affordable way for these people to maintain a sustainable calorie intake.

The Indian subcontinent has a tradition which is inseparably mingled with rice, is more than mere livelihood and has shaped the history, culture, art, and lifestyle of its population in many ways. It is regarded as a sign of fortune and well-being in many South Asian societies. In weddings, seasonal festivals, and rituals, rice plays an inevitable part. The amount of rice consumed in this region explains how important rice is in their lives (Table 1). The so-called dal-vat-tarkari is an iconic menu seen across the entire Indian subcontinent. Vat (cooked rice) is the term for meal in many South Asian languages. In Hindu rituals, paddy is an indispensible item. Two panicles of paddy appear on many logos, monograms and bank notes in Bangladesh.

Table 1 shows that in terms of per capita consumption, Bangladesh is the largest consumer of rice in South Asia which explains why it depends on exports every year despite having the highest production rate. Nepal and Sri Lanka show a similar rice-eating habit. India, being the largest exporter is the fourth largest rice consuming nation in South Asia. Pakistan is the country least dependent on rice as wheat is the staple food for most of its population. Source: adapted from FAOSTAT database [22].

Future prospects and importance of rice research in South Asia

The population in South Asia is predicted to exceed six billion by the end of the century and may face a severe

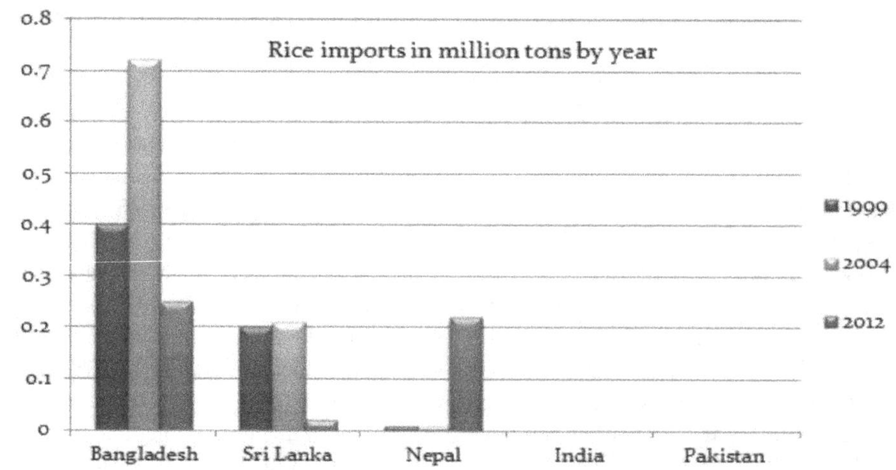

Figure 2 Shows the trend of rice import by Bangladesh, Nepal and Sri Lanka. Bangladesh is highly dependent on import from others countries and the biggest importer of rice in South Asia. India and Pakistan are the only two self-sufficient countries in rice in the region. Source: United States Department of Agriculture [12].

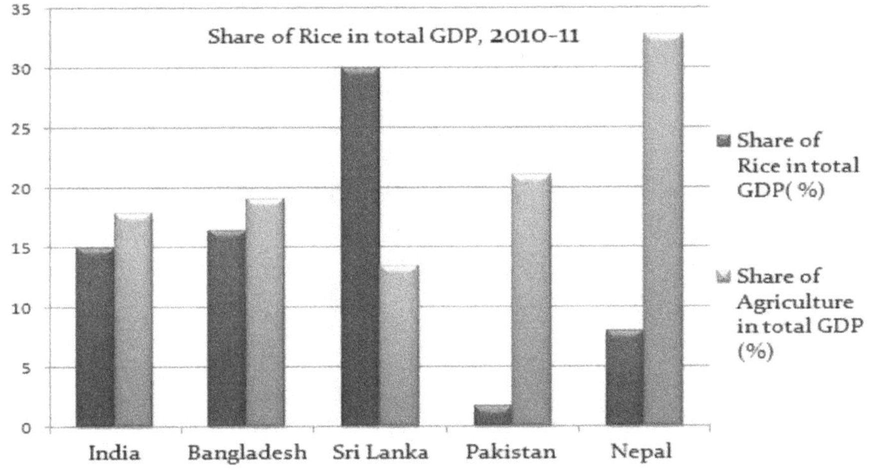

Figure 3 Shows the percentage share of agriculture in total GDP beside the percentage share of rice in total GDP in 2010/2011. Pakistan's agricultural share in total GDP is higher than that of India and Bangladesh but contribution of rice to total GDP remains much lower. Rice contributes the lions share in GDP for Sri Lanka, while India and Bangladesh show a similar pattern of dependency on rice. For Nepal, agriculture has the biggest share in total GDP comparing to the other four countries but shows that contribution of rice to GDP is comparatively lower. Source: [18-21].

food shortage unless production can keep pace with the increased demand. Also, there are many crucial problems including water scarcity, decreasing size of agricultural land, inadequacy of agrochemicals, power shortage and poor infrastructure. Yet there is a ray of hope shed on the future of rice by biotechnology which ensures greater productivity. Additionally, subcontinental soil is suitable for rice and production per unit of land is considerable. Figure 5 shows that there is an increasing trend in average yield of rice in Pakistan, Nepal and Sri Lanka while India and Bangladesh have undergone a sharp decline. However, the demand for food is only rising and is being met by constantly diminishing resources. To boost production, increasing amounts of fertilizer and pesticide are being used and aggravating environmental and ecological problems. Thus the need for a sustainable intervention to these urgencies necessitates innovation in rice technology and development of varieties which are not only high yielding but also can be produced in more eco-friendly ways. The development of high yielding varieties (HYVs) and bio-fortified, disease-, insect-, flood-, drought-, saline-, herbicide- and stress-resistant rice has opened a new era in the history of rice and will boost rice production in near future. Genetic engineering has offered rice cultivars resistant to herbicide, salinity and drought. Malnutrition, being a widespread problem in this region, requires an immediate intervention which may be very effectively addressed by bio-fortified rice. Rice bio-fortified with iron

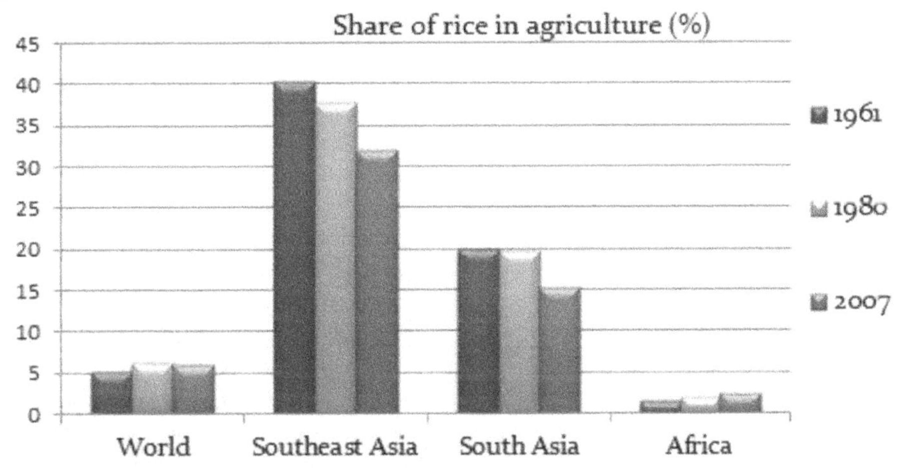

Figure 4 Shows the percentage share of rice in total agricultural output in three different regions of the world. Share of rice is on the decline since 1980 butit is still the chief agricultural product in Southeast and South Asia.

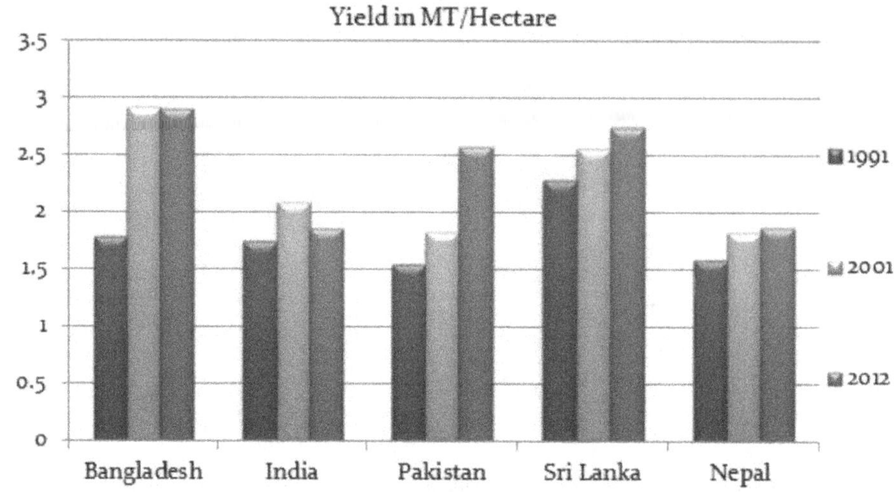

Figure 5 Shows that Bangladesh has been the highest rice yielding nation since 1991 in South Asia. All the countries except India have shown an increasing yield since 2001 while Bangladesh has undergone a sharp decline. Source: United States Department of Agriculture [12].

and zinc can protect against certain deficiency diseases and people who have no access to commercially marketed fortified foods and supplements will greatly benefit from it. Golden rice, which is a genetically modified strain of rice and enriched with beta-carotene, has the potential to save millions of children from blindness and other deadly consequences of acute vitamin A deficiency. More and more biotechnologists around the world are working to develop such micronutrient-rich rice varieties which were not possible by conventional breeding technologies. Field trials are also being undertaken to estimate any potential health impacts of the newly innovated rice cultivars. Gene banks have a major role to play in preserving the native cultivars and the artificially designed ones. From a food security perspective, gene banks are invaluable as they provide all the past cultivar options and if a variety fails to due to certain causes, farmers still can turn to gene banks for the varieties that were previously available.

Regional rice economy

Food security is a multifaceted issue influenced by national and international policy making; social, economic, environmental and demographic variables being the most prominent ones. Rice is not only the staple crop of South Asia, it is also a political crop. Large rice stocks are maintained by many countries and remain untraded until market prices rise abnormally high, and consequently food scarcity remains a common scenario across many poor areas. Though SAARC countries have launched the South Asia Food Security Program, there is still a lot to be done. Lack of mutual cooperation has long debilitated the region's socioeconomic development. Shockingly, in terms of regional trade, South Asia is the world's least integrated territory, trade between the countries accounting for less than two percent of the gross domestic production (GDP) of the region. Promoting regional cooperation and integration with neighboring organizations like ASEAN can prove to be phenomenal in alleviating poverty and achieving food security. Cross-border trading, mutual work on scientific and technological research, and improved communication will greatly facilitate the region's political, economic, and social welfare and would enable the countries to mitigate against food crisis and other related issues more effectively. It is therefore of utmost importance to strengthen regional cooperation and build ties across political borders to increase trade and improve regional food security.

Table 1 Trend of per capita rice consumption

Country	Per capita consumption, kg/person/year		
	1970 to1972	1889 to1891	1999 to 2001
Bangladesh	150	153	155
India	69	79	76
Nepal	82	106	99
Pakistan	29	14	13
Sri Lanka	95	93	94

Conclusion

From this study it is apprehensible that sustained production of rice is central to food security in South Asia. Subcontinental land is quite capable of being self-sufficient in rice production and India and Pakistan have already proved that. Since poverty is another direct cause of food insecurity, people can move out of stark poverty if they can be employed in agricultural activities and many such projects are already in operation in many countries. From the present South Asian perspective, there is no other easier way to promote national food security than through gaining self-sufficiency in rice production. Raising agricultural production is evidently the most direct way to tackle food insecurity in agronomic countries and agronomy is essentially rice economy in South Asia. The fact that most of the poor and undernourished people of South Asia are living in rural areas, and that they are largely dependent on agriculture for their livelihood can be a problem and a solution at the same time. Since agriculture is the mainstay of its economy, and the ratio of rice land to arable land is high in the subcontinent, there remains an opportunity to expand domestic rice production by creating employment and income generating opportunities in the face of ever increasing demand for rice. If self-sufficiency is achieved, it will create scope for surplus production. Also, having a surplus of rice will allow rural people to profit and can lift them above the poverty line. In a strict sense, income generation and food security initiatives go hand-in-hand and this reality makes rice more important to food security in South Asia. This paper thus substantiates that agriculture is the backbone to the South Asian economy and that rice, being the staple agricultural product, holds the capacity to pull people out of stark poverty and ensure sustainable availability of food for the food-insecure population.

Abbreviations

ASEAN: Association of Southeast Asian Nations; FAO: Food and agricultural organization (of the United Nations); HYV: High yielding variety; SAARC: South Asian Association for Regional Cooperation.

Competing interests

The authors have declared that no competing interests exist.

Authors' contributions

GB conceived the study idea; DY, SS, MAK, HG have helped in collecting data and drafting the manuscript. MLJ, DY and SG revised it critically for important intellectual content. All authors read and approved the final manuscript. LGL and GB had primary responsibility for final content.

Acknowledgements

This study could not have been possible without the unconditional help of many of our colleagues who helped to gather data and gave valuable suggestions. Special thanks is given to Mr. Arun Kumar Mondal for his generous contribution in conducting this research and help in revising the manuscript.

Author details

[1]Department of Nutrition and Food Hygiene, School of Public Health, Tong Ji Medical College, Wuhan, China. [2]Department of Epidemiology and Biostatistics, School of Public Health, Tong Ji Medical College, Wuhan, China. [3]Department of Respiration Medicine, Guangzhou First People's Hospital, Guangzhou, China. [4]Department of Journalism, Communication and Media Studies, State University of Bangladesh, Dhaka, Bangladesh. [5]Department of Public Administration, Huazhong University of Science and Technology, Wuhan, China.

References

1. Estudillo JP, Keijiro O: **Rural poverty and income dynamics in Southeast Asia.** *Handb Agric Econ* 2010, **4**:3435. http://www.sciencedirect.com/science/article/pii/S1574007209040675.
2. *The Asia foundation, occasional papernumber 4, October 2010, Food security in Asia and the changing role of rice.* Asia foundation. Australia; 2010.
3. Gumma MK, Nelson A, Thenkabail PS, Singh AN: **Mapping rice areas of South Asia using MODIS multitemporal data.** *J Appl Remote Sens* 2011, **5**:3.
4. *Bangladesh economic review.* Finance Division, Ministry of Finance Economic Adviser's Wing, Bangladesh Government Press; 2011.
5. *Rice situation and outlook yearbook, Market and trade economics division, economic research service.* U.S: Department of Agriculture; 2012:RCS-2012.
6. RICE PRODUCTION TRENDS WORLDWIDE: http://nipunarice.com/rice-o-pedia/rice-production-trends-worldwide.
7. *Top 10 rice consuming countries.* http://www.irri.org.
8. Sustainable Development Networking Program (SDNP): *Agricultural database.* Dhaka, Bangladesh: Sustainable Development Networking Program (SDNP); 2002. http://www.sdnbd.org/sdi/issues/agriculture/index.htm.
9. Handbook of Agricultural Statistics: *FAO/WFP CFSAM 2008 report.* Bangladesh Ministry of agriculture; 2009. http://www.moa.gov.bd/statistics/statistics.htm.
10. *The financial express, volume 20, number 413.* Dhaka; 2012.
11. Mohammad d C: **Sustainability of accelerated rice production in Bangladesh: technological issues and the environment.** *Bangladesh J Agric* 2009, **34**(3):523.
12. *Rice statistics.* USA: United States department of agriculture (USDA). http://www.indexmundi.com/agriculture/.
13. WDR: *Revival of agriculture; south Asia agriculture.* 2013 World Hunger and Poverty Facts and Statistics. World hunger; 2008. http://www.worldhunger.org/articles/Learn/world%20hunger%20facts%202002.htm#Number_of_hungry_people_in_the_world.
14. Alam MS, Alam MR, Islam KK: **Energy flow in agriculture: Bangladesh.** *Am J Environ Sci* 2005, **1**:3. ISSN: 1553-345X.
15. Ministry of Agriculture Development Nepal: (MOAD, http://www.doanepal.gov.np/.
16. Madhab K: **Nepal economic growth assessment.** *Agriculture* 2008.
17. Mendis A: *Sri Lanka GRAIN annual, global agricultural information network, GAIN report number: CE9002.* Sri Lanka: USDA foreign agricultural service; 2009.
18. Department of Agriculture and Cooperation: *Agricultural Census Division.* India: Ministry of Agriculture; 2012. http://pib.nic.in/archieve/others/2012/mar/d2012031302.pdf.
19. Kevin De W: *Gallagher National Agricultural Research Centre (NARC).* Pakistan: FAO Representation; 2011. http://www.fao.org/fileadmin/templates/rap/files/epublications/PakistanedocFINAL.pdf.
20. WWF-ICRISAT: *SRI fact sheet.* INDIA; 2008. http://www.sri-india.net/documents/india.pdf.
21. *World food statistics and graphics.* http://worldfood.apionet.or.jp/.
22. United States: *Department of Agriculture.* http://www.indexmundi.com/agriculture.

8

Potential environmental and population health impacts of local urban food systems under climate change: a life cycle analysis case study of lettuce and chicken

Gillian Hall[1,2], Alison Rothwell[3], Tim Grant[4], Bronwyn Isaacs[1], Laura Ford[1], Jane Dixon[1], Martyn Kirk[1*] and Sharon Friel[1]

Abstract

Background: Climate change is expected to have an impact on food production, processing and transport systems. While food systems have become globalized in recent decades, interest has re-emerged for local production and consumption to contribute to sustainable and secure food systems in an era of increasing urbanization and climate change. To explore environmental health issues related to the production of local food in an urban setting, a life cycle analysis screening study of two food commodities, chicken meat and lettuce, produced at industrial and civic scales was conducted in Sydney, Australia, as well as interviews with consumers and producers to explore their potential motivation to change.

Methods: Determination of environmental impacts was performed using life cycle assessment (LCA) of two civic and one industrial scale producer for each commodity using SimaPro version 7.3.3. Impacts of global warming potential (GWP), land use and water use from the production of these commodities are reported. With a view to producing holistic insights to sustainable practices in Sydney, interviews with producers and consumers were undertaken to assess sociocultural outcomes including views on environmental food sustainability and other motivators of behavioral change.

Results: Local industrial production of chicken meat was found to have a lower carbon footprint than small scale civic production. Small scale civic production of lettuce had a similar carbon footprint to local industrial production. Other environmental health benefits and risks varied across the production scales. Environmental sustainability was not generally a key concern of producers or consumers.

Conclusions: Action can be taken to retain and promote food production in urban settings as a future means of assisting food security. The scale of production can be an important variable in assessing the environmental health impacts of food production in an urban setting. Currently neither producers nor consumers appear motivated to change practices to promote environmental sustainability.

Keywords: Urbanization, Scale of food production, Carbon footprint, Life cycle assessment

* Correspondence: martyn.kirk@anu.edu.au
[1]National Centre for Epidemiology and Population Health, The Australian National University, Canberra, ACT 0200, Australia
Full list of author information is available at the end of the article

Background

Environmental factors can impact directly and indirectly on human health, a concept encapsulated in the term 'environmental health'. The loss of biodiversity and human-induced climate change are environmental factors that are likely to create major pressures on human health outcomes [1]. Food production is one functional aspect of society that is under significant stress from such environmental change. This stress threatens food yields, environmental sustainability, producer livelihoods, regional economies and community cohesion, and food security and equity, which in turn threatens human health [1,2]. With a world population of around seven billion and the level of man-made greenhouse gases in the atmosphere continuing to rise [3,4], the need for resilient and environmentally sustainable food systems capable of producing nutritious food available to everyone becomes ever more pressing [5]. The objective of keeping carbon emissions to a minimum, as well as conserving limited water and land resources, is of importance for improving food security and protecting human health. Producing food at lower relative emission rates is both a local and global priority.

Food insecurity, or when people have inadequate physical and economic access to sufficient, safe and nutritious food to meet their dietary needs, affects one billion people around the world [6,7]. Urban communities may be at risk of increased inaccessibility in the future due to declining local production and reduced security of importing food. Access and cost are important factors in people's ability to obtain food, and as local civic or community production operates outside market structures, it has the added potential to improve accessibility by avoiding market food price fluctuations. In addition, while food supply systems have become globalized in recent decades and food is currently transported around the world [8], these global supply lines may become vulnerable to environmental changes in the future resulting in insufficient food reaching populations that rely on an imported food supply [9]. Areas for food production may shift or yields may be significantly reduced due to land degradation, water shortages and climate change [1]. The dual pressure from the need to control carbon emissions and the impending scarcity of fossil fuels means that eventually mechanized production and long distance transport of food is likely to become more difficult and costly in the future [10], making it harder to supply food to urban populations from distant food production areas.

Under these pressures, it is possible that some countries will find the global food system they rely on will become unreliable and they will need to reinvigorate local capacity to produce food. This may cause a shift of food production close to urban centers [11]. Local urban food systems are already common in some areas of the world, such as in parts of Africa [12], Cuba [13] and in developed countries like the UK [14]. Local urban food systems may contribute to security if global food supply lines fail, but with ever growing urban populations, balancing the use of water and land for food production versus the needs for drinking water and land for housing are other important issues for urban food systems. These issues must be taken into account when attempting to find the most efficient and secure food supply systems in the face of global environmental change.

Environmental change can threaten the security of both local and global food production, and there are potential advantages and disadvantages to food production systems at both sites. Some 'perceived' benefits of locally grown food to consumers include a belief in better freshness and taste, lower use of pesticides and chemicals, and the social interaction that can occur between consumers and local producers [15,16]. Locally produced food is also often perceived by consumers to have a reduced ecological footprint, as the distance of production to the consumer is smaller which reduces the energy requirements used in transport. It is not always recognised that energy for other less obvious aspects of production may vary independently from energy used for transport and may contribute a significant component to total production emissions [17]. On the other hand foods imported over longer distances are perceived as providing greater quantities and greater variety in both the range of products available and the degree of processing, including all kinds of fresh fruit and vegetables all year round and ready-prepared meals. To prevent food spoilage, however, transport over longer distances is likely to require energy-intensive logistics systems [18,19]. Research on 'food miles' is contentious depending on study boundaries such as geographical location, food product studied, seasonality and transport scope [17,20-22]. How this contentious debate on 'food miles' relates to Sydney's food supply chain is not clear.

Consumer views form an important potential source of pressure for food producers to change their practices. How consumers view sustainability issues and whether their views are likely to shape their food provisioning practices is often unknown. While a narrowly defined life cycle assessment (LCA) study can offer a sustainability measurement of particular practices, taking a more holistic approach to include interviews and other qualitative material broadens the insights and predictions that can be made regarding the likelihood of future behavioral change and societal willingness to support policy and food system changes in the interest of sustainability.

As well as variation in the locality of food production, there are variations in scale and approach. Apart from the distance between producers and consumers, the food chain from production to plate can take many varied forms [23]. Within a local urban setting, at one end of local food production is a small 'civic' food production system, which can be commercial or non-commercial, but is small scale or backyard. At the other end is a fully industrialized system with a complex and high intensity production-processing-retail chain, which is located in and servicing the local area and often areas many hundreds of kilometers distant. How these small civic systems compare with the larger more complex industrial systems regarding issues other than profit is largely unknown.

A case study of local food subsystems in Sydney, Australia

In order to further explore some of the environmental advantages and disadvantages of local urban food production, and of consumer and producer perceptions which are likely to be critical to motivate changes in production, we undertook a project in Sydney, Australia, as a case example of local food subsystems in an urban population in a developed country. Such information is relevant to policy makers as it is only by assessing the potential importance of local production in urban settings that resources such as land will be made available.

Sydney sits on the eastern coast of Australia in the Hawkesbury Nepean river basin in New South Wales (Figure 1). The Sydney Basin has been a major source of food for Sydney's population since white settlement [23]; however, with vast population increases and rapid development projects [23,24], prime agricultural land has been reduced. In the Sydney area during the period from 1997 to 2006, there was a decrease in the number of businesses across a broad range of agricultural commodities including eggs, poultry meat, pigs, mushrooms, stone fruit, honey and grain crops. While citrus fruit production has actually increased during this period (from 14.8 million kg to 16.1 million kg), the total value of fruit decreased dramatically from AUD$2.5 million in 1997 to AUD$1.4 million in 2006 [25]. There was a dramatic drop in the number of dairy cows from over 12,000 in 1997 to about 2,500 in 2006 [25]. Nonetheless, the Sydney Basin remains an important source of food production for the local urban population, with agricultural land covering approximately 77,000 hectares [25,26].

By evaluating different food production systems in the Sydney Basin, environmental health impacts can be compared. Local urban civic and industrial food production systems were evaluated using one meat (chicken) and one vegetable (lettuce), as a case study in Western Sydney. The selection of these commodities was guided by: 1) high importance in the Australian diet in terms of quantity consumed; 2) nutritional value; and 3) significant production and consumption in Sydney via different food systems.

Chicken meat is an important source of protein with less saturated fats, compared to other meats [27], and in 2010 was consumed more than any other meat in Australia at an average of 41.7 kg per person per year [28]. There are approximately 12 million broilers located in the Sydney region in industrial production, which is about 15% of the national total [29]. In addition, it is known that there are small scale 'civic producers', but the number is unknown.

Lettuce is consumed in Australia at the rate of 9.3 kg of lettuce per capita per year [30]. It has reasonable nutritional value, providing fiber and various vitamins and minerals [31]. Sydney is largely self-sufficient, producing an equivalent of 88% of the total quantity consumed [32]. Lettuce is also a common civic crop. A survey of 137 backyards in Sydney, Wollongong and Alice Springs found that vegetables were grown in 52% of backyards, although a high migrant population in the sample may have inflated this proportion [33].

The main environmental health outcomes of interest for chicken and lettuce production in this project were global warming potential (GWP), land use and water use. Producer and consumer perceptions of these foods were explored, including interest in environmental impact, and considered in relation to other social factors that may impact on health. Both physical and social outcomes were sought regarding different food subsystems in order to contribute to thinking about possible pathways to enhance future environmental sustainability, food security and human health.

Methods

Two main methods were employed to gather information on potential environmental health impacts related to different food subsystems. First, a life cycle approach was used to undertake a screening assessment of three indicators: GWP (global warming potential or warming of the atmosphere due to human induced emissions of specific greenhouse gases, in CO_2 equivalents), land use and water use. LCA has been an accepted approach to estimating environmental impacts of products since the 1990s [34,35]. In this study, we followed the internationally standardized techniques and approaches outlined in International Organization for Standardization (ISO) documents [36,37] and used SimaPro version 7.3.3. We considered the three LCA indicators selected to be important to the setting of urban food production where, in addition to the overarching concern about GWP, contestation for land and water are key issues. We were confident about collecting reasonable primary foreground data for the three LCA indicators selected and did not want to cloud these results with other indicators (such as eutrophication and summary score values) that would have relied heavily on background data. The use

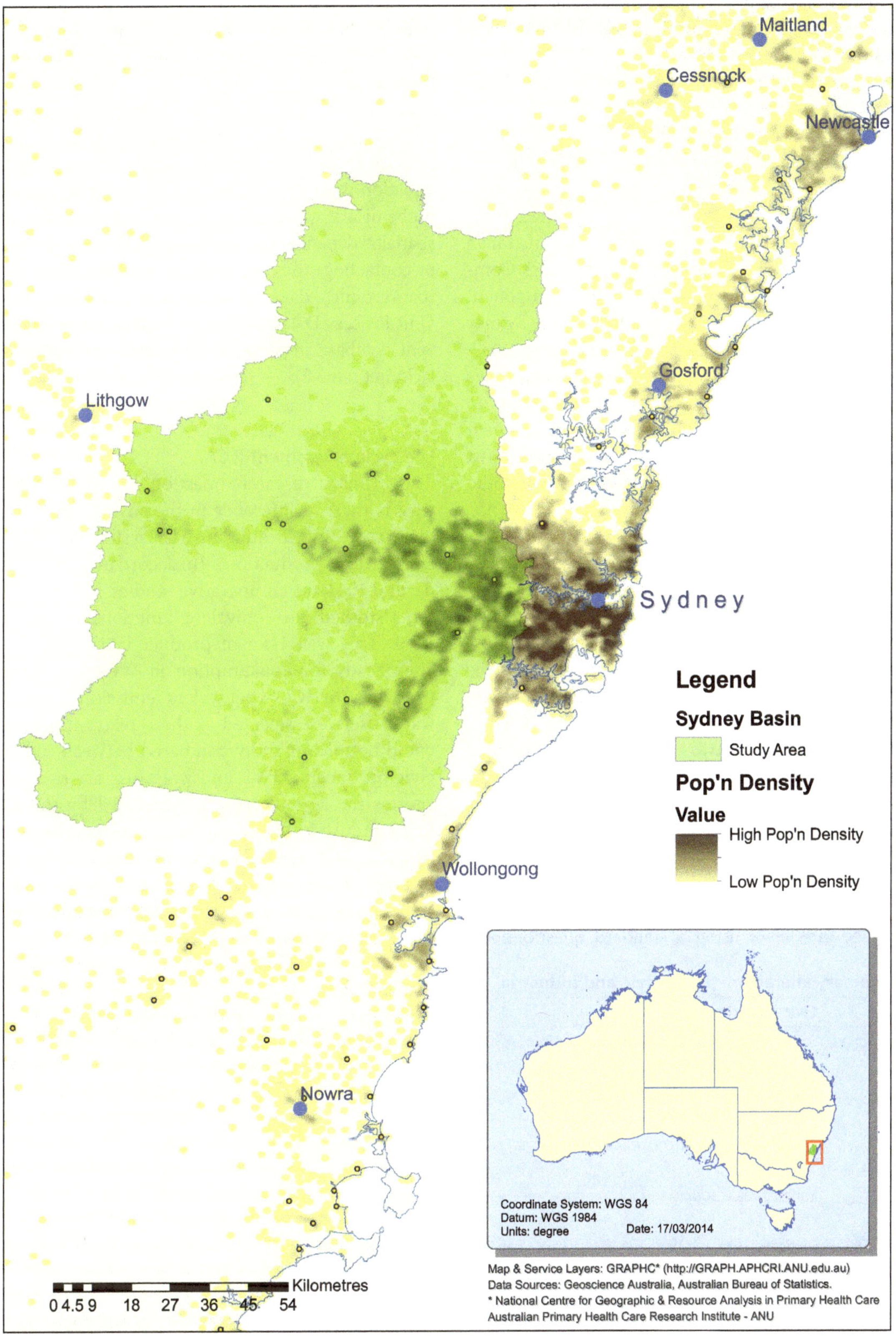

Figure 1 Map of the case study area within the Sydney Basin, Australia.

of quality data that is appropriate to the setting has been emphasized as important to produce reliable life cycle analyses [38].

Second, to complement the LCAs, interviews and focus groups were used to explore views from producers and consumers regarding social and cultural factors relating to environmental sustainability and food provisioning. In particular, motivation to improve environmental outcomes was investigated. Informed consent was obtained from each participant. The study was approved by the Australian National University (ANU) Human Research Ethics Committee, and the industrial lettuce production component was also approved by the University of Western Sydney Ethics Committee.

Criteria for local civic and industrial food subsystems were characterized as shown in Table 1, similar to schema devised by others [39-41]. A food subsystem could be classified as 'civic' or 'industrial' if it met most of the criteria of one or the other. Hybrid systems would contain elements of both.

Participants and data collection

One local medium sized industrial chicken producer agreed to take part in the study and two civic chicken growers provided information suitable for analysis. Industrial lettuce producers were contacted by visiting the Sydney central market and civic lettuce producers were identified via an active gardening group based in Western Sydney. One industrial and two civic producers provided information suitable for analysis. Consumers were contacted by door knocking from three socioeconomic areas. Consumer focus groups were made up of parents of young children who were recruited from local pre-school and day care centers.

Primary (foreground) data were collected from producers in face-to-face interviews using a standard questionnaire. Collecting data at the producer level minimized the use of 'average' background data from publically available databases that may not be appropriate to individual settings and provided a better representation of the different production systems being contrasted with potentially widely different inventories. Using producer-specific data is particularly relevant to land use and transportation, where spatial differentiation is recommended [42-44]. Farm inputs such as agrochemicals, water and transport to the Sydney market are spatially dependent. Verification of data was obtained by re-contacting the producers at a later date. Secondary (background) data were obtained from the Australasian Unit Process Database where possible, and from the ecoinvent database in SimaPro [45] when Australian data were lacking (Table 2).

Life cycle assessment (LCA)

The primary research question for the screening LCA was: Is there a difference in the environmental impact of local 'civic' and local 'industrial' systems of chicken meat and lettuce production? In accordance with ISO standards 14044 on comparative studies [37], for all subsystems studied, an equivalent functional output unit was used. This was 1 kg of product in the consumer's kitchen ready for consumption in Western Sydney. GWP (measured as CO_2 equivalent emissions), land use and water use were selected as three relevant indicators that could be meaningfully compared between industrial and civic producers. GWP and water use are two impact categories for agriculture recommended by the Australian Rural Industries Research and Development Corporation (RIRDC) [46]. GWP is necessary within this comparison if emissions reductions possibilities are to be suggested. Water in Australia is a constrained resource and land in the Sydney region is similarly constrained.

Table 1 Summary characteristics of civic and industrial food subsystems

Criteria	Civic food chain	Industrial food chain
Scale of production	Small scale production with local distribution	Large scale, widespread, high volume production/processing/distribution
Business model	Independent decision making, production highly variable. Probably no business plan, do not make an income or pay tax	Large capital investment for production (hundreds of millions of Australian dollars)
Employees	Likely no paid labour, no workers compensation, no OHS plan or HACCP plan	Paid workforce, worker education, safety plans in place and managed
Value adding	Minimal value adding to food product	Value adding on some products. Use of cooking and freezers is likely
Motivation	Motivation likely to be highly diverse and may include desire for healthy food, preference for organic foods, food taste, saving money, environmental sustainability, enjoyable lifestyle, cultural acceptance	Motivation is primarily monetary profit but other values will also be present
Consumer access	Availability to consumers at same location as production/processing	Available at large retail supermarkets in addition to other food outlets
Location	Location co-exists with house garden or community plot. Urban or residential-rural location	Multiple production and growing sites are likely. Both rural and urban locations

HACCP, hazard analysis and critical control point; OHS, occupational health and safety.

Table 2 Primary (foreground) and secondary (background) data items

Foreground data	Background data (from databases[a])
Product produced: description and amount	Electricity: production and distribution
Co-products: description and amount	Cars and trucks for transport
Land use: area of land used for production, co-uses, provision of any cultural services	Tractors and other machinery in production
Energy use: amount and source of electricity, petrol fuels, gas	Fertilizers and pesticides: manufacture of
Water use: amount and source of water	Wheat, canola and soya: production of
Material use: type, amount and source of other materials, such as pesticides, fertilizers, feed	City water supply
Wastes: type of waste and method of disposal	Waste to landfill
Application of food and worker safety measures	Manure management
Transport used in production	Other

[a]Australasian Unit Process Database and databases available in SimaPro version 7.3.3 software.

As the aim of the study was to conduct a comparative screening assessment of the significance of the potential environmental impacts between two production systems, more certain midpoint modeling was considered appropriate, as opposed to endpoint modeling which would require introduction of additional modeling assumptions [47]. No weighting was performed, as weighting would introduce value choices, as per the recommendation of ISO 14040 [36], ISO 14044 [37] and RIRDC [46]. For the method within SimaPro, the Australian indicator set version 3.01 was used. Greenhouse impacts are 100-year impacts based on the 2007 Intergovernmental Panel on Climate Change (IPCC) figures [48]. Factors for calculating emissions of CO_2 and N_2O in agriculture were taken from the *Australian Methodology for the Estimation of Greenhouse Gas Emissions and Sinks 2006* [49]. Land use and water use were each additive.

The LCA models encompassed the stages of production, processing, retail, consumer acquisition and preparation (Figures 2 and 3). The boundary did not include capital works onsite, but did include waste disposal, and inputs into machinery were generally incorporated through secondary databases. Allocation of inputs and outputs was applied for any secondary co-products.

Five stages were modeled in SimaPro for production of chicken meat: 1) production of eggs; 2) hatching of eggs; 3) growing chickens; 4) harvesting chickens; and 5) preparing for consumption. For both industrial and civic producers, one week of chicken life to produce the egg that led to the actual broiler chicken was the first step modeled. For industrial chicken production, there is an earlier stage of developing fertile elite breeder eggs overseas but this stage was not included in the industrial model. Although this involves an intensive scientific breeding program, one great-great-grandparent accounts for many birds for consumption, in excess of 1,000,000 birds [50]. The production stage focused on inputs for the raising of chickens including feed production, litter production, fertilizer production and its transport, and the raising of chickens on the industrial broiler farm or civic location. The processing stage focused on the slaughtering of chickens and the manufacture of chicken meat products. It included the production and transport of necessary inputs, such as cleaning products and packaging, as well as direct inputs, such as electricity and water and dealing with waste products. The focus for the consumer stage was on storage prior to sale and transport to acquire the product. Cooking the chicken was not included in the models.

Four stages were modeled in SimaPro for production of lettuce: 1) providing seed and growing seedlings; 2) growing and harvesting crop; 3) transporting to market; and 4) preparing for consumption. Seed production and raising seedlings applied to both civic and industrial systems. The growing stage focused on inputs of fertilizer, water and chemicals. Retail did not involve refrigeration, while consumer preparation included acquisition and washing.

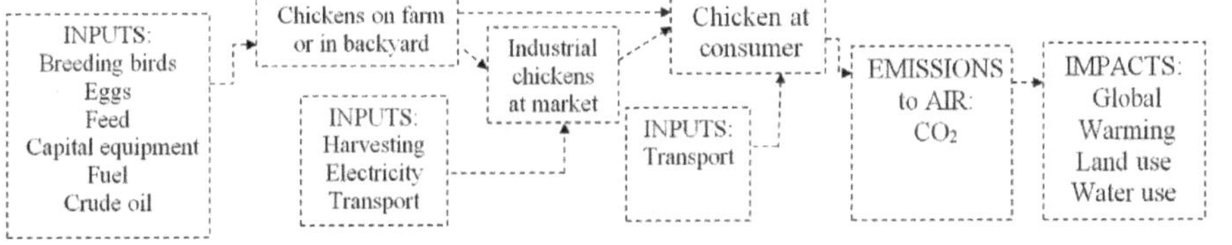

Figure 2 Inputs and boundaries for industrial and civic chicken production.

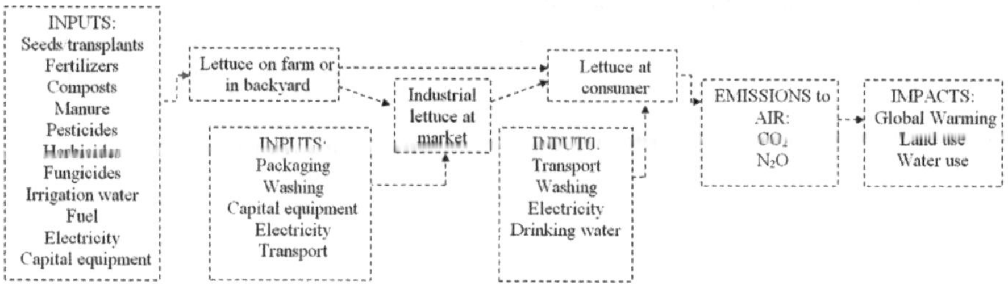

Figure 3 Inputs and boundaries for industrial and civic lettuce production.

Interviews and focus groups on environmental health and sustainability

Our primary research question for producers and consumers was: Are there environmental health motivators and behaviors related to the production and consumption of chicken and lettuce in Western Sydney? Producers were asked about their views on environmental health and sustainability and their motivation for being producers of chicken meat or lettuce. Consumer views related to buying and consuming chicken and lettuce, and were explored through focus groups and door-knock interviews with people living in purposively selected streets in areas of high, middle and low socioeconomic advantage across Western Sydney. Other information from consumers was also collected for a companion project as described elsewhere [51]. The three focus groups each included three to twelve people and in each location at least eight door knock interviews took place. The researcher also accompanied consumers on grocery shopping trips and participants were asked to document their eating and shopping practices in diaries and photo journals. The researcher, a trained anthropologist, analyzed the data with a food sociologist, sharing interpretations of common themes and patterns between various factors and re-reading the transcripts until consensus was reached. This is usual practice in small scale, qualitative research [52].

Results

The LCA models show that industrially produced chicken has less impact on CO_2 equivalent emissions than the civic produced chicken. Industrial chicken produced only 2.6 kg of CO_2 equivalent emissions per 1 kg of chicken meat, compared to the 7.7 kg and 4.0 kg of CO_2 equivalent emissions produced by civic producer 1 and civic producer 2, respectively. In addition, the industrial chicken producer in this study used less land, and less or the same amount of water as the two civic producers (Table 3). The chicken production networks for industrial and civic producers (Additional file 1) show that in both food subsystems, the main contribution to CO_2 emissions was made during the growing stage.

Industrial lettuce production had a similar impact on CO_2 equivalent emissions to one of the civic lettuce producers in this study, with 0.25 kg of CO_2 equivalent emissions for 1 kg of civic produced lettuce, and 0.32 kg of CO_2 equivalent emission produced by the industrially produced lettuce. Civic lettuce producer 2 produced 0.08 kg CO_2 equivalent emissions because of little car transport and use of home chicken manure instead of synthetic fertilizers, which are a large contributor to GWP for industrial lettuce production. The civic producer 2 model does not take account of the chickens to provide manure, as they produce a by-product of eggs. Civic lettuce producer 1 used the least land and water, with civic producer 2 and the industrial lettuce producer exhibiting similar land and usage (Table 3). The lettuce production networks for industrial and civic producers (Additional file 1) show that the main contribution to CO_2 emissions was made during the growing stage.

Producer and consumer views on sustainability and environmental health

Chicken

Interviews with chicken meat producers indicated that views on environmental health and sustainability were not primary motivations for choosing to farm chickens or for selecting particular methods. The two civic producers

Table 3 CO_2 equivalent emissions, land use and water use for local civic and industrial production in Western Sydney per 1 kg of chicken or lettuce production

Production	CO_2 equivalent (kg)	Land use (hectare-years)	Water use (m^3)
Chicken			
Civic producer 1	7.7	0.005	0.17
Civic producer 2	4.0	0.003	0.32
Industrial	2.6	0.001	0.18
Lettuce			
Civic producer 1	0.25	4.0E-06	0.004
Civic producer 2	0.08	1.4E-05	0.004
Industrial	0.32	2.0E-05	0.072

who participated were animal lovers who enjoyed hobby farming as an activity that allowed them to express independence and 'backyard innovation'. However, despite environmental sustainability not being of the first concern, the civic producers employed a number of methods in order to ensure the environmental health of their operations and their land. Civic producer 1 sourced all his necessary water from his own dam and poultry litter was delivered in the same delivery for horse supplies, thus negating the need for extra transport. All egg and chicken wastes were fed to pet dogs. Civic producer 2 used rain water and rarely relied on tap water for feeding chickens, while feed was purchased locally and chickens were also fed kitchen waste.

Views concerning environmental health held by those working in industrial chicken production varied widely across different company employees and contract farmers interviewed. At the company level, environmental health typically emerged in interviews as an important priority, relevant to company efficiency, responsibility and best practice. Cost considerations, however, were primary, as evidenced in encouraging farmers to replace conventional sheds with tunnel sheds due to their greater efficiency, despite the high electricity costs. Health and sanitary regulations also led to environmental costs, such as the use of cheap, polyester protective garments disposed of daily by hatchery employees. Growers did not usually measure value in terms of sustainability outcomes; however, primary motivations, such as making an efficient use of resources and caring for the welfare of the flocks, had significant positive environmental benefit. For some of these growers their most immediate concern was the future security of their livelihood due to increasing council regulation.

Chicken meat consumers discussed purchasing and eating chicken primarily in terms of convenience. Chicken was described as convenient because of its popularity among the family and its ready availability at the supermarket. With very few exceptions, participants purchased their chickens from large supermarkets, although some participants did mention going to the butcher, and one mentioned going directly to a chicken producer (factory outlet). The importance of convenience was seen clearly in the fact that half of the focus group participants (who were exclusively parents with young children) only bought pre-cooked barbecue chicken. Barbecue chickens were firstly described as 'easy' and 'convenient', and secondly as 'tasty'. Those who only bought barbecue chickens often described purchasing raw chickens as inconvenient and/or complicated. When discussing their chicken purchases and eating habits almost none of the participants mentioned the environment as important, although when asked about their waste methods, there was a trend among some consumers to minimize waste by using leftovers in new meals and the bones for creating stock or feeding to pets.

Lettuce

Civic lettuce producers showed the greatest ideological commitment and motivation to environmental health and sustainability. The participating lettuce growers grew a range of fruit and vegetable crops and spoke of their gardening as a positive environmental project. The civic producers had significant rain water and waste management systems, and all spoke of their gardening as having educational benefits for the community by demonstrating the possibilities in backyard gardening or by sharing knowledge in a community garden space. One of the civic producers was particularly careful to limit reliance on cars as a form of transport, including when they were sourcing necessities for the garden. Civic producers were also able to create environmental benefit by picking lettuce just before eating it and not relying on refrigeration. The interviews suggested that currently lack of access and high cost are not strong reasons for people to engage in civic production in the setting of Sydney.

Among industrial lettuce producers, care for the environment was conceived of as an ongoing responsibility necessary to 'good farming'. While environmental health and sustainability were never articulated as primary concerns, farmers expressed a responsibility to the health of their land and to Australian consumers. This was sometimes expressed in monetary terms, as farming practice, or with an intimacy of knowing the land, as farmers who had worked with the land for generations they might know when to use less (or more) fertilizers than those recommended by agronomists. Environmental concerns fit into the larger picture of farmer practice and identity, but were not presented as the most pressing of concerns. The most immediate concern to lettuce producers was that all of their product would be sold 'and not wasted'. On a longer and ongoing basis, the lettuce farmers were highly concerned with the lack of understanding and appreciation of farming by the wider Australian culture without a framework to appreciate their hard work and detailed knowledge.

Lettuce consumer responses to issues of environmental sustainability were more diverse on the topic of lettuce than they were in regards to whole chicken products. As with chicken, the majority of consumers did not consider environmental concerns as their top priority. Across geographic locations and socioeconomic backgrounds, 'price' and 'convenience', followed by 'taste' were considered the key factors when making consumer choices. When asked about the environment, many participants reflected the sentiment that it was not a high enough priority to merit time and effort, such as one focus group member who said, 'I know it probably should be important but honestly I just don't think about it'. However, a minority of participants did express environmental concerns, particularly concerning fruit and vegetable products. Among some of these consumers there was distrust expressed towards food traveling long distances impacting upon the quality of lettuce, such as its 'freshness'. For this reason some consumers chose to purchase lettuce and other vegetables from the market and independent grocers, or grow their own. However, people who grew their own lettuce or used markets and independent growers were just as likely, if not more likely, to do so for reasons that were not expressed as environmental motivations, such as 'supporting the local grocers' or 'a hobby for the kids'. Some backyard lettuce growers explicitly stated that they were not interested in the environment. Lettuce products were typically kept in the fridge for between 2 to 14 days. Wasted lettuce amounts were reported as quite high ranging between 'nothing wasted' to 'about half wasted'. Very few consumers composted their leftover vegetable product, although in one site the council provided 'food waste bins' which were used by the majority of people interviewed.

In summary, environmental concerns were not a strong factor motivating most producers' and consumers' behavioral patterns. However, some motivating factors, such as efficient use of resources and desire for freshness, may have an unintended bonus of being beneficial for environmental sustainability.

Discussion

Chicken meat and lettuce have high consumption and production rates in Australia and in the Sydney Basin [28,30]. In terms of GWP, this study found that local industrial chicken production is more efficient than civic production, and local industrial and civic lettuce production exhibit similar environmental efficiency. Taking into account current and future environmental change, improving the viability and efficiency of all kinds of local urban food production is likely to be fundamental to increasing food security for urban communities. However, our results suggest that generally neither producers nor consumers of chicken and lettuce consider environmental sustainability as a primary motivating factor for their behaviors. Conscious application of environmental sustainability by those who are in the best position to influence better environmental outcomes seems to be still in its infancy.

Although comparison of the civic and industrial food production systems highlighted the areas in both systems that can be improved in order to reduce the impact on GWP, land use and water use, changes are only likely to occur if both producers and consumers consider this a priority. In particular, significantly increased urban production at a civic level is only likely to occur if people are motivated and supported to try this. Currently there is a high reliance on, and acceptance of, the industrial scale producer for both chicken and lettuce. The motivations of producers at the civic and industrial scale were generally both dominated by reasons other than a drive to reduce environmental impacts, illustrating the market nature of society and Australia's long period of reliance being fed by industrial producers [53]. Even among the civic poultry producers, where there was evidence of the adoption of closed loop practices, such as waste refuse for chickens or animals, environmental impact reduction was not necessarily the primary target. Similarly, even without the conscious adoption of sustainability imperatives, one of our case studies found that the business as usual industrial approach to commodity production may have environmental benefits.

Our LCA results indicated that industrial chicken meat production is highly efficient. Supporting these results, average values recently reported for another Australian industrial chicken producer were comparable, at 2.6 kg CO_2 equivalent emissions per 1 kg of chicken meat [54]. Another large Australian LCA of the chicken meat industry found that industrial chicken production was highly efficient, with the industrial average GWP ranging from 1.89 ± 0.15 kg CO_2 equivalent emissions to 2.38 ± 0.16 kg CO_2 equivalent emissions [55]. In our models, chicken meat production was significantly influenced by commercial chicken feed production, which has the greatest impact on global warming potential. The low feed-to-product ratio used by industrial producers contributes to their higher efficiency. Industrial producers used relatively less food mainly due to selection of the breeding stock for feed efficiency, a scientific approach to food composition and partly due to temperature control, reduced movement by chickens in growing sheds and a short lifespan of broiler chickens. Interviews with civic producers showed that they feed their chickens more per day and for longer periods before slaughter. Civic chicken production could improve efficiency by reducing the input of commercial feed through controlling pests and maximizing the use of other feed, such as kitchen scraps. There would be considerable benefit in a system where manure produced on site was used to produce crops that could in turn be fed to the chickens.

The lower use of feed in the industrial setting reflects greater efficiency of the system as a whole, but this could still be improved if electricity usage could be reduced. In a recent Australian LCA of industrial chicken production, the authors recommend utilizing by-products, such as spent litter, as an energy source, in order to decrease reliance on electricity [55]. Renewable energy could also be a cost-effective option for industrial producers, who in interviews describe 'high electricity costs' as a concern. Reducing the feed and electricity inputs would improve efficiency in civic and industrial chicken production, respectively.

Other lettuce LCA studies in Australia show GWP of similar magnitude to our study. Although other studies have not included the consumer contribution to GWP, our study showed that on-farm impacts were dominant and values of similar orders of magnitude were found in Queensland [56] and in an assessment of the Australian vegetables industry [57]. In our LCA study of lettuce, the impact of lettuce production on GWP was most influenced by the use fuel (tractor- and transport-related) and fertilizer (manure and synthetic fertilizer). For industrial lettuce, a combination of poultry litter and synthetic fertilizers were used. Fertilizers are known to make large contributions to GWP due to N_2O emissions [58,59]. Sensitivity to the type of fertilizer was investigated by replacing the poultry litter in the model with the equivalent quantity of synthetic fertilizer, which caused the GWP to increase, primarily due to the increased fossil fuel requirements to produce synthetic fertilizers. The civic producers typically used manures and the same sensitivity was displayed when manures were replaced by synthetic fertilizers. By increasing the efficiency of fertilizer use, lettuce production could become more environmentally efficient. Using home produced composts would further improve environmental efficiencies for civic producers, as fuel for transport to collect manures also contributed to GWP in civic production.

There are a series of issues to consider when evaluating a food commodity in terms of its production and security. While the environmental efficiency of a food production system is one of the most important considerations, a commodity's ability to meet population nutritional needs through availability and resilience is also very important. Food supplies must be sufficient to meet daily requirements, reliable and stable, resilient and diverse on a local scale to ensure healthy diets and food security. Having a range of food available may require production of different forms (from backyard to industrial producer). The industrial production system allows for the efficient production of large amounts of food, approximately 114 g of chicken meat per person per day is already produced in Australia countrywide [28], which is more than a recommended global consumption target of 90 g of all kinds of meat per day [60]. A US study found that industrially produced chicken travels an average of 1,400 metric ton-kilometers per household, which is significantly less than red meat and cereals [17]. Local civic food production adds to the resilience for urban food supplies, in the case of a break down in these food supply lines due to conflict or lack of fuel for transport under environmental pressures. Wartime England promoted home production of all foods where possible [61] and chickens were fed significantly less grain and their feed was supplemented by kitchen scraps. This is one way to maximize the efficiency of civic chicken production and improve the resilience of urban food supplies.

Also for sufficient and resilient food supplies, physical supplies of land and water must be available for production. Civic food production is not possible if people do not live in households with enough land to have chickens or grow lettuce, and local industrial production is not possible if agricultural lands are overrun by urbanization. Most importantly, it is critical to not only have physical results about the most sustainable options but also to have community motivation to change. Our results do not support the concept of strong community concerns about environmental sustainability regarding food. An example of high community motivation to undertake civic food production was wartime England, when many people used available land for local production. However, the current perception of environmental issues does not appear to have reached a critical level that is perhaps required before it might become a strong motivator in current day Sydney. In Sydney's urban environment, there is a limited amount of land, which must be shared between agriculture and housing development [23]. As more land is developed to accommodate a growing urban population, less land is available for agriculture. The 2006 and 2011 Australian censuses show that the proportion of people living in separate houses has declined and the proportion of people living in semi-detached dwellings, flats, units and apartments has increased [62,63], meaning that fewer people have backyards in which they could grow lettuce or raise chickens or other food commodities. Moreover, those in employment in the area under study can have long commutes to work, denying them time to grow food and increasing the acceptability of 'convenience' or shop-bought foods [51]. Increasing land prices also makes it difficult for agricultural businesses to expand and there is increased pressure to sell for both chicken and lettuce farmers [2]. In addition, the local industrial chicken producer faced increasing difficulties of sustaining a local industrial company due to residential complaints about the smell of chicken farms.

Water is another contested resource needed for urban food production, as well as for urban development and drinking. The lettuce producers in this study were stressed by increasing water regulations put in place by councils which favored housing over farming. Under future climate scenarios annual rainfall may decrease and there may be more frequent and severe droughts and storms in New South Wales [64,65], and water may become even more contested, so food production systems using less water will be favored. Grafton and Kompas [66] performed simulations and found that during low rainfall periods, Sydney's current system is already insufficient to prevent water levels from reaching critically low thresholds, as there is an imbalance between supply and demand.

Besides environmental efficiency and resilience, other issues to consider when evaluating the production of a food commodity include food safety, animal welfare, social and cultural acceptance, and cost. As chicken meat can carry pathogens, such as *Campylobacter* and *Salmonella* [67,68], biosecurity is important for evaluating civic and industrial chicken production. Graham *et al.* [69] show how modern methods of poultry production are a threat to biosecurity through pathogen transfer from animals to humans in industrial settings. However, Australian chicken meat companies have reached a high level of adoption of different biosecurity practices to protect their farms against infections [70]. There is limited data on the biosecurity risks of civic chicken production. Pathogen transmission from backyard chickens appears to be low but understanding and adherence to proper hygiene and animal husbandry would be needed to mitigate risks [71]. In addition to biosecurity, social issues and animal welfare are other considerations. Civic chicken production involves chicken owners killing the chickens themselves, which may not be a socially popular practice, although skills could be learnt if motivation is sufficient. Industrial chicken production has been the subject of animal welfare concerns including food restrictions, fast growth requirements and physiological stress during transit [72-74].

As well as being efficient, there are other potential benefits to civic lettuce production. Urban gardens can contribute to a sense of community and add beauty to city environments [75]. In addition, members of the focus groups liked 'supporting the local grocers' and some saw growing lettuce in the backyards as a good 'hobby for the kids' and a good way to exercise. However, there may be some safety issues with civic lettuce production to consider, as soils in urban areas can contain trace elements, such as copper, zinc and lead, and eating lettuce grown in urban soils could increase the amount of trace elements being ingested [75].

For the GWP, land use and water use endpoints examined in this study, it is possible that the estimates for GWP of 1 kg of simple whole chicken meat for the industrial model may be an overestimation. This is because in reality the industrial production of '1 kg of chicken meat' included some whole chicken and some processed product. Some resources in the model would have been used to 'value add' to the carcass, for example, by chopping it up. Another limitation is that there is some uncertainty in the inputs. For example, the chicken production models had to rely on some overseas data for some feeds, as there is limited Australian feed data. Capital goods were only assessed for industrial lettuce production, for cool rooms and irrigation infrastructure. However, inclusion of capital goods into the industrial lettuce model did not alter the top contributors to environmental impact results. There are mixed views about the importance of capital infrastructure [76,77].

It is important to note that the study of chicken and lettuce in Sydney is a case study, and may not be representative of all civic and industrial meat and vegetable production. Such case studies are, however, a good place to start examining the viability of civic and industrial food production systems in an urban setting. The efficiency of civic and industrial chicken and lettuce production in our study were based on GWP, land use and water use; there are other endpoints for environmental health outcomes and food security that could be examined.

Conclusions

We developed models of food production systems for industrial and civic production of chicken meat and lettuce that allowed for assessment of the contribution of these foods to GWP, land use and water use. We also explored producer and consumer views and attitudes to these foods as these are important to facilitating improved production practices. Per kg of product, industrial chicken meat production had less environmental impact than civic production of chicken meat, primarily due to the fact that industrial producers had very low land use and slaughtered chickens much younger than civic producers. This led to a consequent reduced amount of feed per kg of chicken meat produced and feed is a key component of GWP for this system. In terms of lettuce production, carbon use was similar between civic and industrial producers, with the main influences on carbon efficiency being the use of fertilizers and manures, and tractor use in the industrial system. Using composts and manures instead of synthetic fertilizers appears to benefit GWP results for both civic and industrial producers. Maximizing environmental resource use efficiency in production systems at both the industrial and civic scale will be needed in order to ensure food security and thereby safeguard human health and well-being in the face of environmental change.

Tools such as LCA can serve to measure environmental impacts of different production systems and provide evidence that may be used to support the drive for improved environmental outcomes such as emissions reductions. However, interviews with producers and consumers showed that generally both were more influenced by reasons other than a drive to adopt practices that better support sustainability and environmental health. The results from measurements need to be aligned with motivation in a population for whom sustainability is important in order for change to be enacted. Given the current apparent lack of interest in environmental sustainability by both producers and consumers, steps may be required to encourage this. These might include a shift in market mechanisms to make resource-hungry foods more expensive as well as new regulations to reward the conscious application of sustainability practices.

Abbreviations

ANU: Australian National University; GWP: Global warming potential; HACCP: Hazard analysis and critical control point; IPCC: Intergovernmental Panel on Climate Change; ISO: International Organization for Standardization; LCA: Life cycle assessment; OHS: Occupational health and safety; RIRDC: Rural Industries Research and Development Corporation.

Competing interests

The authors declare that they have no competing interests.

Authors' contributions

The whole team contributed to all aspects of the study. Members led the following components as follows: conception of the idea by GH, JD and SH; overview and design by GH; data collection by BI and AR; data analysis by TG, AR and BI; and interpretation and report writing by GH, LF, AR and MK.

Acknowledgements

We would like to thank Paul Konings (APHCRI, ANU) for the map of the case study area.

This research received support from the Climate and Health Cluster, which is funded by the Commonwealth Scientific and Industrial Research Organisation (CSIRO) Flagship Collaboration Fund.

Author details

[1]National Centre for Epidemiology and Population Health, The Australian National University, Canberra, ACT 0200, Australia. [2]Medical School, The Australian National University, Canberra, ACT 0200, Australia. [3]School of Science and Health, University of Western Sydney, Locked Bag 1797, Penrith, NSW 1797, Australia. [4]Life Cycle Strategies, 29 Forest Street, Collingwood, VIC 3066, Australia.

References

1. McMichael AJ: **Globalization, climate change, and human health.** *N Engl J Med* 2013, **368:**1335–1343.
2. Mason D, Dixon J, Isaacs B, Edwards F, Hall G: *Briefing Paper: The Dynamic Situation of Urban Agriculture in the Sydney Basin*, Report for the Urbanism, Climate Adaptation and Health CSIRO Climate Adaptation Flagship. Canberra: National Centre for Epidemiology and Population Health; 2011.
3. Core Writing Team, Pachauri RK, Reisinger A: *Climate Change 2007: Synthesis Report. Contribution of Working Groups I, II and III to the Fourth Assessment Report of the Intergovernmental Panel on Climate Change.* Geneva: Intergovernmental Panel on Climate Change; 2007:104.
4. United Nations: *World Population Prospects: The 2010 Revision, CD-ROM Edition.* New York, NY: United Nations, Department of Economic and Social Affairs, Population Division; 2011.
5. Friel S, Barosh L, Lawrence M: **Towards healthy and sustainable food consumption: An Australian case study.** *Public Health Nutrition* 2013:1–11. doi:10.1017/S1368980013001523.
6. Friel S: **Climate change, food insecurity and chronic diseases: sustainable and healthy policy opportunities for Australia.** *N S W Public Health Bull* 2010, **21:**129–133.
7. Food and Agriculture Organization (FAO): *World Food Security: A Reappraisal of the Concepts and Approaches. Director General's Report.* Rome: FAO; 1983.
8. Satterthwaite D, McGranahan G, Tacoli C: **Urbanization and its implications for food and farming.** *Phil Trans R Soc B* 2010, **365:**2809–2820.
9. Edwards F, Dixon J, Friel S, Hall G, Larsen K, Lockie S, Wood B, Lawrence M, Hanigan I, Hogan A, Hattersley L: **Climate change adaptation at the intersection of food and health.** *Asia Pac J Public Health* 2011, **23:**94–104.
10. Graham PW, Smart A: *Possible Futures: Scenario Modelling of Australian Alternative Transport Fuels to 2050.* Newcastle: CSIRO Energy Transformed Flagship; 2011.
11. Deelstra T, Girardet H: **Urban agriculture and sustainable cities.** In *Growing Cities Growing Food: Urban Agriculture on the Policy Agenda: A Reader on Urban Agriculture.* Edited by Bakker N, Dubbeling M, Guendel S, Sabel Koschella U, de Zeeuw H. Feldafing: German Foundation for International Development; 2001:43–65.
12. Gabel S: **Exploring the gender dimensions of urban open-space cultivation in Harare, Zimbabwe.** In *Agropolis: The Social, Political, and Environmental Dimensions of Urban Agriculture.* Edited by Mougeot LJA. London: Earthscan; 2005:107–130.
13. Premat A: **Moving between the plan and the ground: shifting perspectives on urban agriculture in Havana, Cuba.** In *Agropolis: The Social, Political, and Environmental Dimensions of Urban Agriculture.* Edited by Mougeot LJA. London: Earthscan; 2005:153–179.
14. Perez-Vazquez A, Anderson S, Rogers A: **Assessing benefits from allotments as a component of urban agriculture in England.** In *Agropolis: The Social, Political, and Environmental Dimensions of Urban Agriculture.* Edited by Mougeot LJA. London: Earthscan; 2005:239–266.
15. La Trobe H: **Farmers' markets: consuming local rural produce.** *Int J Consum Stud* 2001, **25:**181–192.
16. Kneafsey M, Cox R, Holloway L, Dowler E, Venn L, Tuomainen H: *Reconnecting Consumers, Producers and Food: Exploring Alternatives.* Oxford: Berg; 2008.
17. Weber CL, Matthews HS: **Food-miles and the relative climate impacts of food choices in the United States.** *Environ Sci Technol* 2008, **42:**3508–3513.
18. Harvey M: **The rise of supermarkets and asymmetries of economic power.** In *Supermarkets and Agri-Food Supply Chains: Transformations in the Production and Consumption of Foods.* Edited by Burch D, Lawrence G. Cheltenham: Edward Elgar; 2007:51–73.
19. Estrada-Flores S, Larsen K: *Best Practice Food Distribution Systems.* Melbourne: Victorian Eco-Innovation Lab (VEIL) and Food Chain Intelligence (FCI); 2010.
20. Milà I, Canals L, Muñoz I, Hospido A, Plassman K, McLaren S: *Life Cycle Assessment (LCA) of Domestic vs. Imported Vegetables. Case Studies on Broccoli, Salad Crops and Green Beans.* Guilford: Centre for Environmental Strategy, University of Surrey; 2008.
21. Hospido A, Milà I, Canal L, McLaren S, Truninger M, Edwards-Jones G, Clift R: **The role of seasonality in lettuce consumption: a case study of environmental and social aspects.** *Int J Life Cycle Assess* 2009, **14:**381–391.
22. Mundler P, Rumpus L: **The energy efficiency of local systems: a comparison between different modes of distribution.** *Food Policy* 2012, **37:**609–615.
23. Mason D, Knowd I: **The emergence of urban agriculture: Sydney, Australia.** *Int J Agric Sustain* 2010, **8:**62–71.
24. Western Sydney Regional Organisation of Councils Ltd (WSROC): *Greater Western Sydney Regional Organisation of Councils: Community Profile: 2001 and 2006 Enumerated Census Information.* Melbourne: ID Consulting Pty Ltd; 2009.
25. Australian Bureau of Statistics (ABS): *7125.0 - Agricultural Commodities: Small Area Data, Australia, 2005-06 (Reissue).* Canberra: ABS; 2008.
26. New South Wales Department of Agriculture: *Regional Review: Sydney and South East Region.* Orange: New South Wales Department of Agriculture; 2003.
27. Farrell D: **Poultry development review.** In *Poultry and Human Nutrition.* Rome: Food and Agriculture Organization of the United Nations; 2013 [www.fao.org/docrep/019/i3531e/i3531e.pdf]
28. Australian Bureau of Agricultural and Resource Economics and Sciences (ABARES): *Agricultural Commodity Statistics 2011*, Edited by Department of Agriculture, Fisheries and Forestry. Canberra: ABARES; 2011.
29. Australian Bureau of Statistics (ABS): *7121.0 - Agricultural Commodities, Australia, 2008-9*, Table 2 Agricultural Commodities, State and NRM Region - New South Wales 2008-09. 7120DO001_200809. Canberra: ABS; 2010.
30. AUSVEG: *Vegetable Spotlight: Lettuce.* Camberwell: AUSVEG; 2014 [http://ausveg.com.au/resources/statistics/vegetable-spotlight/lettuce.htm]
31. Food Standards Australia New Zealand (FSANZ): *Australian Food Composition Tables.* Canberra: FSANZ; 2006.
32. Malcolm P, Fahd R: *Ground Truthing of Sydney Vegetable Industry in 2008.* Orange: New South Wales Department of Primary Industries; 2009.
33. Head L, Muir P, Hampel E: **Australian backyard gardens and the journey of migration.** *Geogr Rev* 2004, **94:**326–347.

34. Natural Resource Management Ministerial Council (NRMMC): *National Agriculture and Climate Change Action Plan 2006-2009.* Canberra: Department of Agriculture, Fisheries and Forestry; 2006.
35. Chesson J, Morgan L, Whitworth B: *Assessing the Environmental Performance of the Food Value Chain: An extension of the Signposts for Australian Agriculture Framework.* Canberra: Department of Agriculture, Fisheries and Forestry, Bureau of Rural Sciences; 2006.
36. International Organization for Standardization (ISO): *ISO 14040 - Environmental Management - Life Cycle Assessment - Principles and Framework.* Geneva: ISO; 2006.
37. International Organization for Standardization (ISO): *ISO 14044 - Environmental Management - Life Cycle Assessment - Requirements and Guidelines.* Geneva: ISO; 2006.
38. Reap J, Roman F, Duncan S, Bras B: **A survey of unresolved problems in life cycle assessment. Part 1: goal and scope and inventory analysis.** *Int J Life Cycle Assess* 2008, **13:**290–300.
39. Lyson TA: *Civic Agriculture: Reconnecting Farm, Food, and Community.* Lebanon, NH: University Press of New England; 2004.
40. Kloppenburg J, Lezberg S, De Master K, Stevenson GW, Hendrickson J: **Tasting food, tasting sustainability: defining the attributes of an alternative food system with competent, ordinary people.** *Hum Organ* 2000, **59:**177–186.
41. Marsden T, Smith E: **Ecological entrepreneurship: sustainable development in local communities through quality food production and local branding.** *Geoforum* 2005, **36:**440–451.
42. Brentrup F, Kusters J, Lammel J, Kuhlmann H: **Life cycle impact assessment of land use based on the hemeroby concept.** *Int J Life Cycle Assess* 2002, **7:**339–348.
43. Reap J, Roman F, Duncan S, Bras B: **A survey of unresolved problems in life cycle assessment. Part 2: impact assessment and interpretation.** *Int J Life Cycle Assess* 2008, **13:**374–388.
44. Milà I, Canals L, Bauer C, Depestele J, Dubreuil A, Knuchel R, Gaillard G, Michelson O, Muller-Wenk R, Rydgren B: **Land use in LCA: key elements in a framework for land use impact assessment within LCA.** *Int J Life Cycle Assess* 2007, **12:**5–15.
45. PRé Consultants: *SimaPro 7.3.3 (software).* Amersfoort: PRé Consultants; 2012.
46. Harris S, Narayanaswamy V: *Life Cycle Assessment Methodology for Australian Rural Industries.* Canberra: Rural Industries Research and Development Corporation; 2009.
47. Goedkoop MJ, Heijungs R, Huijbregts M, De Schryver A, Struijs J, van Zelm R: *ReCiPe 2008: A Life Cycle Impact Assessment Method Which Comprises Harmonised Category Indicators at the Midpoint and the Endpoint Level.* The Hague: Ministry of Housing, Spatial Planning and the Environment; 2009.
48. Solomon S, Qin D, Manning M, Chen Z, Marquis M, Averyt KB, Tignor M, Miller HL: *Climate Change 2007: The Physical Science Basis. Contribution of Working Group I to the Fourth Assessment Report of the Intergovernmental Panel on Climate Change.* Geneva: Intergovernmental Panel on Climate Change; 2007.
49. National Greenhouse Gas Inventory Committee: *Australian Methodology for the Estimation of Greenhouse Gas Emissions and Sinks 2006: Land Use, Land Use Change, and Forestry.* Canberra: Department of Climate Change; 2008.
50. Benjamin Wells BVSc: *Personal communication.* 2013.
51. Dixon J, Isaacs B: **Why sustainable and 'nutritionally correct' food is not on the local agenda: Western Sydney, the moral arts of everyday life and public policy.** *Food Policy* 2013, **43:**67–76.
52. Miles M, Huberman M, Saldaña J: *Qualitative Data Analysis: A Sourcebook of New Methods.* Thousand Oaks, CA: Sage Publications; 1984.
53. Symons M: *On Continuous Picnic: A History of Eating in Australia.* Adelaide: Duck Press; 1982.
54. Bengtsson J, Seddon J: **Cradle to retailer or quick service restaurant gate life cycle assessment of chicken products in Australia.** *J Clean Prod* 2013, **41:**291–300.
55. Wiedemann S, McGahan E, Poad G: *Using Life Cycle Assessment to Quantify the Environmental Impact of Chicken Meat Production.* Canberra: Rural Industries Research and Development Corporation; 2012.
56. Maraseni T, Mushtag S, Reardon-Smith K: **Integrated analysis for a carbon- and water-constrained future: an assessment of drip irrigation in a lettuce production system in eastern Australia.** *J Environ Manage* 2012, **111:**220–226.
57. Maraseni T, Cockfield G, Maroulis J, Chen G: **An assessment of greenhouse gas emissions from the Australian vegetables industry.** *J Environ Sci Health B* 2010, **45:**578–588.
58. Dalal RC, Wang W, Robertson GP, Parton WJ: **Nitrous oxide emission from Australian agricultural lands and mitigation options: a review.** *Soil Res* 2003, **41:**165–195.
59. Department of Environment and Primary Industries (DEPI): *Nitrogen Fertilisers - Improving Efficiency and Saving Money.* Melbourne: DEPI; 2014 [http://www.depi.vic.gov.au/agriculture-and-food/farm-management/weather-and-climate/understanding-carbon-and-emissions/nitrogen-fertilisers]
60. McMichael AJ, Powles JW, Butler CD, Uauy R: **Food, livestock production, energy, climate change, and health.** *Lancet* 2007, **370:**1253–1263.
61. Goodchild C, Thompson A: *Keeping Poultry and Rabbits on Scraps.* Harmonsworth: Penguin Books; 1941.
62. Australian Bureau of Statistics (ABS): **B31 Dwelling structure.** In *Sydney 2006 Census Data*, Community Profile Series: Basic Community Profile (catalogue number 2001.0). Latest issue edition. Canberra: ABS; 2008.
63. Australian Bureau of Statistics (ABS): **B31 Dwelling structure.** In *Greater Sydney 2011 Census of Population and Housing*, Basic Community Profile (catalogue number 2001.0). Latest issue edition. Canberra: ABS; 2013.
64. Commonwealth Scientific and Industrial Research Organisation (CSIRO): *Climate Change in the Hawkesbury-Nepean Catchment.* Clayton South: New South Wales Government, CSIRO; 2007.
65. Commonwealth Scientific and Industrial Research Organisation (CSIRO): *Climate Change in the Sydney Metropolitan Catchments.* Clayton South: New South Wales Government: CSIRO; 2007.
66. Grafton RQ, Kompas T: **Pricing Sydney water.** *Aust J Agric Resor Econ* 2007, **51:**227–241.
67. Mullner P, Jones G, Noble A, Spencer SE, Hathaway S, French NP: **Source attribution of food-borne zoonoses in New Zealand: a modified Hald model.** *Risk Anal* 2009, **29:**970–984.
68. Pires SM, Vigre H, Makela P, Hald T: **Using outbreak data for source attribution of human salmonellosis and campylobacteriosis in Europe.** *Foodborne Pathog Dis* 2010, **7:**1351–1361.
69. Graham JP, Leibler JH, Price LB, Otte JM, Pfeiffer DU, Tiensin T, Silbergeld EK: **The animal-human interface and infectious disease in industrial food animal production: rethinking biosecurity and biocontainment.** *Public Health Rep* 2008, **123:**282–299.
70. East IJ: **Adoption of biosecurity practices in the Australian poultry industries.** *Aust Vet J* 2007, **85:**107–112.
71. Pollock SL, Stephen C, Skuridina N, Kosatsky T: **Raising chickens in city backyards: the public health role.** *J Community Health* 2012, **37:**734–742.
72. Mitchell MA, Kettlewell PJ: **Physiological stress and welfare of broiler chickens in transit: solutions not problems!** *Poult Sci* 1998, **77:**1803–1814.
73. Duncan IJH: **Animal welfare issues in the poultry industry: is there a lesson to be learned?** *J Appl Anim Welf Sci* 2001, **4:**207–221.
74. RSPCA Australia: *Meat Chickens.* Deakin West: RSPCA; 2014 [http://www.rspca.org.au/how-you-can-help/campaigns/meat-chickens.html]
75. Hendershot W, Turmel P: **Is food grown in urban gardens safe?** *Integr Environ Assess Manag* 2007, **3:**463–464.
76. Frischknecht R, Jungbluth N, Althaus HJ, Bauer C, Doka G, Dones R, Hischier R, Hellweg S, Humbert S, Köllner T, Loerincik Y, Margni M, Nemecek T: *Implementation of Life Cycle Impact Assessment Methods. ecoinvent report No. 3, v2.0.* Dübendorf: Swiss Centre for Life Cycle Inventories; 2007.
77. Moller H: *Environmental Impacts From Capital Goods in LCA of Meat Products.* Bari: LCA Food; 2010.

9

Contribution of wetland resources to household food security in Uganda

Nelson Turyahabwe[1*], Willy Kakuru[2], Mnason Tweheyo[2] and David Mwesigye Tumusiime[3,4]

Abstract

Background: In Uganda, nearly 1.4 million people are currently food insecure, with the prevalence of food energy deficiency at the country level standing at 37%. Local farmers are vulnerable to starvation in times of environmental stress, drought and floods because of dependence on rain-fed agriculture. Accordingly, the farmer's means of increasing food production has always been an expansion of area under cultivation from virgin and fragile areas, especially wetlands. Consequently, Uganda has lost about 11,268 km^2 of wetland, representing a loss of 30% of the country's wetlands from 1994 to 2009. While the environmental importance of wetland ecosystems is widely recognized, their contribution to household food security is still hardly explored. In this paper an assessment of the contribution of wetland resources to household food security and factors influencing use of wetland resources in Uganda are reported.

Methods: A number of livelihood tools in food security assessment including focus group discussions, key informant interviews, direct observations and a household questionnaire survey, were used to collect the data. A total of 247 respondents from areas adjacent to wetlands were involved in the household questionnaire survey conducted in three agro-ecological zones that are frequently characterized as food insecure.

Results: The findings indicate that about 83% of the households experienced food insecurity. The main indicators of food insecurity were low harvest (30.9%) and when people buy locally grown food items (18%). Most households felt food secure when they had perennial crops (43.2%) in their gardens, or adequate money to buy food (23.9%). The prevalence of food insecurity was significantly lower among households with older and better educated household heads, but also among households located in Lake Victoria Crescent and South western farmlands agro-ecological zones, but significantly higher among households that were female headed, larger and participate in collection of wetland resources. Over 80% of the respondents reported that wetland resources provide products and services that contribute enormously to their household food security. Besides, they also indirectly contribute to food security by providing services that foster food production such as weather modifications and nutrient retention. Households with older heads and those that reside in the Lake Victoria Crescent agro-ecological zone when compared to counterparts in the Lake Kyoga agro-ecological zone are more likely to have a higher dependence on wetlands for food security.

Conclusions: With increasing population around the wetlands, coupled with land shortage and weather variations, households with limited options will continue to generally rely on wetlands for food security and income for sustaining their livelihoods unless alternative livelihood options are provided. There is thus a need to design appropriate food production technologies that ensure sustainable use of wetland resources for food security.

Keywords: Food security, Food insecurity, Wetland management, Wetland resources, Uganda

* Correspondence: turyahabwe@forest.mak.ac.ug
[1]Department of Extension and Innovation Studies, College of Agricultural and Environmental Sciences, Makerere University, P.O. Box 7062, Kampala, Uganda

Introduction

Food is the basic human need for survival, health and productivity. It is the foundation for human and economic development [1]. In a broad sense, food security exists when "all people, at all times, have physical, social and economic access to sufficient, safe and nutritious food that meets their dietary needs and food preferences to live an active and healthy life" [2]. On the other hand, food insecurity occurs when food systems are stressed such that food is not available, accessible or utilized properly.

Most households especially in developing nations are food insecure partly due to the rapid increase in human population, weather and climatic variability, and environmental degradation. For instance in 2009, it was estimated that 102 billion people were undernourished worldwide, which is about 37% higher than 20 years ago [3]. Sub-Saharan Africa and South Asia are the regions most affected by food insecurity, being home to 60% of the world's food insecure people. In the case of sub-Saharan Africa, the food situation is further aggravated by low per capita food availability, high fluctuations in food supply and lack of innovative ideas as well as responsive policies for sustainable use and management of natural resources. In sub-Saharan Africa, the predominance of rain-fed agriculture often results in food systems that are highly sensitive to rainfall variability [4,5]. The region thus remains susceptible to frequent food crises and famines. Sub-Saharan Africa is the only region of the world where hunger is projected to worsen over the next two decades unless drastic measures are taken to improve governance of its natural resources and achieve the economic development required to reverse the current trend [6].

Over, the past decades, there has been an increasing influx of people into wetland areas as a coping strategy, especially in areas where uplands are predominantly characterized by low agricultural potential, dominated by poor soils and low unpredictable rainfall [7-11]. This is further due to the presence of water in wetlands during the dry season, combined with their natural fertility and irrigation potential [12-14]. Wetlands are also used to secure food not only directly through dry season subsistence cultivation but also indirectly through income generation from cash crops, the production of clay for pottery, reed and palm mats, baskets and beehives, and the sale of collected items, thus acting as safety nets for most adjacent communities [15-18]. In the drier regions, wetlands are the only sites where people can get water, varieties of food and other basic supplies [19-21].

The National Development Plan (NDP) indicated that the food security situation in Uganda has been unsatisfactory [22]. Nearly 1.4 million people are currently food insecure despite the country's abundant resources [23], with the prevalence of food energy deficiency at the country level standing at 37% [1]. According to a report by the World Food Programme (WFP) [24], about 6.1 million (21%) people in Uganda are undernourished. The report further identifies that at the household level, about 6.3% of the households in Uganda are food insecure and that food insecurity is most common among the natural resources dependant households. About 86% of Uganda's population live in rural areas and are predominantly rural farmers and agricultural practice is predominantly rain-fed, characterized by low levels of crop productivity. The people are generally cash-poor, with over 40% living below the poverty line, on less than a dollar a day. Most of these people are perennially food insecure and are thus vulnerable to starvation in times of environmental stress, drought and floods [25].

Accordingly, the farmer's means of increasing agricultural output has always been an expansion of area under cultivation. Additional land is often brought under agriculture either through reduction in fallow periods or cultivation of virgin areas, especially forests and wetlands [25,26]. As noted by Nyakana [27], and Mwakubo and Obare [21], an increasing number of marginalized people are moving into fragile wetland areas, in search of new means of livelihood, including crop farming, fishing and livestock grazing. Crops commonly grown on the wetland periphery include: *Dioscorea* spp (yams), beans, *Zea mays* (maize), *Ipomoea batatas* (L.) Lam. (sweet potatoes), *Manihot esculenta* Crantz (cassava), *Brassica oleracea* var. *capitata* (cabbages), *Saccharum officinale* (sugar cane) and low land rice [28]. Consequently, Uganda has lost about 11,268 km^2 of wetland, down from 37,575 km^2 (15.6%) in 1994 to about 26,308 km^2 (10.9%) in 2009. This represents a loss of 30% of the country's wetlands [28]. To date, this loss is expected to be even higher.

Despite their contribution to rural livelihoods, wetland resources have been overlooked in national economic development planning. Thus, the current development pathways will continue to underestimate the significance of these resources, and in so doing, miss opportunities for reducing food insecurity and sustainable management of wetland systems. While the environmental importance of wetland ecosystems is widely recognized, the potential contribution of wetland resources to household food security is still hardly explored. Understanding the degree to which wetlands contribute to people's food security may be vital in steering decisions that minimize negative impacts or enhance the benefits that wetlands have for communities. As such, this study endeavored to provide this information. Such information is fundamental in developing interventions for sustainable wetland management in order to achieve the National Food Security targets and the Millennium Development Goal (MDG) 1 of halving the proportion of people who suffer from hunger by 2015, and MDG 7 on environmental sustainability. The objectives of this study were thus threefold, namely to: i) map the experiences and perceptions of the local people on the food security situation; ii) assess the contribution of wetlands to food security; and iii) determine factors that influence dependence on wetlands for food production in Uganda.

Methods

Study area

Out of the ten Ugandan agro-ecological zones, three were randomly selected (Lake Victoria Crescent, Kyoga plains and South western farmlands) (Figure 1). A National Wetland Inventory of 1999 [29] identifies four key factors by which to categories wetlands. These include agro-ecological factors of the wetland system and the level of food security, population density and farming systems of the local communities in proximity to the wetland system. Based on these characteristics, wetlands in each of the three zones were categorized into strata. Random selection was used to sample Munyere and Mabamba Bay wetlands in Wakiso district to represent the Lake Victoria Crescent agro-ecological zone. This zone is characterized by medium level of food security, and a high population density of 484 persons per km^2. Lake Opeta in Pallisa district and Limoto wetland in Kibuku district represented Kyoga

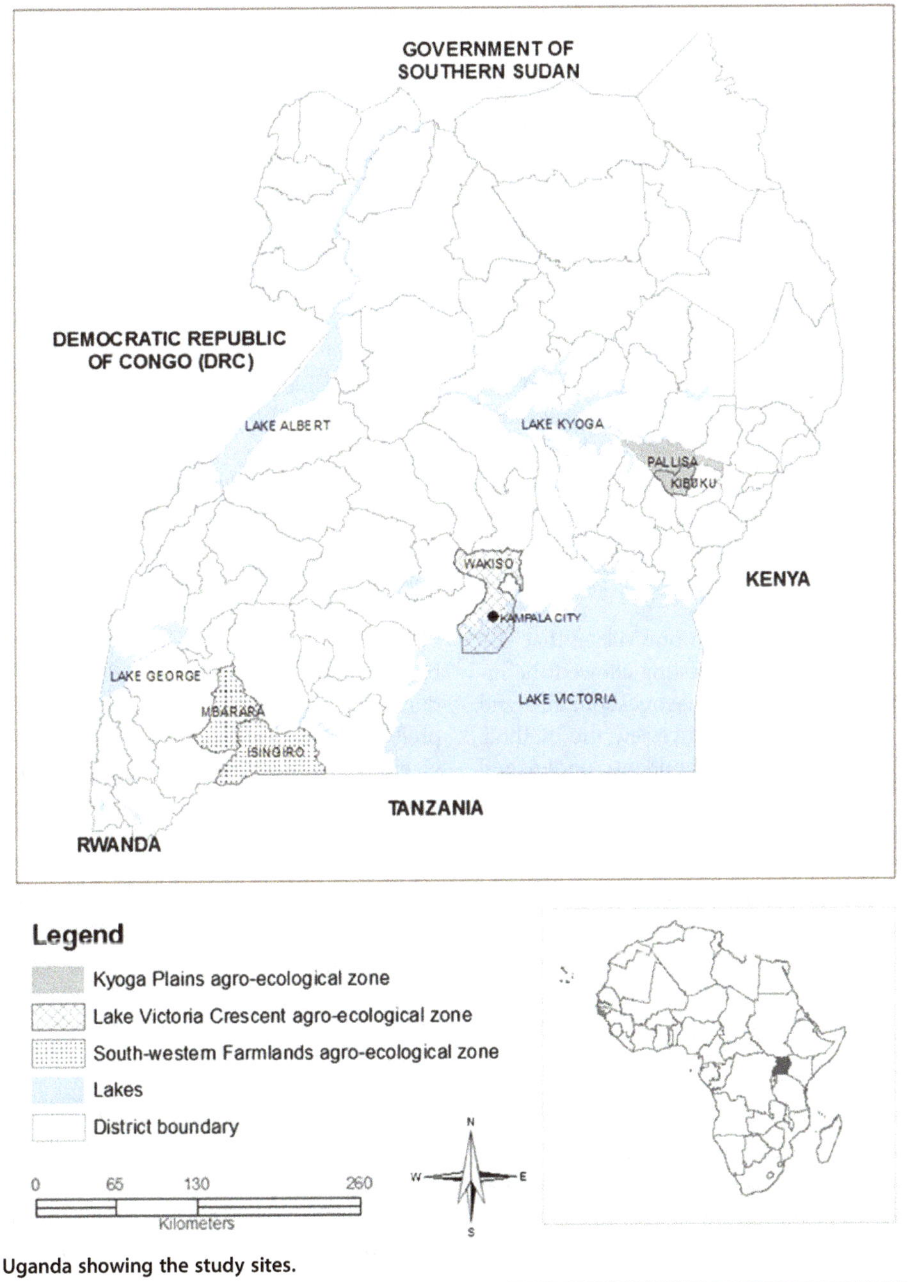

Figure 1 Map of Uganda showing the study sites.

plains agro-ecological zone characterized by small-scale subsistence, mainly annual crops with some pastoralism and with a high level of food insecurity and a moderate population of 252 persons per km^2. While Lake Nakivale in Isingiro district and Rucece in Mbarara district represented the south-western farmlands agro-ecological zone with moderate level of food security and a moderate population density of about 247 persons per km^2. The three zones represent three regions in Uganda which are often regarded as food insecure [24].

Data collection

We employed data collection methods and tools for assessing food security following Lisa *et al.* [1] and Young *et al.* [30], by taking a qualitative measure to capture people's own perceptions of the extent to which they suffer from food shortage [31] and their understanding of the causes of food insecurity; and information on what resources local people directly and indirectly obtain from wetlands for food security. Semi-structured interviews were administered to two hundred and forty seven (247) households randomly selected from 5 km of the wetland edges (*i.e.*, 82 from south-western farmlands agro-ecological zone, 81 from Lake Victoria Crescent and 84 from Kyoga plains). We targeted household heads as our interviewees, but in a few instances where the heads were absent, selected the most senior and knowledgeable of the adults present. We carried out the interviews in the common local language. Before conducting each individual interview, it was made clear that the purpose of the study was purely scientific and academic. It was emphasized that the study had no legal implications whatsoever, and the respondents were also assured of confidentiality and anonymity.

The questionnaire was pre-tested in one village that was not part of the selected sample. Pre-testing allowed the interviewers to gain familiarity with the questionnaire and provided an opportunity to apply and review the method. The focus was on assessing how respondents understood our questions and identifying any problems encountered in providing answers. Changes were proposed, reviewed and incorporated into our final questionnaire. The questionnaire focused on respondents' understanding of food security/insecurity issues, main wetland products harvested for consumption or sale and other activities undertaken that directly or indirectly contribute to household food security.

Further, one focus group discussion (FGD) was conducted in each sample parish to generate information on key wetland resources used by surrounding communities. Particular attention was given to women due to their key role in household food security [28]. Key informant interviews were held with staff from National Environmental Management Authority (NEMA), Wetlands Management Department, Wetland and Production units of the Respective District and sub-county Local Governments, Natural Resources Committees at the Districts, production and environmental Committees, National Agricultural Advisory Services (NAADS), and representatives of different wetland user groups to ascertain: the kind of wetland resources directly harvested for food security, the status of food security/insecurity in the areas and means of deriving livelihoods. Also, direct field observations were made on activities undertaken in the sample wetlands.

Data analysis

Questionnaire responses were edited, coded and analyzed using SPSS version 18.0 for Windows. Spatial differences were captured through the construction of three dummy variables based on location of the sample household in one of the three agro-ecological zones. In all cases, a dummy variable was coded 1 for a household located in that agro-ecological zone and 0 otherwise. Frequencies were generated to capture local experiences and perceptions of the food security situation. A binary variable depicting whether or not a household had experienced food insecurity in the last five years was regressed against household socioeconomic factors to determine factors that influence susceptibility to food insecurity [32]. The contribution of wetlands to food security was ascertained by examining the frequencies of the wetland products harvested for food security. Binary variables depicting whether or not a household i) had experienced food insecurity, and ii) depended on wetland resources for food security, were regressed against household socioeconomic factors to determine the latter's influence over these two outcomes. Marginal effects of each of these variables were examined using Stata Margeff program [33]. All categorical variables were specified as dummies using the dummies option and the agro-ecological zones were specified as dummies with Kyoga plains as the base. Following Wooldridge's [34] suggestion, for location dummies, we report exact percentage differences in the predicted dependent variable when a household is resident in either zone compared to Kyoga plains. The percentage difference is computed as $100[e^{(\beta\wedge)} - 1]$, where $\beta\wedge$, is the coefficient on the respective dummy variable; χ^2 tests were used to examine the existence of associations between the contribution of wetlands to food security and age, landholding, occupation, household size, gender and education level of the respondent. All statistical tests were performed at 5% significance level.

Results

Household characteristics

The majority of respondents (65%) were males, with an average age of 40 years. Most respondents (90%) had formal education. The main occupation of respondents was farming. The other sources of income included tailoring, petty trade, brick making, fishing and working as drivers, tourist guides, mechanics, builders and causal laborers.

The average household size was eight persons with some households having up to 21 members. On average, most respondents had stayed in the area for at least 24 years, and within a distance of 5 km to the wetland edge. Most respondents earned less than UGX 100,000 (approx. USD 40) per month with average monthly income of UGX 97,000 (approx. USD 39) indicating that they were generally poor. Most of the respondents owned less than five acres of land with average land owned at 3.8 acres, indicating a low per capita land holding.

Local experiences and perception of food security situation

Over 80% of the households reported experiences of food insecurity during the past five years and attribute it to different causes. The most frequently mentioned cause was climate change. Included in this category is the explicit mention of climate change (50%), mentions of the related causes: prolonged drought (42%) as in the quote below, floods (2%), and reduction in quantity of water in wetland (1%). Other relatively pervasive causes included limited access to land, labor and the sale of food crops to raise cash income (Figure 2).

> *In my opinion, the single most important cause of food insecurity in our villages are prolonged droughts which do not only scorch our crops in the fields, but also limit our abilities to plant new crops. Droughts affect crop and livestock farmers alike* (Participant in a focus group discussion in Isingiro District).

A number of factors are perceived as indicators of food insecurity, but main ones are situations of low household food harvest and when people buy locally grown food items (Figure 3).

However, prevalence of food insecurity is significantly lower among households with older and better educated household heads. However, the marginal effect of age is low (Table 1). Keeping other things equal, the likelihood of being food insecure reduces by only 0.3% for an additional year in age of household head. Education has a greater effect, and the likelihood of household experiencing food insecurity reduces by about 12% for every extra year its head spent in school. In terms of location, keeping other things equal, the difference in likelihood of experiencing food insecurity reduces by nearly 17% ($100[e^{(0.155)}-1]$) if a household is resident in south western farmlands as compared to Kyoga plains, and about 14% ($100[e^{(0.130)}-1]$) if resident in Lake Victoria Crescent.

On the other hand, keeping other variables equal, the likelihood of experiencing food insecurity increases by 13% if a household is female headed, by 1.3% for each extra member to a household, and 23% if a household participates in collection of wetland products for food security.

Contribution of wetlands to household food security

Wetlands were reported to contribute to household food security through provision of wetland products and services.

Wetland products

Over 75% of the respondents acknowledged that wetlands directly contribute to their household food security. They do so in a number of ways, but the most pervasive are three; direct consumption of wetland products, wetlands providing space for growing crops, and the sale of wetland products to raise cash income that is then used to purchase food (Figure 4). As one farmer noted in a focus group discussion in the Kyoga plains "wetlands are the lifeline of many a local farmer who feed on and sometimes sell wetland products".

Direct consumption of wetland products

Wetlands are reported to be a source of a variety of resources that are directly consumed among 86% of the sample households (Figure 4). Water was identified by up to 60% of the sample households as the most important product directly obtained for domestic and livestock use.

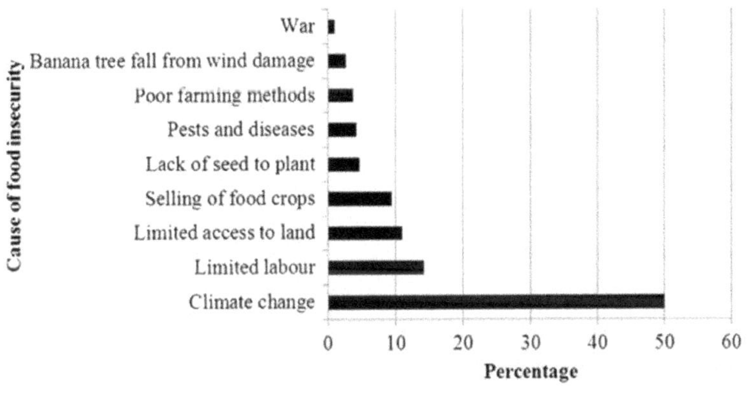

Figure 2 Frequency of mention of the different causes of food insecurity among households adjacent to wetlands in Uganda.

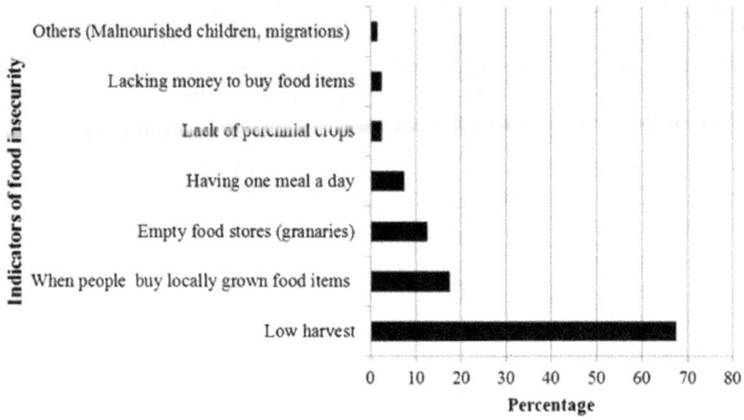

Figure 3 Perceived indicators of food insecurity among households adjacent to Wetlands in Uganda.

Other important products included fish and bush meat (mainly Sitatunga and wild rat). The main indigenous fruits harvested from wetlands were *Afromomum spp* (*A. angustfolium,* and *A.mildbraedi)* and *Physalis micrantha* (Entutu), while main vegetables included *Amaranthus spp* (dodo) *and Solanum nigrum* (Eswiga).

There was a significant association between collection of wetland products for direct home use and the income level of the respondents (χ^2 = 6.858, df = 3, P = 0.001). Generally, respondents earning less than UGX 100,000 (approx. USD 40) per month were more likely to collect products for direct home use. Other noticeable variations included variation in harvesting of fish with gender, age and household size. Fish collection is closely associated with males (χ^2 = 5.755, df = 1, P = 0.016), younger individuals (<45 years) (χ^2 = 5.307, df = 2, P = 0.042), and members of larger households (> 7 members) (χ^2 = 24.984, df = 2, P = 0.000). The latter were also more associated with collection of fruits and vegetables (14.1%) (χ^2 = 6.901, df = 2, P = 0.032), papyrus materials (χ^2 = 10.405, df = 2, P = 0.006), and

Table 1 Household characteristics that influence susceptibility to food insecurity among households adjacent to wetlands in Uganda

Dependent variable*	Units	Coef.	Std. Err.	z	P >z
Household characteristics					
Age	Years	−0.003	0.002	−1.80	0.071
Household female headed	0/1	0.129	0.045	2.85	0.004
Education level	Years	−0.117	0.028	−4.14	0.000
Household size	Number	0.013	0.006	2.09	0.037
Length of stay in the area	Years	0.001	0.001	0.66	0.508
Distance to wetland	km	0.061	0.043	1.40	0.160
Location in south-western farmlands	0/1	−0.155	0.083	−1.87	0.062
Location in Lake Victoria Crescent	0/1	−0.130	0.078	−1.66	0.097
Primary occupation is farming	0/1	−0.029	0.068	−0.43	0.669
Primary occupation is formal employment	0/1	0.038	0.105	0.36	0.719
Household collects wetland products	0/1	0.229	0.117	1.97	0.049
Number of observations					236
Log likelihood					−83.1
LR χ^2 (11)					42.04
Prob >χ^2					0.000
Pseudo R2					0.202

*Dependent variable is a binary variable depicting whether or not a household had experienced food insecurity (1 = Household had experienced food insecurity in the last five years; 0 = Otherwise).

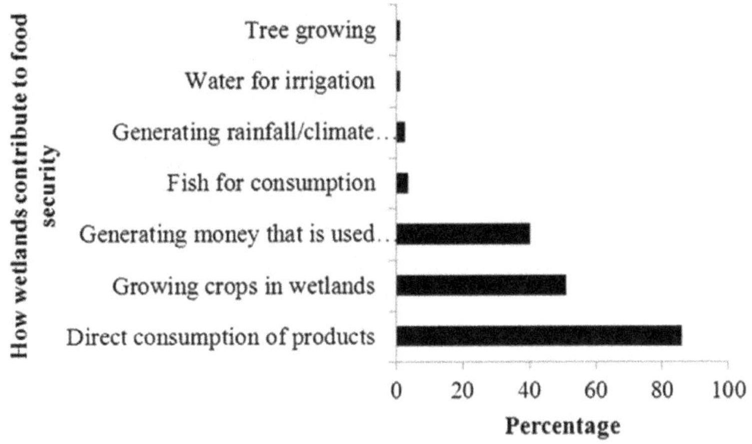

Figure 4 Ways through which wetlands contribute directly to household food security in areas adjacent to wetlands in Uganda.

medicinal herbs (χ^2 = 11.760, df = 2, P = 0.003). The herbs particularly played an important role in maintaining the health of local people.

Growing of crops in wetlands

Wetlands also contributed to food security through providing space for growing crops, and the sale of wetland products to raise cash income that is then used to purchase food (Figure 4). The main crops grown by respondents in wetlands were vegetables, sugarcane, coco yams and paddy rice. Paddy rice was the main crop grown in Kyoga plains, and vegetables in south-western farmlands and Lake Victoria Crescent. Vegetable production was common to all wetlands. The main vegetables grown were *Amaranthus dubius* (Dodo), *Amaranthus lividus* (*A. blitum*) Ebugga, *Solanum aethiopicum* (Nakati), *Brassica olerace* (cabbages), tomatoes (*Solanum lycopersicum*), *Solanum melongena L.* (egg plants), pumpkins (*Cucurbita maxima* Lam.), watermelons *Citrullus lanatus* (Thunb.) Matsum and Entura (*Solanum aethiopicum gilo.*). The other crops grown were maize, sweet potatoes, fruits, ground nuts and cotton. There are observable significant differences in the extent of use of wetlands for growing crops across the agro-ecological zones (χ^2 = 33.34, df = 2, P = 0.000). The practice is widespread in Kyoga plains where households report to derive about half of the household food from the wetlands particularly through cultivation of paddy rice, and an average household derives more than half of its cash income from the sale of crops grown in the wetlands.

Sale of wetland products for cash

Wetlands also indirectly contributed to food security through provision of resources sold for cash that was used to buy food. About 60% of the respondents harvested and sold wetland resources for cash to purchase food. Of these, about 38.4% spent one quarter of the income generated from sale of wetland products to buy food, 28.5% one quarter to half, 20.1% spent a half to three quarters and 12.9% three quarters. Main food items bought with such cash were mainly food security crops such as millet, maize flour and cassava.

There was a wide variation across agro-ecological zones in participation in these sales. Highest records are in the south-western farmlands where up to 86% of the households participate as compared to 46% and 38% in Kyoga plains and Lake Victoria Crescent, respectively. The main items sold were fish, fruit and vegetables, papyrus, medicinal herbs and craft materials, and these jointly generated a monthly income of UGX 99,208 (approx. USD 36) to an average household. However, the income accruing to an average household differed significantly between the agro-ecological zones (F = 3.89, df = 2, P = 0.022). Households in Kyoga plains obtained a significantly higher amount of cash from such sales (Figure 5). Fish was generally ranked as the most frequently sold resource, followed by poles which were either sold for cash or processed into different items such as hoe handles, and mortars and pestles for local processing of food.

There was an association between dependence on sales of wetland products for purchasing household food and income level (χ^2 = 21.742, df =3, P = 0.000), proximity to the wetland (χ^2 = 7.291, df =1, P = 0.000) and size of land owned (χ^2 = 9.508, df =1, P = 0.000). The practice was most frequent among households with monthly income less than UGX 200,000 (approx. USD 80.), living within a distance of less than <1 km of the wetland edge, and owning less than one acre of land.

Wetland services

Wetlands provided services to local people that enhanced their livelihoods in meeting their food security requirements. The most pervasive of these services include weather modification, cleaning water before local use, acting

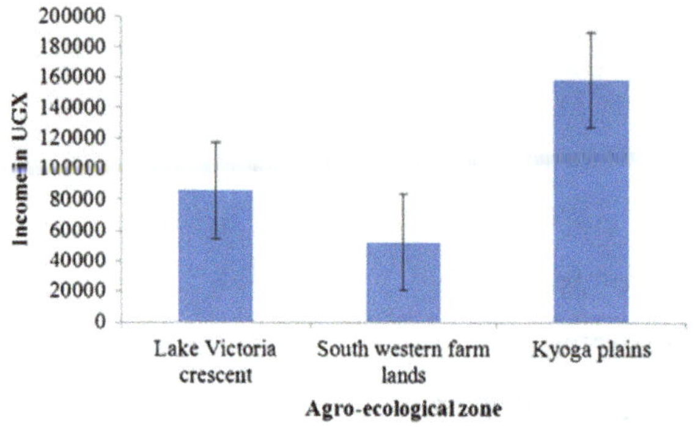

Figure 5 Monthly cash income from sale of wetlands products in three agro-ecological zones of Uganda.

as breeding grounds for fish, and provision of water transport and tourism (Figure 6).

Factors that influence dependence on wetlands for food production

There is a wide variation in dependence on wetlands for food security (Table 2). Keeping other things equal, seniority in age of the household head increases dependence on wetlands for food security, although only marginally. For every additional year the likelihood increases by 0.3%. On the other hand, residence in the Lake Victoria Crescent increases this likelihood by about 6% ($100[e^{(0.056)}-1]$) compared to residence in Kyoga plains. On the other hand, keeping other things equal, a female headed household is 15% less likely to depend on wetlands for food security.

Discussion

People's experiences of food insecurity

The poor socio-economic status of the households, particularly the low incomes, suggests susceptibility to food insecurity. As noted by Kydd *et al.* [35] and DFID [36], food insecurity is closely related to poverty and an inability to purchase food, especially in the agriculture-based rural areas. Uganda is frequently considered food secure, especially when compared to the other eastern African countries, but this study reveals high incidences of food insecurity. While there is increasing recognition that situations of food insecurity are brought about by a complex and dynamic set of causes [37], food insecurity in Uganda is increasingly reported to coincide with harsh weather conditions especially during the prolonged droughts and periods of heavy rain. Changes in the patterns of extreme weather events such as floods and droughts affect food production as well as stability of and access to food supplies. This is a general trend in crop production, where huge losses due to crop failures, arising from droughts and flooding, are being experienced more frequently than ever before in Africa, causing famines and economic hardships [38]. Local people's linkage of food insecurity to climate change is supported by, among others, Vlassenroot *et al.* [38], who assert that food insecurity is a direct consequence of food shortages caused by climatic variables or demographic pressures. Perhaps this could be the reason why local people use wetlands as safety nets during the drought since wetlands are able to store more water and moisture during the dry periods. Taking a single meal a day seems to be a commonly agreed indicator of food insecurity, but also adapted as a coping strategy across sub-Saharan Africa [39].

Prevalence of food insecurity is significantly lower among households with older and better educated household heads because ability to access assets needed to secure livelihoods increase with seniority and, as noted by Muchapodwa [40], education makes it easier for households to comprehend negative externalities and be able to generate cash for buying food. Households located in Lake Victoria Crescent and south-western farmlands are less susceptible to food insecurity because both zones have better access to perennial crops, particularly bananas. The south-western farmlands are the main areas for banana production in Uganda and have a vibrant livestock-based economy that can complement crop production for both cash and subsistence

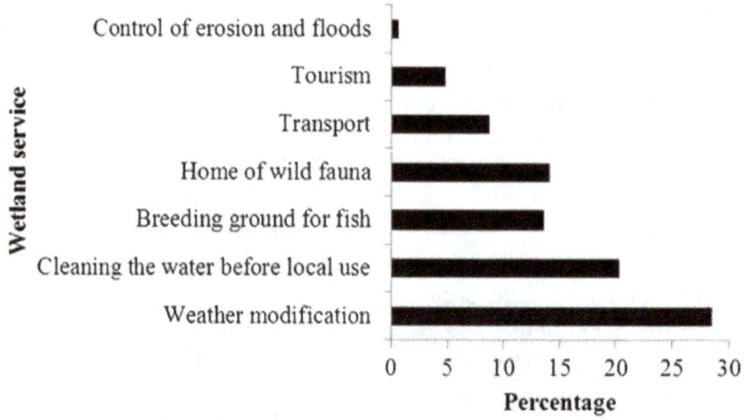

Figure 6 Wetland services indirectly contributing to household food security in Uganda.

Table 2 Factors influencing dependence on wetlands for food security among households adjacent to wetlands in Uganda

Dependent variable*	Units	Coef.	Std. Err.	z	P >z
Household characteristics					
Age	Years	0.003	0.001	2.12	0.034
Household female headed	0/1	−0.150	0.080	−1.86	0.062
Education level	Years	−0.008	0.020	−0.40	0.693
Household size	Number	−0.001	0.003	−0.17	0.868
Length of stay in the area	Years	−0.001	0.001	−0.66	0.512
Distance to wetland	km	0.003	0.030	0.09	0.931
Location in South western farmlands	0/1	−0.041	0.039	−1.05	0.295
Location in Lake Victoria Crescent	0/1	0.056	0.031	1.83	0.067
Primary occupation is farming	0/1	−0.033	0.038	−0.85	0.394
Primary occupation is formal employment	0/1	−0.208	0.188	−1.11	0.268
Household suffers food insecurity	0/1	0.087	0.062	1.39	0.165
Number of observations					236
Log likelihood					−41.79
LR $\chi^2(11)$					22.67
Prob $>\chi^2$					0.019
Pseudo R2					0.213

*Dependent variable is a binary variable depicting whether or not a household depends on wetlands for food security (1 = Household is dependent; 0 = Otherwise).

incomes. On the other hand, the Lake Victoria Crescent is adjacent to Kampala, Uganda's capital city and the industrial town of Jinja, and the local people thus have better access to alternative sources of cash income that can then be used to purchase food.

On the other hand, the prevalence of food insecurity is higher among female headed households, possibly because they frequently have less access to adult labor [41], and may lack the means to seek employment away from their families [42]. Generally, female headed households are more vulnerable to food insecurity because of their tighter time schedules and income constraints [43-45].

Larger households were more food insecure because these typically have a greater dependence and consumer burden [46]. This then becomes a problem in areas where production is already low, typical of most rural areas in Uganda.

Increased susceptibility to food insecurity for a household that utilizes wetlands for food security is a relationship that can be interpreted in two ways: i) households collected wetland resources were susceptible to food insecurity because of their dependence on wetlands; or ii) households collected wetland resources because they were food insecure. The latter seems to be more likely as food insecure households turn to collecting wetland resources as safety net or gap-filler. This is typical in sub-Saharan countries where households with limited livelihood options utilize common pool resources, including wetlands [11,47] and forest resources [41,48] as a means of meeting household food and income.

Contribution of wetlands to household food security

A vast majority of people were directly dependent on wetland resources for food security similar to previous studies and reports in Uganda, particularly among resource poor and larger households [28,49,50], for whom wetlands are then a major source of cash and subsistence income that supplements other sources.

The practice of growing crops in wetlands may be the preponderance of rural households with limited access to other productive assets (such as land) or sources of income. This is evident among the households studied with average land size of 3.8 acres and household size of eight persons. Studies conducted elsewhere in Uganda indicated that in situations of inadequate land, people resort to natural resources including wetlands as alternative sources of land for crop and livestock farming [28,51,52]. Studies carried out elsewhere in Africa also report rural dependence on wetlands for food production [20]. This is mainly because wetlands have some relatively higher levels of water/moisture, particularly in the dry season, compared to the surrounding catchment areas. In some cases, wetlands have some relatively high levels of fertility due to the silt accumulated by run-off from surrounding catchments. Given the current unpredictable rainfall, it is inevitable that local communities will largely rely on wetlands for food security, especially during prolonged dry periods, a situation also noted by Grimble *et al.* [53], which justifies the need for research in more technological and social innovations to improve sustainable use of wetlands for improved food security.

Wetlands take a more pronounced role in Kyoga plains where rice is grown as a main source of both food and cash. Similar levels of reliance on rice growing were reported by Karanja *et al.* [49]. In Rwanda's wetlands of Cyabayaga, rice was also reported as the largest contributor to household income providing on average $1,045 (approx. UGX 2,612,500) per household per season [14]. Wetlands in the two districts are also one of the most productive and resourceful areas, which provide food from aquatic ecosystems such as fish for the local people.

The local people recognize wetlands to provide a variety of services, which reinforces local valuation of wetlands and, by extension, willingness to participate in wetland conservation.

Factors influencing dependence on wetlands for food production

Wetland utilization for food security is higher among households with older heads possibly because these have better access to social networks through which to access common pool resources including wetlands. On the other hand, residence in the Lake Victoria Crescent is associated with higher dependence on wetlands because the zone has many urban poor households that at best have very limited access to land and thus end up encroaching on wetlands which are still regarded as common pool resources. Lower utilization of wetlands among households headed by females is possibly because of lower access to social networks crucial for access to such common pool resources.

Conclusions and recommendation

Over 80% of all households had experienced food insecurity, characterized by low harvest and households having a single meal in a day, and the main causes being unpredictable weather and inadequate arable land. The prevalence of food insecurity is higher in large sized, female headed and uneducated households. Land shortage, household size coupled with limited off-farm opportunities to generate cash for purchase of food drive local people into utilization of wetland resources as an alternative source of household food.

Wetlands are the basis of food security, directly providing resources for consumption, indirectly supporting crop and livestock production, materials that are sold for purchasing food in emergency situations and services that support food production. The most interesting feature of wetlands is that they provide conditions that enable a wider range of crops than dry lands, and therefore provide ready food supplies to wetland adjacent communities during unfavorable conditions that are otherwise unavailable for the traditional crops grown in the uplands. Beyond subsistence agriculture, wetlands are increasingly offering products for additional income through cultivation of locally marketable crops such as rice, sugar cane and vegetables. These products are sold and income is used to buy household food supplements.

With increasing population around the wetlands, coupled with land shortage and weather variations, the poor people, especially in rural areas, will continue to generally rely on wetland ecosystem services directly for subsistence and income generating activities for sustaining their livelihoods unless alternative livelihood options are provided. With rain-fed agriculture being the primary food production option for people living adjacent to wetlands in Uganda, there are risks of many people being vulnerable, and who could see their food security seriously limited. Thus, there is a need to design food production technologies and alternative income generating activities that ensure sustainable use of wetland resources for food security.

Abbreviations

FAO: Food and Agriculture Organization of the United Nations; MDG: Millennium Development Goal; NDP: National Development Plan; NEMA: National Management Authority; WFP: World Food Programme.

Competing interests

The authors declare that they have no competing interests.

Authors' contribution

NT designed the study, collected data and coordinated the write up. WK participated in collection and analysis of data and write up. MT participated in literature search and write up. DMT participated in data analysis and write up. All authors read and approved the manuscript.

Authors' information

NT is an Associate Professor in the Department of Extension and Innovation Studies, College of Agricultural and Environmental Sciences, Makerere University.
WK is a part time lecturer in the School of Forestry, Environmental and Geographical Sciences, Makerere University and a Consultant on Natural Resources Management.
MT is an Ecologist and a Professor in the Department of Forestry, Biodiversity and Tourism, School of Forestry, Environmental and Geographical Sciences, College of Agriculture and Environmental Sciences, Makerere University.
DMT is a Research Fellow at the Norwegian University of Life Sciences and lectures at the Department of Environmental Management, School of Forestry, Environmental and Geographical Sciences, Makerere University (Uganda).

Acknowledgements

The authors would like to thank the International Development Research Centre (IDRC) for funding this study in partnership with Makerere University and Wetlands Management Department, Ministry of Water and Environment and National Agricultural Research Organisation (NARO). We would like to thank Fred Yikii, Teddy Tindamanyire and Vincent Barugahare for participation in data collection and preparation of the draft report. We would also like to thank Lucy Iyango, Gilbert Ituka, Euzobia Arinaitwe and Asiimwe Joseph for participation in data collection. We would like to thank communities around wetlands in Wakiso, Pallisa, Kibuku, Isingiro and Mbarara Districts for their cooperation during data collection.

Author details

[1]Department of Extension and Innovation Studies, College of Agricultural and Environmental Sciences, Makerere University, P.O. Box 7062, Kampala,

Uganda. [2]Department of Forestry, Biodiversity and Tourism, College of Agricultural and Environmental Sciences, Makerere University, P.O. Box 7062, Kampala, Uganda. [3]Department of Environmental Management, School of Forestry, Environment and Geographical Sciences, Makerere University, P.O. Box 7062, Kampala, Uganda. [4]Department of International Environment and Development Studies (Noragric), Norwegian University of Life Sciences (UMB), P.O. Box 1432, Aas, Norway.

References

1. Lisa LC, Alderman H, Aduayom D: *Food Insecurity in Sub-Saharan Africa: New Estimates from Household Expenditure Surveys. Research Report, 146. Volume Research Report.* Washington, DC: 146, International Food Policy Research Institute (IFPRI); 2006.
2. Food and Agriculture Organization of the United Nations: *The State of Food Insecurity in the World.* Rome: FAO; 2001.
3. Food and Agriculture Organization of the United Nations: *The State of Food Insecurity in the World.* Rome: FAO; 2009.
4. Cooper PJM, Dimes J, Rao K, Shapiro B, Twomlow S: **Coping better with current climatic variability in the rain-fed farming systems of Sub-Saharan Africa: An essential first step in adapting to future climate change?** *Agric Ecosystem Environ* 2008, **126**:24–35.
5. Jones PG, Thornton PK: **Croppers to livestock keepers: Livelihood transitions to 2050 in Africa due to climate change.** *Environ Sc. Policy* 2009, **12**:427–437.
6. Food and Agriculture Organization of the United Nations: *Food Security and Agricultural Development in Sub-Saharan Africa: Building a Case for More Public Support.* Rome: FAO; 2006.
7. Ellis-Jones J, Mudhara M: **Factors affecting the adoption of soil and water conservation technologies in semi-arid Zimbabwe.** In *Soil and Water Conservation for Small Farmers in Semi Arid Zimbabwe, Transfers between Research and Extension. Proceedings of Technical Workshop, 3–7 April 1995, Masvingo Zimbabwe.* Edited by Twomlow S, Ellis-Jones J, Hagmann J, Loos H. Masvingo: Silsoe Research Institute Report OD/95/16; 1995.
8. Mutambikwa A, Muzenda S, Chivinge O, Ellis-Jones J, Riches C, Twomlow S: *Participatory Evaluation of Vlei Utilization and Weed Problems in Communal, Resettlement and Small Scale Commercial Farming Systems of Masvingo and Gutu districts; Report IDG/00/8.* Silsoe Bedford: Silsoe Research Institute; 2000.
9. Dixon AB: **The hydrological impacts and sustainability of wetland drainage cultivation in Illubabor, Ethiopia.** *Land Degr Develop* 2002, **13**:17–31.
10. Dixon AB, Wood AP: **Wetland cultivation and hydrological management in eastern Africa: matching community and hydrological needs through sustainable wetland use.** *Nat Resources Forum* 2003, **27**:117–129.
11. Halima KH, Munishi KTP: **Contribution of wetlands to household income and food security in the Nyumba Ya Mungu wetland system, northern Tanzania.** *Tanzania J Forestry Nat Conserv* 2009, **79**(2):99–108.
12. Adams WM: **Indigenous use of wetlands and sustainable development in West Africa.** *Geogr J* 1993, **159**:209–218.
13. Abbot P, Afework H: *Report on the PRA Training Workshop for Sustainable Wetland Management.* EWRP, Metu. Working Paper; 1997.
14. Nabahungu NL, Visser SM: **Contribution of wetland agriculture to farmers' livelihood in Rwanda.** *Ecol Econ* 2011, **7**:4–12.
15. Maclean I, Tinch R, Hassall M, Boar R: *Social and Economic Use of Wetland Resources: A Case Study from Lake Bunyonyi, Uganda. Working Paper ECM 03-09.* Norwich, United Kingdom: University of East Anglia, Centre for Social and Economic Research on the Global Environment; 2003.
16. Schuyt KD: **Economic consequences of wetland degradation for local populations in Africa.** *Ecol Econ* 2005, **53**:177–190.
17. Turpie J, Barnes J, Arntzen J, Nherera B, Lange GM, Buzwani B: *Economic Value of the Okavango Delta, Botswana and Implications for Management.* Okavango Delta Management Plan; 2006.
18. Shackleton CM, Shackleton S: **Household wealth status and natural resource use in the Kat River valley, South Africa.** *Ecol Econ* 2006, **57**:306–317.
19. Masiyandima M, McCartney MP, Van Koppen B: *Wetland Contributions to Livelihoods in Zambia FAO Netherlands Partnership Program: Sustainable Development and Management of Wetlands* Rome: FAO; 2004:50.
20. McCartney MP, Van Koppen B: *Wetland Contributions to Livelihoods in United Republic of Tanzania. FAO Netherlands Partnership Program: Sustainable Development and Management of Wetlands.* Rome: FAO; 2004:42.
21. Mwakubo SM, Obare GA: **Vulnerability, livelihood assets and institutional dynamics in the management of wetlands in Lake Victoria watershed basin.** *Wetl Ecol Manag* 2009, **17**:613–626.
22. The Republic of Uganda: *National Development Plan for Uganda. 2010/11-2014/15.* Kampala; 2010.
23. Famine Early Warning Systems Network: *Uganda Food Security Outlook.* Kampala: FEWS-Net; 2010.
24. United Nations World Food Programme: *Comprehensive Food Security and Vulnerability Analysis (CFSVA) in Uganda.* UNWFP; 2009.
25. International Food Policy Research Centre: *IFPRI Strategy Towards Food and Nutrition Security.* Washington, DC: International Food Institute; 2003.
26. International Fund for Agricultural Development: *Rural Poverty Report.* London: Oxford University Press; 2001.
27. Nyakana JB: **Sustainable wetland resource utilization of Sango Bay through eco-tourism development.** *Afr J Environ Sci Technol* 2008, **2**:326–335.
28. Wetlands Management Department: *Ministry of Water and Environment, Uganda Bureau of Statistics, International Livestock Research Institute, World Resources Institute: Mapping a Better Future: How Spatial Analysis can Benefit Wetlands and Reduce Poverty in Uganda.* Washington DC and Kampala: World Resources Institute; 2009.
29. Ministry of Water: *Lands and Environment: National Wetlands Programme - End of Phase III report.* Kampala: Wetlands Inspection Division; 1999.
30. Young H, Jaspars S, Brown R, Frize J, Khogali H: *Food-Security Assessments in Emergencies: A Livelihoods Approach.* London: Humanitarian Practice Network Paper No.36, Overseas Development Institute; 2001.
31. Kennedy E: **Qualitative measures of food insecurity and hunger.** In *Measurement and Assessment of Food Deprivation and Under Nutrition.* Rome: FAO; 2003:165–180.
32. Agresti A: *Categorical Data Analysis.* 2nd edition. Hoboken, New Jersey: Wiley Series in Probability and Statistics; 2002.
33. Bartus T: **Estimation of marginal effects using Margeff.** *Stata J* 2005, **5**:309–329.
34. Wooldridge JM: *Introductory Econometrics: A Modern Approach.* UK: South-Western; 2009.
35. Kydd J, Dorward A, Morrison J, Cadisch G: *Agricultural Development and Pro-Poor Economic Growth in Sub-Saharan Africa: Potential and Policy. ADU Working Paper 02/04.* Imperial College: Wye; 2002.
36. Department for International Development: *Agriculture, Growth and Poverty Reduction.* London: DFID; 2004.
37. Devereux S: *Famine in the Twentieth Century. IDS Working Paper No. 105.* Brighton: Institute of Development Studies; 2000.
38. Mwendera E: *Situation Analysis for Water and Wetlands Sector in Eastern and Southern Africa: Draft Report, International Union for Conservation of Nature (IUCN).* Nairobi: Eastern and Southern Africa Regional Office (ESARO); 2010.
39. Vlassenroot K, Salomé Ntububa S, Raeymaekers T: *Food Security Responses to the Protracted Crisis Context of the Democratic Republic of the Congo.* 2006.
40. Muchapodwa E: *The Economics of Community-Based Wildlife Conservation in Zimbabwe. PhD Thesis, Department of Economics.* Harare: Göteborg University, Sweden and University of Zimbabwe; 2003.
41. Vedeld P, Angelsen A, Sjaastad E, Kobugabe BG: *Counting on the Environment: Forest Incomes and the Rural Poor, Environment Economics Series Paper 98.* Washington DC: World Bank; 2004.
42. Shackleton C, Shackleton S: **The importance of non-timber forest products in rural livelihood security and as safety nets: a review of evidence from South Africa.** *South African J Sci* 2004, **100**:658–664.
43. Caldwell R, Seroussi M, Hagens C, Huddle J, Mbizule C, Uny K: *Malawi Baseline Survey: Report Findings Prepared for C-SAFE Tueson.* Ariz, USA: Tango International; 2003.
44. Food and Agriculture Organisation of United Nations: *Women and Food Security.* Rome: FAO; 1996.
45. Gladwin C, Thomson A: *Food or Cash Crops? Which is the Key to Food Security: Prepared for the University of Florida's Gender and Soil Fertility in Africa.* 2004.
46. Godoy R, O'neill K, Groff S, Kostishack P, Cubas A, Demmer J, McSweeney K, Overman J, Wilkie D, Brokaw N: **Household determinants of deforestation by Amerindians in Honduras.** *World Dev* 1997, **25**:977–987.
47. MEA (Millennium Ecosystem Assessment): *Ecosystems and Human Well-Being: Wetlands and Water, Synthesis.* Washington, DC, USA: World Resources Institute; 2005.
48. Tumusiime DM, Vedeld P, Gombya-Ssembajjwe W: **Breaking the law? Illegal livelihoods from a protected area in Uganda.** *Forest Policy Econ* 2011, **13**:273–283.

49. Karanja F, Emerton L, Mafumbo J, Kakuru W: *Assessment of the Economic Value of Pallisa District Wetlands.* Kampala, Uganda: Biodiversity Economics Programme for Eastern Africa, IUCN-The World Conservation Union and Uganda National Wetlands Programme; 2001.
50. National Environment Management Authority (NEMA): *State of the Environmental Report for Uganda, 2008.* Kampala; NEMA; 2008
51. National Environment Management Authority (NEMA): *Environmental Sensitivity Atlas for the Albert Graben.* Secondth edition. Kampala: NEMA; 2010.
52. Rwanda Environmental Management Authority: *Rwanda State of Environment and Outlook: Our Environment for Economic Development.* Kigali: REMA; 2009.
53. Grimble R, Cardoso C, Omar-Chowdhury S: *Poor People and the Environment: Issues and Linkages.* 16th edition. Greenwich: Natural Resources Institute; 2002.

Developing GM super cassava for improved health and food security: future challenges in Africa

Ademola A Adenle[1,2*], Ogugua C Aworh[3], Richard Akromah[4] and Govindan Parayil[1]

Abstract

Background: There is an urgent need to solve the problem of micronutrient malnutrition that is prevalent among young children and women in Africa. Genetically modified (GM) biofortified cassava has great potential to solve part of this problem, but controversy surrounding GM technology and lack of awareness, limited facilities, biased news and other factors may hinder the adoption of GM cassava in the future.

Method: Using semi-structured interviews in Ghana and Nigeria, this paper examines the perspectives of scientists, including the BioCassava Plus (BC+) team, on the potential adoption of GM cassava for improving health and food security in Africa. The article also examines issues around the regulatory system and transfer and acceptance of GM cassava among scientists.

Results and discussion: The result suggests that an overwhelming majority of scientists agree that GM biofortified cassava will benefit the health of millions in Africa, and that GM cassava conferred with disease and pest resistance will increase cassava production as it is currently plagued by cassava mosaic diseases (CMD). However, respondents are wary of long-term effects of GM cassava on the environment and lack of a regulatory framework to facilitate the adoption of GM cassava. Even though scientists expressed little or no concern about health risks of GM cassava, they were concerned that consumers may express such concerns given limited understanding of GM technology.

Conclusion: The article concludes with a summary of priorities for policy development with regard to adopting biofortified food products.

Keywords: GM biofortified cassava, Health, Food security, Policy, Scientists, Africa, BioCassava Plus (BC+)

Introduction

Cassava (*Manihot esculenta* Crantz) is a staple food crop grown by poor farmers on marginal lands, particularly in Africa, Asia and Latin America. Around 200 million tons of cassava are currently produced globally, of which 51%, 34% and 15% are produced by Africa, Asia and Latin America respectively [1]. Cassava is one of the major sources of farm income and is an important food security crop for the people of Africa. It is the third African major staple food crop after maize and rice, and contributes about 40% of the food calories consumed in tropical Africa [2]. Despite the fact that cassava is very limited in nutritional value, particularly essential micronutrients such as vitamin A, iron and zinc, it remains an important diet for over 250 million people in Africa. The provision of adequate essential micronutrients is critical to health improvement in Africa, particularly in Sub-Saharan Africa (SSA). These essential micronutrients play a significant role in metabolic activities and energy storage in the human body, particularly in women during menstruation and pregnancy, and in children during normal growth and development [3].

Despite the understanding of the importance of micronutrients, micronutrient deficiencies remain a huge problem among young children and women in SSA. More than 800 million people suffer from micronutrient

* Correspondence: adenle@ias.unu.edu
[1]United Nations University-Institute of Advanced Studies (UNU-IAS), 6F International Organisations Centre, Pacifico Yokohama, 1-1-1 Minato Mirai, Nishi-ku, Yokohama 220-8502, Japan
[2]National Graduate Institute for Policy Studies (GRIPS), 7-22-1 Roppongi, Minato-ku, Tokyo 106-8677, Japan

malnutrition in developing countries with Africa accounting for almost 50% of the children who are clinically or sub-clinically deficient in vitamin A, particularly under five years of age [4,5]. This deficiency causes (preventable) blindness in children and confers an increased risk of infection due to immune system impairment. Iron deficiency, which causes anemia, is also a problem. According to the United Nations [6], the prevalence of anemia in preschool children ranges from 42% to 53% in East and West Africa, respectively. Additionally, it is estimated that in West Africa (Benin and Burkina Faso) the prevalence of anemia (Hb < 110 g/l) in children under five years old is more than 80% [7]. Lastly, in 2004, 453,207 deaths were recorded as a result of zinc deficiency (a major cause of child death due to severe diarrhea) worldwide; 260502 (57%) of these deaths were in Africa, mainly among children between six months and five years of age [8].

Clearly, the prevalence of diseases caused by micronutrient deficiencies is a reflection of inadequate diets or poor diets in this region. One reason for these poor diets is that the majority of people living in this region are extremely poor and can not afford micronutrient-rich foods such as milk, meat, eggs and fish that normally form part of essential balanced diets. This is simply because these foods are too expensive. Therefore, people in this region need a diverse diet including vegetables, fruits, animal-sourced foods and fortified staple foods at low-cost. Although there have been efforts to alleviate micronutrient malnutrition in this region, the diseases remain highly prevalent. Past efforts have included: promoting dietary diversification, promoting improved locally available resources, improving nutritional knowledge, providing pharmaceutical products and/or dietary fortified food supplements. For example, West African countries, such as Burkina Faso and Mali have promoted the increased production of Vitamin A rich-foods in school and local communities through the support of international non-governmental organizations (NGOs) such as Helen Keller International (HKI) in collaboration with national governments [9]. In another example provided by Birner *et al.*, [9], enriched local products, such as infant porridge (mother-and-child clinics), and locally processed pre-prepared flours, such as Misola (millet, soya and peanut), have been promoted in a National Plan of Action for Nutrition and National Program for Food Fortification in Burkina Faso and Mali, respectively. As mentioned, despite some efforts by a few national governments, the majority of people in this region still lack access to a diversified diet due to a high level of poverty, inefficient and poor control of locally available resources, and an inability of the African governments to provide adequate support to, or otherwise prioritize alleviating micronutrient deficiencies as part of their public health programs.

We believe that the development of foods with increased nutritional value and/or content of vitamins and minerals, a process termed biofortification, is required to solve part of the problem of micronutrient malnutrition. Biofortification of staple food crops has been advocated by others as one of the possible solutions to combat the scourge of micronutrient malnutrition in developing countries including Africa [10]. According to the Millennium Development Goal One (MDG-1), "the prevalence of underweight children under five years of age is an indicator where malnutrition contributes in turn to poverty and hunger"[11]. The use of biofortification in reducing child mortality is fundamental to achieving MDG-1by 2015.

For example, reduction of the child-mortality rate of under five-year-olds by two thirds and the maternal mortality rate by three-quarters will require additional technology such as biofortification for nutrition improvement, particularly in isolated areas of Africa and South Asia [12]. Biofortification is believed to be a cost effective means of affecting millions of lives on a sustainable basis in developing countries [10,13], particularly in isolated areas of Africa and South Asia. Production of such biofortified food products can be achieved either through conventional crop breeding or genetic modification as described in section 2. HarvestPlus is one of the programs of the Consultative Group for International Agricultural Research (CGIAR) that was established in 2004 and the first recipient of funding for biofortification research by the Bill and Melinda Gates Foundation [14]. Since its establishment, HarvestPlus has been playing a leading role in biofortification projects with a focus on three previously discussed micronutrients that pose a major health problem in Africa: vitamin A, iron, and zinc. This project is targeted towards improving the nutritional content of seven staple food crops that include sweet potato, pearl millet, beans, maize, wheat and cassava. The goal is improving health among young children and women in developing countries. Also, in 2004, as part of Grand Challenges in Global Health Programs, the Bill and Melinda Gates Foundation made a $7.5 million commitment to the BioCassava Plus (BC+) project under the leadership of the Donald Danforth Plant Science Center (DDPSC), St. Louis, MO, USA. This project has two key African partners: Nigerian based National Root Crop Research Institute (NRCRI) and Kenya Agricultural Research Institute (KARI) and also includes other consortium partners from different part of the world (Table 1). The overall aim of BC + is to develop improved cassava of different varieties particularly in terms of biofortifying cassava with iron, zinc, protein and the vitamin A precursor, β-carotene. Again, the belief is that biofortified cassava has the ability to improve health and quality of life among poor farmers, including women, and young children in Africa.

Table 1 BioCassava Plus Project consortium partners

Project consortium partners	Roles	Country
Major consortium partners		
Donald Danforth Plant Science Center (DDPSC)	Overall project management and scientific leadership, product development, communication, biosafety and regulatory affairs, intellectual property rights (IPRs), technical affairs and finance	USA
National Root Crop Research Institute (NRCRI)	Finance, crop breeding, laboratory, field trials and logistics	Nigeria
Kenya Agricultural Research Institute (KARI)	Crop breeding, laboratory and field trials	Kenya
Other consortium partners		
International Center for Tropical Agriculture (CIAT)	Cassava breeding and genetic transformation, field trials	Colombia
International Institute for Tropical Agriculture (IITA)	Cassava breeding, nutrition assessment	Nigeria
Ohio State University	Cassava genetic transformation	USA
University of Bath	Cassava genetic transformation	UK
Eidgenossische Technische Hochschule (ETH)	Cassava genetic transformation	Switzerland
Bill and Melinda Gates Foundation	Donor	USA
Monsanto	Donor	USA

While this article provides background on biofortification for nutrition enhancement through either conventional breeding or genetic modification (GM), the main purpose of this article is to examine the perceptions of scientists regarding the potential adoption of GM technology for increasing the nutritional content of cassava and improving the agronomic traits of cassava production in general. This study relies on semi-structured interviews with some key members of the BC + project and relevant scientists in biotechnology and agricultural science at research institutes and universities in Ghana and Nigeria. This research was carried out in the context of a wider study of biotechnology which focuses on GM cassava technology as a relatively new technique which touches upon the vital aspects of benefits and risks that are associated with potential adoption of new technologies. This study concludes with a summary of findings and a list of policy considerations for adoption of biofortified products.

Biofortification for nutritional enhancement in Africa: Is it conventional breeding or genetic modification?

Given the prevalence of micronutrient malnutrition among young children and women, the development of foods with enhanced nutrients is crucial in Africa, particularly SSA. The technology advancement in this era and subsequent adoption of innovative tools has the potential to pave the way for better crop productivity and higher quality food at lower cost in order to solve the micronutrient problem in this region. The science of biotechnology, either through conventional breeding (often in conjunction with marker assisted selection) or genetic modification approaches has great potential to achieve biofortification for nutritional benefits in SSA. However, this raises important questions as to which biotechnological approach can be used to achieve biofortification in the most efficient, cost-effective and sustainable manner; both methods have unique capabilities and constraints.

Conventional breeding is achieved by identifying parents with traits that complement each other and which have the ability or capacity to compensate the shortcomings of each other through sexual recombination. This requires selection of offspring with desired traits over a long period of time. For example, production of improved crop lines can take as long as 10 years from the first parental crossing to generation and distribution of selected improved crops. The conventional breeding technique can be used in staple food crops such as rice, wheat, beans, maize, banana and cassava to study the genetics of trace minerals (for example, iron, zinc, provitamin-A carotenoids) inheritance and, in theory, to improve the nutritional value of such trace minerals in staple crops. This technique has been used for many years in crop production and has created several desired varieties. For example, using conventional breeding, Nigeria-based International Institute for Tropical Agriculture (IITA) has developed several cassava varieties and germplasm with improved yield, quality, disease and pest resistance; more than 200 of these varieties have been distributed and released to African farmers [15]. Similarly, Colombia-based International Center for Tropical Agriculture (CIAT) has also distributed advanced cassava breeding materials to many national programs in Asian and African countries through conventional breeding programs [16]. Other than using conventional breeding for cassava, several papers have reported the use of conventional breeding for improving other staple crop varieties in increasing micronutrient density [17-22]. For example, using conventional breeding, a rare variation of a gene called *crtRB1* (normally occurring in temperate maize) has resulted in higher production of ß-carotene (18-fold) [20], and these traits are being introduced into tropical maize strains which are common in Africa. Another example is orange-fleshed sweet potato and orange maize which were biofortified using conventional breeding. These biofortified products, with provitamin-A, have proven to be successful as they are well accepted by young children in rural Mozambique [21,22].

In spite of the potential of conventional breeding, irregular flowering, high degrees of heterozygosity, low seedling number, and genetic variability represent significant drawbacks for cassava breeders [13,23]. The ability of cassava roots to produce enough ß-carotene to meet the daily requirements of adults (that is., 6 mg/d ß-carotene) is due to the presence of genetic variability in germplasm [24]. Unfortunately, the use of conventional breeding will not be an option where genetic variability for a given trait is too low to achieve the target levels [24,25]. Thus, even though desirable varieties of cassava can be conventionally bred, to some level, conventional breeding may prove useless due to problems with other germplasm of cassava. As a result, small-scale farmers are unlikely to benefit from conventional breeding as it might not serve the purpose of improving crop production and increasing the opportunity for large scale farming in tropical areas.

Transgenic methods, otherwise known as genetic modification, are performed by inserting only a small additional piece of information (foreign DNA) into locally adopted and farmer preferred cultivars to control a specific trait within the selected cultivar. This process is achieved through the splicing of a well characterized chunk of foreign DNA containing a particular known gene within the chromosome of a host organism using a pair of molecular scissors called 'restriction enzymes'. Restriction enzymes cut DNA strands at specific restriction sites into specific DNA fragments so that they can be inserted and integrated into the host chromosome at the restriction sites. All this can take place within a relatively short period of time when all the resources required to perform the tasks are available. Given the inability of conventional breeding to perform this specific task in cassava, genetic engineering can be used to fill the gaps. It was 15 years ago, precisely in 1996, when the genetic engineering technique was first developed for cassava transformation. This was carried out by a research group at the Eidgenössische Technische Hochschule (ETH), Switzerland, that produced transgenic cassava via regeneration from embryogenic systems [26], and by research groups at the International Laboratory for Tropical Agricultural Biotechnology (ILTAB) and the University of Bath, the UK, that produced transgenic cassava via friable embryogenic callus (FEC)[27]. Since the first breakthrough, a growing body of literature [28-31] has reported genetic transformation of cassava, particularly for biofortification, with a focus on nutritional enhancement such as elevating the vitamin, protein and mineral content levels. Other genetic transformations of cassava have also focused on important areas such as reducing cyanogenic content, creating disease and pest resistant varieties, delaying post-harvest physiology deterioration in storage roots, increasing starch production, and improving the overall quality of cassava in general terms [31,32]. The application of transgenic methods has led to a wide range of germplasm with more than 20 cultivars from Africa, Asia and the Americas [33], and has demonstrated a new and highly efficient, robust and reproducible cassava transformation protocol [34].

These transgenic methods have also been applied in biofortification of other staple food crops such as potato, maize, bean and rice [30,35]. A similar approach can be applied to genetic improvement in cassava. For example, successful studies with transgenic potatoes suggests that an increase in total protein and most of the essential amino acids can be achieved [35]. Similarly, in rice, transgenic rice has been developed with increased ß-carotene levels. First, rice called 'Golden Rice 1' (GR1) with 1.6 μg/g total carotenoids (mainly ß-carotene) was developed, [36] and later boosted to 37 μg/g total carotenoids for 'Golden Rice 2' (GR2) [37]. The benefit of GR2 for poor and malnourished countries including those in Africa, particularly among young children between one- and three-years-old in terms of recommended dietary allowance (300 μg vitamin A) can be as high as 25 μg/g ß-carotene of 72 g GR2 based on a retinol equivalence ratio for ß-carotene of 12:1. Based on the same ratio (12:1), Bouis *et al.,* [30] reported that total provitamin-A intakes would be increased by 25% for adult women and preschool children, if GR1 were to replace non-transgenic rice. All of these improvements of various traits represent a significant major trend that could be applied to cassava transformation and clearly can benefit the poor in countries with micronutrient malnutrition.

The use of conventional breeding, marker assisted selection and transgenic methods will largely depend on the regulation, production cost, and suitability required to produce a desired result in terms of achieving nutritional quality. These must be balanced against cost effectiveness in poor countries such as those in SSA. According to Hunt [38], the undernourished may enjoy biofortified varieties for a cost of only US $0.02-0.03 for a single person per year, compared with commercial iron fortification for a cost of $0.12, and iron supplement exceeding the cost of $ 3.00 per person based on a World Bank report [39]. This analysis suggests that the cost of iron supplementation is 100 times more than the cost of biofortified rice, indicating how biofortification might save considerable money when it becomes available. While the cost of conventional breeding can be moderately low, for example, $4 million per variety over a given period of 10 years (equivalent to 0.2% of global vitamin A supplement) [12], it will cost between five and eight fold more to produce and regulate a GM, or transgenic, crop [40,41]. However, a previous study by Hunt [38] suggests that development of an iron or zinc-dense variety by conventional method will each cost $10 million, which sharply contrasts the subsequent study [12]. In addition, another report [30] argues that higher costs of meeting regulatory approval remain unknown, making it difficult to conclude that transgenic methods will be more or less expensive than conventional methods. Bouis *et al.,* [30] suggests that the higher regulatory costs of transgenics may be offset by greater costs and the lengthy period of time required of conventional methods. In the case of the former, future introduction of comprehensive regulatory procedures based on robust scientific findings may minimize the effect of higher costs, thereby making it more affordable.

In summary, conventional breeding including marker assisted selection [42] and genetic engineering have both been described in terms of their important roles toward crop improvement, particularly biofortification. Even though they are different and complementary ways of improving crops, both are still questionable, particularly in terms of limitations and constraints. While conventional breeding is still a better choice in Africa, due to its technical simplicity, low cost (depending on breeding procedures) and lack of government regulation, this technique is limited due to long breeding cycles and the sexual process itself [25], therefore making it less suitable for biofortification. Genetic engineering has great potential in making conventional breeding faster and more efficient as well as complementing its limitations, particularly in targeting a specific gene trait that requires less need for a large population and multiple generations of selection. In the past 15 years, cassava transgenic technology has made significant progress in terms of enhanced nutritional values (for example., GM biofortified cassava), increased disease and pest resistance, improved starch yield and quality [31,32]. However, transgenic technology is hotly debated around the world with several factors (for example., biosafety regulation, low level of awareness, facilities, infrastructures, and so on) hindering the adoption of new technology in developing countries, including those in Africa. Moreover, the dispute between the US and the European Union (EU) over adoption of GM technology, particularly with regards to the precautionary principle as mandated through the Cartagena Protocol on Biosafety (CPB)[43], has become a political tussle which is one of the major obstacles to potential adoption of GM crops in Africa; this tussle has a serious impact on the African continent as to whether to adopt GM technology or not [44,45]. The tough EU regulation through a highly precautionary approach represents a significant challenge to the development of GMO policies in many African countries, and also hinders the opportunity of donating new agricultural technologies that may come under humanitarian purposes by multinational firms and other organizations.

Methodology

Scientists were interviewed as some of the key stakeholders that play an important role in adopting new agricultural technology. Their perceptions around the potential adoption of GM technology were collected in two West African countries, Ghana and Nigeria, in January, 2011 through a qualitative approach. A qualitative approach was chosen for this study because it offers the opportunity to explore a wider range of issues in much greater detail and it also uncovers new areas that are not anticipated at the outset of the research [46,47]. Accordingly, the research questions posed in the study are not amenable to a quantitative approach such as is commonly employed in many large-scale investigations (which are necessarily constrained by decisions on which highly structured questions will be used). Amongst GM crops that are likely to be cultivated in both countries is GM cassava given the ongoing field trial in Nigeria. The West African countries were chosen as they are part of the only region in Africa with a large production of cassava and are therefore two important countries in this region which could benefit from adoption of GM technology for improving cassava production. Nigeria has the largest production of cassava in the world with approximately 34 million tonnes of cassava produced in 2002 [48], Ghana is also among the top ten producers of cassava, occupying the third and sixth position in Africa and the world, respectively, in 2002.

Interviews were conducted using a semi-structured interview style. In most cases some directors of relevant institutes were included in the group of interviewees. Individual interviews and focus groups were employed and comprised groups of six to nine scientists. While several relevant questions were asked regarding adoption of other GM crops such as GM cotton, GM sorghum, GM maize, specific questions on GM cassava were asked given the importance of cassava as a food security crop in the West African region. Potential adoption of GM cassava technology was chosen as an example to elicit relevant opinions from respondents. Specific questions included health improvement and food security, regulatory issues, and transfer and acceptance of new technology. A key aspect of the interviews was to ensure that experts selected for interview were relevant in the field and associated with institutions that can influence agricultural or national policy; this approach has previously been adopted in the selection of important stakeholders for policy development [49]. Given the nature of the topic and the possible role of scientists in decision-making, we tried as much as possible to keep the questions neutral to reduce any potential bias in response to our question. For example, what are the risk assessments of GM cassava? (see the Additional file 1 for the details). We identified scientists, in both countries, in the field of agriculture and natural sciences, including biotechnology, at research institutes and universities including the ministry of agriculture. These scientists were identified with the help of each country's government. The Nigeria-based NRCRI was selected due to the ongoing GM cassava field trials and the presence of members of the BC + project at this institute. The selection also included scientists at the two best universities in the field of agricultural science in the two countries. The total number of respondents for this research is 58: 43 scientists at research institutes and 15 scientists at universities as described in Table 2.

Prior to the interviews, permission was sought from the individual country governments. We explained the objectives of our meetings with the key stakeholders, particularly with regards to potential adoption of GM technology in each country. After the approval from the governments, all the institutes and universities that were contacted agreed to speak with our team on the topic, except the University of Ghana and Nigeria-based IITA that did not participate in the interviews after a formal approach due to some undisclosed reasons.The interviews and discussions were conducted in English, tape recorded and notes were simultaneously taken during the discussions. Following the interviews, audio recordings and notes were immediately transcribed and coded for the analysis of the results. All pieces of key information and important quotes from the interviews were put together for this article.

Results and Discussion

Scientists' perspectives on GM cassava technology

Scientists' views were sought with regards to GM cassava for improving health and food security in Africa. In Ghana and Nigeria, scientists agreed that GM technology can be important in improving the quality of root and tuber crop production (for example, cassava which is widely consumed in SSA and almost every household in West Africa). Emphasis was laid on nutritional benefits, increased yields and disease resistance in cassava through the application of GM technology. Many scientists stressed that GM technology can be more efficient where traditional plant breeding techniques have historically performed poorly and that the new technology can be useful in solving the micronutrient deficiencies that lead to the death of young children due to chronic malnutrition and poor-quality diets. The BC + team placed much emphasis on the potential of GM cassava in solving nutritional problems such as iron, zinc, vitamin A, and protein deficiencies. This group also stated that the BC + project's focus on these important components of nutrition has led to the production of more than 10 times the levels of provitamin A in GM cassava compared with non-GM cassava. It is expected that GM cassava will make a significant impact in terms of improving the quality of life and health among poor farmers and consumers in Africa. According to a scientist; *'BC + project wants to deliver nutrient dense cassava that has potential to solve malnutrition problems among young children and women'* Another scientist from the BC + team added: *'GM technology offers a much more effective approach to improve other traits that are present in germplasm 60444 apart from β-carotene because we don't*

Table 2 Interviews conducted at various research institutes and universities in Ghana and Nigeria

Country	Institutions	Number
Ghana		
	Biotechnology and Nuclear Agriculture Research Institute (BNARI)	9
	Kwame Nkrumah University of Science and Technology (KNUST)	6
	Crop Research Institute (CRI)	6
	Regional Agricultural Development Unit (RADU)	8
	Savannah Agriculture Research Institute (SARI)	3
Nigeria		
	National Root Crops Research Institute (NRCRI)	6
	University of Ibadan (UI)	9
	Lagos State Agricultural Development Authority (LSADA)	9
	National Biotechnology Development Agency (NABDA)	2
	Total (including individual interviews)	58

have answers to those traits in the germplasm even with application of traditional plant breeding in the past'. According to the BC + scientists, the Phase I biofortified germplasm 60444 which contains β-carotene (Figure 1) was first tested in Puerto Rico before the confined field trials in Nigeria which were recently completed at NRCRI (the confined field trial took place between 9 and 12 months). This particular GM cassava trait has been harvested. Currently, there is also an ongoing confined field trial of 60444 iron and zinc nutrient enhanced cassava (Figure 1).

An overwhelming majority of scientists mentioned that creating GM cassava varieties that are resistant to diseases (especially, cassava mosaic disease: CMD) and pests so as to increase the yields of cassava production will be very important. According to the BC + team, the recently completed field trial of β-carotene nutrient enhanced cassava did not involve cassava that was genetically modified to be disease resistant. However, there is ongoing transformation of cassava for both disease resistant and nutrient enhancement traits that will be part of Phase II trials. The current 60444 may be susceptible to CMD in Nigeria but this is not a major problem in Puerto Rico due to the lack of widespread CMD in that country. In epidemic areas, the impact of CMD, and other diseases, can be devastating with great losses in yield particularly at early stages. For example, as a result of CMD attack, the yields can be as low as 4 ton/ha or less compared with the normal average yield of 8 ton/ha in Africa which is still below the potential yield of cassava elsewhere. Even in some improved cassava varieties the yield can be very low due to an attack of CMD depending on the degree of severity and time period during which the crop is planted. Moreover, some scientists argued that good yields can be obtained from improved cassava varieties despite an attack of CMD, as the improved cultivars are resistant to some strains of CMD. As described by one of the scientists, an increase in yield by controlling disease attack, particularly in a major staple crop such as cassava is paramount to ensuring food security in Africa, and governments must step up action toward rising agricultural challenges and must embrace innovative tools such as GM technology to solve the food problem.

Figure 1 Confined field trials of germplasm 60444 genetically modified (GM) cassava (biofortified). **A**- Germplasm 60444 GM cassava with provitamin A (β-carotene) that has been harvested. **B**- Germplasm 60444 GM cassava with high iron at early stage. **C**- Germplasm 60444 GM cassava with high zinc at early stage.

Regulatory issues

The regulation of GMO products is a key factor for successful adoption of GM technology in Africa. Specific questions asked included risk assessment in terms of health and environmental safety of GM cassava. Despite the fact that GM technology is said to have the potential to improve crop productivity and improve a variety of traits in cassava, concerns were expressed amongst the scientists about the regulatory process and potential risks that may be associated with GMO products. Some scientists were wary of the long-term effects of GMO products, particularly with regard to the environmental impact which could spell doom for their indigenous crops. Scientists agreed that the large majority of African governments are not well equipped in terms of regulatory systems for controlling possible adverse effects of GMO products and the need for proper regulation is fundamental to the release of GM cassava. However, there is no evidence yet with regard to a negative impact of GMOs on the environment [50], whereas evidence in terms of benefit to the environment has continued to grow [51]. Currently, only Nigeria and Kenya are conducting confined field trials of GM cassava. However, some Ghanaian scientists mentioned that there may not be a need for confined trials of GM cassava in Ghana due to the ongoing GM cassava trials in Nigeria, but regulatory procedures must be in place to ensure the safety of human health and the environment when it becomes available in Ghana. The regulation of GM cassava could present a significant challenge due to limited facilities and lack of scientific expertise. It is a long-term process to access GM food as such food will have to go through research, testing, safety assessment, screening and market evaluation before being released to farmers and consumers [52].

When asked about the safety evaluation of GM cassava no clear cut strategy or well defined guidelines on the risk assessment process were mentioned for releasing GM cassava in Africa even amongst the BC + team that participated in the focus group discussion. An attempt to get the majority view from the experts involved in the BC + project at DDPSC, proved equally difficult. However, an African scientist, and one of the key team members of the BC + project from the same institute in the US that participated in the interview, said that BC + has spent considerable time and energy, as a group, to provide answers on the safety evaluation of GM cassava. Biotechnology-derived products are subjected to various analyses on a case-by-case study basis, and a key principle is comparative safety assessment versus a conventional counterpart [53-55]. Thus, GM cassava will have to undergo various standard tests before it can be released in Africa. Whether this will happen in the next five to ten years or less depends largely upon African governments, international regulatory bodies and the BC + project consortium.

By way of comparison, Golden Rice was developed over ten years ago at a cost of more than $100 M USD [56] and, despite having gone through some safety assessment tests [53,57], the so called 'Humanitarian Golden Rice' is yet to be released to the beneficiaries in developing countries. This lack of release is attributed to the cumbersome process of regulatory clearance and extreme application of the precautionary principle [58]. Potrykus [58] argues that the rules and regulations established around the world for control and handling of genetically engineered crops can be demanding, as reflected in 10 years spent on preparation of a regulatory dossier for the Golden Rice. In addition to these regulatory hurdles, farmers may be unwilling to adopt Golden Rice and consumers may not accept Golden Rice given the fact that they are used to white rice as opposed to a rice with yellow or golden color. However, a recent study suggests that existing preference for white maize did not affect the acceptance of orange maize (biofortified food product) in Mozambique [22]. Recent opposition to Golden Rice by a scientist who claims that there is no 'Gold' in 'Golden Rice' may also threaten the release of Golden Rice [59]. Currently, golden rice is scheduled to be officially released to farmers in Asian countries by the year 2012 [58], provided all the necessary procedures for release go well.

GM cassava may not escape a tough, high cost, regulatory process similar to the process to which Golden Rice has been subjected. According to the ETH, the review of an application for field testing CMD-resistant cassava in Nigeria would have far exceeded the actual cost of the field test itself (Personal Communication). GM cassava is only at the initial stages of confined field trials in only two African countries. More confined field trials may have to take place in other African countries under stringent and tight regulatory control systems and compulsory tests in line with international safety assessments that may be required before GM cassava is deemed fit for release. A recent report suggests the same [60], specifically noting that more extensive field tests and resistance stability evaluations of GM cassava are required before improved lines can be released for cultivation. It was pointed out among scientists that Africa is a different terrain where there is a history of things often not working out easily, particularly when new tools or techniques are introduced. Whether adequate attention is being paid to developing, maintaining and adapting local capacity to regulate GMOs is deemed very unclear. This view appears to be supported by a recent article [61] which shows that it can be cumbersome to set up and run a well-functioning laboratory for plant biotechnology in Africa. Of course, in order to ensure an effective regulatory process of confined field trials, testing procedures and smooth delivery of GM cassava, there must be functioning laboratories, regulatory authority, enabling agricultural policy and scientific expertise which many African countries lack. Moreover, controversy over potential adoption of GM cassava coupled with NGOs' campaigns against GMOs might present a significant setback for the release of GM cassava in the future (Table 3). While a food crop such as cassava is not usually consumed in Europe, it is not clear whether the presence of GM cassava in this region will affect trade relationships with the EU countries given the strict EU precautionary principle on GMOs. Realistically, these are the hurdles that GM cassava may have to face and overcome before it becomes available to the farmers and consumers in Africa.

Can GM cassava technology be effectively transferred to African farmers?

Technology assessment, transfer processes and the right of farmers to have free access to the technology will be fundamental to successful adoption of GM cassava technology in Africa. When participants were asked to give their opinions or evaluate methods that are put in place to ensure that GM cassava technology reaches farmers in a sustainable manner, not many of them were able to shed light on a successful technology delivery program and/or the future impact of intellectual property rights (IPRs) on GM cassava. However, some participants from the BC + project described the first attempt taken by local scientists to solve the problem of pests and diseases of germplasm 60444 through breeding. Unfortunately, this attempt was abandoned by Nigerian farmers due to germplasm 60444's susceptibility to diseases and pests as the traditional plant breeding approach failed to solve this problem. They also explained that germplasm 60444 has been sent to laboratories in the US and around the world, and that scientists in these labs have been working on improving the root nutritional and quality traits.

Table 3 Controversies over potential adoption of GM cassava in Africa

Controversial statements about GM cassava		References
1.	'A team led by Richard Sayre, a professor of plant cellular and molecular biology at Ohio State University, had inserted into the cassava's DNA a bacterial gene that affects starch production. The modified plants' roots were said to be more numerous and up to 2.6 times larger than in normal plants.'	[62]
2.	Nagib Nassar, the Brazilian Professor of genetics at the University of Brasilia, responded to 'GM cassava has 'super size' roots' that 'My colleagues and I have produced cassava with roots that are ten times the normal size without resorting to genetic modification.'	[63]
3.	According to DDPSC 'virus-resistance technology will initially be deployed in the East African region's most popular cassava cultivar (Ebwanatareka) for adoption by the 22,000 Kenyan farming families and the project will help 200,000 Kenyan cassava farmers and their families and increase cassava harvests by 50% on a sustainable basis.'	[64]
4.	Less than 9 months later a statement by the DDPSC dated May 26, 2006, says that although resistance to cassava mosaic disease (CMD) had been established through genetic engineering seven years ago, 'the resistance was subsequently lost, and changes to the plant's DNA had taken place.'	[65]
5.	'Many groups oppose the introduction of these seemingly miraculous types of GM cassava. When asked about the health consequences of consuming GM foods, Gareth Jones of Biosafety Africa explained simply, 'Nobody really knows! This is the problem... No long term studies have been done.'	[66]
6.	'The deliberate release of GM into the environment... can lead to irreversible damage to ecosystems,' explained Glen Tyler of Greenpeace International's agricultural campaign. 'Once one of these crops is out of the bag, it's hard to put it back in the bag, so to speak.'	[66]
7.	'A GM crop research project funded by the Bill and Melinda Gates Foundation has developed a highly nutritious strain of cassava sufficient to provide most of a day's nutrients in a single meal'.	[67]
8.	'The world does not need GM cassava and research laboratories should not turn Africans into guinea pigs on the pretext of helping to fight malnutrition. GMOs are not the solution to the hunger and nutrition question in Africa.'	[68]
9.	'Opponents stood in the way of field trials in Nigeria, Uganda and Kenya by spreading unfounded scare stories about 'killer cassava'.'	[69]
10.	According to Dr Claude Fauquet of DDPSC 'acquisition of the cassava genome sequence will provide a platform to explore the vast biodiversity within cassava wild species. Ultimately, these activities will position cassava as a valuable source of renewable bio-energy.'	[70]
11.	Critics say, 'the objectives of the project go beyond food security, and touch on the search by the United States of a cheap source of starch other than maize to manufacture ethanol to help wean it from oil.'	[70]

Further, scientists have been sent to the US and the ETH where they have received training on tissue culture and transformation of cassava, particularly on germplasm 60444 which is the focus of the BC + project's current research. In addition, the respondents said that germplasm 60444 was chosen for the initial tests of GM cassava because it was more amenable to transformation than other available varieties, and on that basis it was accepted for proof of concept. They argued further that this particular cultivar was local to farmers because it has been grown on their farms before, and as a result, the improved new variety through GM technology would be well disseminated and incorporated into their farming systems. These scientists believe that this approach forms part of the technology assessment and transfer process, particularly in modifying and adapting the selected variety or new trait(s) to suit farmers' conditions and environments, thus having a significant impact on the adoption of the new technology when it becomes available.

On the other hand, some scientists argued that technology assessment and transfer processes should go beyond taking a particular variety of cassava for transformation or modification from the farmers, transforming it and then returning it to the farmers. Rather, assessment and transfer must include different types of farm plot trials (for example., initial trials and confirmation trials), extension services, adaptation strategy and other relevant steps that are clearly spelt out as part of a delivery system before the new technology is released to the farmers, and that no technology should be imposed on farmers without following the due process. They are not aware of any established program by the governments or what the future plan is, particularly with regards to reaching out to the farmers while developing biofortified cassava. According to a scientist; *'this is Africa, we don't go through due process because governments, the major player, always thinks it is their right to choose technology for the farmers and impose it on them without assessing the advantages and disadvantages, but the same thing will not happen where the technology is coming from...what we are asking for is type of measures being put in place to introduce GM cassava to our farmers and whether we have it is not clear to us'.*

This group of scientists said further that farmers, as one of the key stakeholders, should be involved from the beginning of the technology development process right through until the last stage. They felt this should be accomplished through participatory approach methods such as seminars, training courses and open debates. Further, for the long term prospects of adopting this new technology successfully, obtaining all the necessary information on technology development, particularly in achieving an understanding of end user perception regarding the potential values and benefits of the technology, coupled with the adequate provision of extension services is vital. While little or no information was provided on the delivery program of GM cassava technology during the interviews, introduction of new agricultural technology should be based on a consultative process with the key stakeholders such as the farmers, in order to identify challenges to successful uptake of new technology [71]. For example, identifying the right germplasm based on consultation with the farmers with regard to suitability of the selected variety and ability to grow in any given region or location will save time and facilitate the speedy adoption of newly transformed traits.

The fact that the selected germplasm 60444 only contains nutrient enhanced performance (β-carotene) without resistance to diseases may suggest that inadequate consultation with the farmers has taken place and/or failure of the participating scientists to reach a consensus on the importance of introducing the two traits at the same time. As it should be possible to confer both traits of nutrient enhancement and disease resistance to germplasm 60444 at the same time, some felt that this should have been considered during the selection processes. It was clear during the interviews that GM cassava where those two traits are present together has not been created, yet transformation in the laboratory is still ongoing as explained by one of the scientists. Surprisingly, a recent report by Sayre [72], states that biofortification of cassava with elevated provitamin A and cassava mosaic disease virus resistance has been developed through a conventional breeding method. Perhaps this approach should have formed the basis for the transformation of GM cassava from the onset. Moreover, the ETH have also developed CMD-resistant 60444 [34,73] but these transformants have yet to be incorporated into the BC + project. The ETH CMD-resistant 60444 has been tested for agronomic performance in Puerto Rico for two years but has yet to undergo field trials in Africa due to lack of funding [74]. A similar issue was raised among scientists during a recent interview in Egypt in which it was indicated that they had not been included in the Water Efficient Maize for Africa (WEMA) breeding project on drought tolerant maize [75]. The sidelining of an important product that could have important uses in Africa would appear to reflect a lack of good teamwork among the scientists and funding agencies and/or an incomplete understanding of African farmer's needs. In this case, it would appear that lack of teamwork and lack of proper consultation and/or inability of the scientists to arrive at proper conclusions on a suitable trait may have resulted in wasted resources and time, particularly as field trials and various stages of regulatory assessment are not yet complete. Therefore, it is important to take necessary measures through a specific approach and integrated action that will facilitate the development, adoption and diffusion of new technology among the farmers.

The transfer of GM cassava technology to African countries may be seriously constrained in the future as the large majority of scientists interviewed are not aware of government structures or systems in place to ensure that farmers have free access to this new technology, although the BC + project members claimed that GM cassava technology would be provided to African farmers on a humanitarian basis. But the critical question remains whether Monsanto, which has invested in the BC + project (over $6 million for CMD-resistant cassava), and which holds the patent on the *Agrobacterium* co-transformation method (the technique used to create transgenic cassava) [76], will keep it free for a short or long period of time. A recent report by Sayre [72] suggests that "these improved varieties will be distributed to farmers for humanitarian purposes either free or at a small cost." The notion of a small cost may suggest that there be might be a twist along the line. It appears for now, provitamin A and other improved nutrient varieties will become available for humanitarian reasons. But what happens when improved agronomic traits (for example., increased yields and starch production) of cassava become available through this transgenic method? Will this become available for humanitarian reasons again in Africa, and what about other developing countries in Asia and Latin America where cassava is used as animal feed and a raw material for starch production and commercial biofuel production (bioethanol)? These IPR issues appear to represent a significant challenge in introducing this new technology into African farming systems, and therefore, should not be considered as a trivial problem despite the so called 'humanitarian way of providing GM cassava technology'. Similar to this point, the technology used to develop drought tolerant maize coordinated under WEMA project is owned by Monsanto [77]. The statement that the drought tolerant maize will be available at a cost within reasonable means of the poor farmers in SSA still lacks clarity and enough information as to how African governments and their farmers benefit from the introduction of these new agricultural technologies despite project collaboration agreement provided on IPRs issues [78]. Given these two examples, some cost may be introduced when this new technology becomes available in the future despite promotion of a humanitarian agenda. African governments should consider that the reality may be that they cannot always get new technology for free and must start negotiating what will benefit their farmers on a long-term basis. Such consideration that includes establishing a proper legal system for IPR so as to facilitate future introduction of new agricultural technologies must be at the heart of national government while working with the African Agriculture Technology Foundation (AATF).

Consumers' acceptance of GM cassava

Respondents also shared their views on the acceptance of GM cassava by the consumer and its adoption by the farmers. Most respondents agreed that, a modified product based upon cooking, palatability, storage quality and improved nutritional values, particularly protein quality, will promote and increase the acceptance of GM cassava among consumers. Increased yield, disease-resistance and profitability were identified as more likely to remain key issues with respect to farmer adoption of GM cassava. Respondents also indicated that the acceptance of GM cassava would depend on how much information is made available to the public. Issues related to ethics and health safety were suggested as perhaps representing the largest concerns which could be a potential barrier to the acceptance of GM cassava. The media can play an important role in disseminating information on health and safety issues as, in the past, the media have played a significant role in encouraging most people to avoid so-called "contaminated" food products due to health concerns. Regardless, once attention is brought to health risk concerns by the media people will start avoiding it. Additionally, media reports on GM can be negative as they tend to rely on unsubstantiated information without proper investigation [79]. In particular, the information provided tends to come from various sources that are pushing an agenda that this is mostly reported without balanced views.

As described by the scientists (albeit, not necessarily representing the views of consumers), the enhanced nutritional attribute(s) of GM cassava is fundamental to its acceptance by consumers. This view is consistent with similar findings [80,81] that show consumers are more likely to accept biofortified foods including GM cassava due to its improved nutritional values. Wolson [81], however, reports that consumer acceptance will be difficult to win if biofortified products with altered taste, color and texture are introduced to the public. The view of the respondents underlines the fact that, given the health concerns that arose from improved crop varieties in the past, the acceptance of GM cassava will require public education and awareness, particularly with regard to potential benefits and risks associated with GM crops. While emphasis was laid on an effective marketing campaign to inform the public on the benefits of biofortification (for example., GM crops), [80-83] acknowledged a low level of public education on GM crops while assessing their attitudes towards GM crops and foods. These authors emphasized that the public must be educated and informed about the potential benefits and risks of GM crops.

Conclusion and policy options

This paper sheds new light on the potential adoption of GM cassava technology in Africa by asking scientists, as one of the key stakeholders, about their perspectives on the new technology. To our knowledge, this is the first study to report scientists' opinions on GM cassava in Africa. This study will add more knowledge and understanding on the potential adoption of this new technology, particularly towards developing appropriate policy to adopt GM technology that will benefit agricultural development in Africa.

In this article, we have only focused on scientists' perspectives, perhaps one of the limitations of this study. However, these scientists agreed that a key need is to also seek opinions of consumers and farmers as other key stakeholders in the future potential adoption of GM cassava. The views of these scientists are fundamental to the successful adoption of GM technology and their advice on potential adoption of this new technology can be a determinant as to whether farmers will adopt the technology or not. While the opinions of scientists, farmers and consumers matter in adopting biofortified food products, so too, does risk assessment. Given the high cost of GMO regulatory processes including risk assessment, there is a need for a cost-effective approach. A recent article proposes a semi-quantitative risk-benefit analysis methodology as a potential tool that requires limited expertise and facilities and can be useful for GMO risk assessment in developing countries [84]. South Africa is likely to take the lead to demonstrate how this innovative tool can balance risks and benefits of GMOs with a view to facilitating a cost effective risk assessment. Given the potential benefits of this new method, national governments should encourage GMO assessments that will facilitate adoption of this kind of accessible tool for better decision making for the regulators in the light of the controversial precautionary principle

Moreover, it also should be stressed that economic studies are needed to analyze the potential benefits of GM cassava, particularly the biofortified products with iron, zinc and provitamin A as there is little or no information on ex-ante socio-economic impact analyses in Africa. The methods and key assumptions for assessing the impacts and cost-effectiveness of biofortification have previously been set out in detail [85,86] and this methodology has been applied in Asia (for example., India, Bangladesh and the Philippine) for calculation based upon optimistic and pessimistic scenarios required for adoption of biofortified foods. This kind of information will be useful for empirical analysis of biofortification in Africa.

The scientists' perceptions about GM cassava technology suggest that this technology has great potential to increase yields, to create disease and pest resistant cultivars, and to increase the nutritional values of cassava. While much emphasis was placed on the human health benefits of GM cassava, little or no concern was raised about health risks. This was in contrast to scientists' beliefs that, based upon previous experience, consumers might express concern about the health risks of any unfamiliar crop product. It was suggested that media reports on GMO products can be very biased and can result in loss of consumer confidence in the introduction of new improved crop varieties. Hence, the scientists felt that GMO products going through proper safety assessments coupled with increased educational efforts targeting the media and the public (for example., potential benefits and risks associated with GMOs) would increase the chance of adoption or acceptance among consumers and farmers. Some concerns were also raised. Uncertainty about the effects of GMO products on the environment remained a serious concern among some scientists, particular concerns were raised about long-term damage to ecosystems and indigenous crops.

Given the perspective of scientists on the potential benefits and concerns in adopting GM cassava in Africa, there is a need for African governments to ensure that an enabling policy environment is created while working together with international partners. Based upon the

Table 4 Summary of policy priority for present and future development of GM biofortified foods

Summary points
1. Engage local institutions and communities in introduction and adoption of new technology
2. Educate the public, farmers, government institutions, the media and private companies to increase understanding of GM technology through communication
3. Establish biosafety cost effective risk-benefit assessments of GMOs
4. Adapt local capacity to assess and safeguard human health and the environment upon the release of GMO products
5. Aligned support for national plans on nutrition and health issues must be clearly spelled out in government policy by the Ministry of Agriculture and Ministry of Health
6. Identify and support national agricultural research institutes and industries that focus on nutrition enhancement and improvement (for example, biofortified food)
7. Build national data resources to monitor the progress and impact of biofortified products
8. Conduct an *ex-ante* cost benefit analysis to identify and assess potential market opportunities for biofortified products in different regions including rural areas
9. Improve infrastructure, processing and marketing sectors for biofortified products
10.Set up well defined terms, agreements and legal structures on IPR between local seed sectors and multinational companies
11.Introduce policies that specifically target and address the underlying social, economic and political causes of undernutrition as technological solutions alone may not address the root causes of these problems.

opinions of scientists during the interviews and the current status of achieving biofortification, particularly through genetic modification, this study highlights areas that require policy priority for present and future development of GM biofortified foods as shown in Table 4.

The importance of biofortification in fighting malnutrition can not be overemphasised, but at the same time, it cannot solve the underlying problems responsible for micronutrient malnutrition or food insecurity in Africa. The international community and African governments must increase their efforts in tackling problems such as conflicts, diseases, poverty, climate change, lack of access to food and safe drinking water that contribute largely to the malnutrition problem. In SSA, the number of people affected by extreme poverty doubled from 164 million in 1981 to 313 million in 2002 [87], which contributed significantly to higher malnutrition rates. For example, major malnutrition is not a result of lack of food, but of the high level of poverty, as the majority of these people are extremely poor, and therefore cannot afford to buy food that contains essential nutrients. Added to this, lack of potable water and unacceptable current practices of hygiene, which are very common in Africa, increase the vulnerability to contagious water-borne diseases as one of the main sources of acute malnutrition. Given the existing limited access to food and potable water due to the above factors, governments need to focus on social policy issues that are targeted at solving these problems coupled with implementing appropriate policy as mentioned in Table 4 before biofortification can be a huge success in Africa.

Competing interests

The authors declare that they have no competing interests.

Authors' contributions

AAA collected the data, designed, coordinated and led the study, developed the concepts and the arguments, and wrote the biggest part of the manuscript. OCA and RA contributed to the data collection and wrote part of the manuscript. GP helped design the study and in incorporating the arguments. All authors have read and approved the final manuscript.

Authors' information

AAA is a Research Fellow in Science and Technology for Sustainable Societies at the UNU-IAS. OCA is a Professor of Food Science at the UI. RA is a Professor of Agricultural Science at the KNUST. GP is a Professor of Science and Technology and the Vice-Rector of the United Nations University and the Director of the UNU-IAS.

Acknowledgments

AAA would like to thank the Japan Society for the Promotion of Science for support. All interview participants at the research institutes and the universities in Ghana and Nigeria, the Ministry of Agriculture, Ghana and Lagos State Ministry of Agriculture, Nigeria are gratefully acknowledged for their valuable time and contributions. Grateful thanks to three anonymous reviewers for their comments on the article. Special thanks to Nathaniel Szewczyk, John Beeching, Alex Winter-Nelson and Jane Morris for their constructive comments, discussion and helpful review on earlier versions of the manuscript.

Author details

[1]United Nations University-Institute of Advanced Studies (UNU-IAS), 6F International Organisations Centre, Pacifico Yokohama, 1-1-1 Minato Mirai, Nishi-ku, Yokohama 220-8502, Japan. [2]National Graduate Institute for Policy Studies (GRIPS), 7-22-1 Roppongi, Minato-ku, Tokyo 106-8677, Japan. [3]Department of Food Technology, University of Ibadan (UI), Ibadan, Nigeria. [4]Kwame Nkrumah University of Science and Technology (KNUST), Faculty of Agriculture, Kumasi, Ghana.

References

1. FAO: *FAOSTAT.* 2010. http//:faostat.fao.org
2. IITA: *Cassava in Tropical Africa. a reference manual. International Institute of Tropical Agriculture (IITA).* Ibandan, Nigeria: 1990.
3. FAO: *Food and Agricultural Organization of the United Nations (FAO), State of Food Insecurity in World 2009.* FAO, Rome, 2010; 2010.
4. FAO/WHO: *Vitamin and mineral requirements in human nutrition. Report of a joint FAO/WHO expert consultation.* Bangkok, Thailand:; 1998. 21–30 September.
5. WHO: *Micronutrient deficiencies: Iron deficiency anaemia.* Geneva: WHO; 2007. http://www.who.int/nutrition/topics/ida/en/. Accessed 10 November 2011.
6. ACC/SCN: *United Nations Administrative Committee on Coordination, Sub-Committee on Nutrition (ACC/SCN). Forth report on the World nutrition situation: Nutrition throughtout the life cycle.* Geneva: ACC/SCN in collaboration with IFPRI; 2000.
7. UNICEF: *United Nations Children's Fund (UNICEF). Vitamin and mineral deficiency. A global progress report.* New York:; 2004.
8. Walker CLF, Ezzati M, Black RE: **Global and regional child mortality and burden of disease attributable to zinc deficiency.** *Eur J Clin Nutr* 2009, **63:**591–597.
9. Birner R, Kone SA, Linacre N, Resnick D: **Biofortified Foods and Crops in West Africa: Mali and Burkina Faso.** *AgBioforum* 2007, **10:**192–200.
10. Bouis HE: **Plant breeding: a new tool for fighting micronutrient malnutrition.** *J Nutr* 2002, **131:**604S–615S.
11. UNSCN: *United Nations-Standing Committee on Nutrition (UNSCN): Nutrition and the MDGs: accelerating progress towards 2015.* 2010. http://www.unscn.org/files/Statements/2010_SCN_MDG_Statement_160910_final.pdf. Accessed 10 May 2012.
12. Nestel P, Bouis HE, Meenakshi JV, Pfeiffer W: **Biofortification of staple food crops.** *J Nutr* 2006, **136:**1064–1067.
13. Beyer P: **Golden Rice and 'Golden' crops for human nutrition.** *New Biotechnology.* 2010, **27:**478–481.
14. HarvestPlus: *Micronutrient malnutrition.* 2004. http://www.harvestplus.org/content/about-harvestplus. Accessed 10 November 2011.
15. Manyong VM, Dixon AGO, Makinde KO, Bokanga M, Whyte J: *The contribution of IITA-improved cassava to food security in sub-Saharan Africa: an impact study.* IITA, Ibadan, Nigeria: IITA report; 2000.
16. Kawano K: **Thirty years of cassava breeding for productivity:biological and social factors for success.** *Crop Sci* 2003, **43:**1325–1335.
17. Graham RD, Welch RM: *Breeding for staple-food crops with high micronutrient density: Agricultural Strategies for Micronutrients.* Washington, D.C: Working Paper No 3; International Food Policy Research Institute; 1996.
18. Graham R, Welch R, Bouis H: **Addressing micronutrient malnutrition through the nutritional quality of staple foods: principles, perspectives, and knowledge gaps.** *Adv Agron* 2001, **70:**77–142.
19. Graham RD, Senadhira D, Beebe SE, Iglesias C, Ortiz-Monasterio I: **Breeding for micronutrient density in edible portions of staple food crops: conventional approaches.** *Field Crops Res* 1999, **60:**57–80.
20. Yan J, Kandianis CB, Harjes CE, Bai L, Kim EH, Yang X, *et al*: **Rare genetic variation at Zea mays crtRB1 increases β-carotene in maize grain.** *Nat Genet* 2010, **42:**322–327.
21. Low JW, Arimond M, Osman N, Cunguara B, Zano F, Tschirley D: **A Food-Based Approach Introducing Orange-Fleshed Sweet Potatoes Increased**

Vitamin A Intake and Serum Retinol Concentrations in Young Children in Rural Mozambique. *J Nutr* 2007, **137:**1320–1327.

22. Steven R, Winter-Nelson A: **Consumer acceptance of provitamin A-biofortified maize in Maputo.** *Mozambique. Food Policy* 2008, **33:**341–351.
23. Jennings DL, Iglesias C: **Breeding for Crop Improvement.** In *Cassava: Biology*. Edited by Hillocks RJ, Thresh JM, Bellotti AC. CAB International; Production and Utilization; 2002:149–166.
24. Iglesias C, Mayer J, Chavez L, Calle F: **Genetic potential and stability of carotene content of cassava roots.** *Euphytica* 1997, **94:**367–373.
25. Ceballos H, Iglesias CA, Perez JC, Dixon AG: **Cassava breeding: opportunities and challenges.** *Plant Mol Biol* 2004, **56**(4):503–516.
26. Li HQ, Sautter C, Potrykus I, Pounti-Kaerlas J: **Genetic transformation of cassava (*Manihot esculenta*Crantz).** *Nat Biotechnol* 1996, **14:**736–740.
27. Schöpke C, Taylor N, Carcamo R, Konan NK, Marmey P, Henshaw GG, *et al*: **Regeneration of transgenic cassava plants (*Manihot esculenta*Crantz) from microbombarded embryogenic suspension cultures.** *Nat Biotechnol* 1996, **14:**731–735.
28. Taylor N, Chavarriaga P, Raemakers K, Siritunga D, Zhang P: **Development and application of transgenic technologies in cassava.** *Plant Mol Biol* 2004, **56**(4):671–688.
29. Stupak M, Vanderschuren H, Gruissem W, Zhang P: **Biotechnological approaches to cassava protein improvement.** *Trends Food Sci Technol* 2006, **17:**634–641.
30. Bouis HE, Chassy M, Ochanda O: **Genetically modified food crops and their contribution to human nutrition and food quality.** *Trends Food Sci Technol* 2003, **14:**191–209.
31. Liu JQ, Zheng Q, Ma Q, Gadidasu KK, Zhang P: **Cassava Genetic Transformation and its Application in Breeding.** *J Integr Plant Biol* 2011, **53**(7):552–569.
32. Sayre R, Beeching JR, Cahoon EB, Egesi C, Fauquet CM, Fellman J, *et al*: **The BioCassava Plus Program: Biofortification of Cassava for Sub-Saharan Africa.** *Annu Rev Plant Biol* 2011, **62:**251–272.
33. Raemakers CJJM, Schreuder M, Pereira I, Munyikwa T, Jacobsen E, Visser RGF: **Progress made in FEC transformation of cassava.** *Euphytica* 2001, **120:**15–24.
34. Bull SE, Owiti JA, Niklaus M, Beeching JR, Gruissem W, Vanderschuren H: ***Agrobacterium*-mediated transformation of friable embryogenic calli and regeneration of transgenic cassava.** *Nat Protoc* 2009, **4:**1845–1854.
35. Chakraborty S, Chakraborty N, Datta A: **Increased nutritive value of transgenic potato by expressing a nonallergenic seed albumin gene from Amaranthus hypochondriacus.** *Proc Natl Acad Sci USA* 2000, **97**(7):3724–3729.
36. Ye X, Al-Babili S, Klöti A, Zhang J, Lucca P, Beyer P, *et al*: **Engineering the provitamin A (beta-carotene) biosynthetic pathway into (carotenoid-free) rice endosperm.** *Science* 2000, **287:**303–305.
37. Paine JA, Shipton CA, Chaggar S, Howells RM, Kennedy MJ, Vernon G, *et al*: **A new version of Golden Rice with increased pro-vitamin A content.** *Nat Biotechnol* 2005, **23:**482–487.
38. Hunt JM: **Reversing productivity losses from iron deficiency: the economic case.** *J Nutr* 2002, **132:**794–801.
39. WB. The World Bank (WB): *Enriching lives: overcoming vitamin and mineral malnutrition in developing countries (Development in Practice Series)*. Washington DC: 1994.
40. Kalaitzandonakes N, Alston JM, Bradford KJ: **Compliance costs for regulatory approval of new biotech crops.** *Nat Biotechnol* 2007, **25:**509–511.
41. Manalo AJ, Ramon GP: **The cost of product development of Bt corn event MON 810 in the Philippines.** *AgBioforum* 2007, **10:**19–32.
42. Fauquet CM, Taylor NJ, Tohme J: **The Global Cassava Partnership for the 21st Century (GCP21).** *Tropical Plant Bio* 2012, **5:**4–8.
43. CBD. Convention on Biological Diversity (CBD): *Cartagena Protocol on Biosafety to the Convention on Biological Diversity: text and annexes. Montreal: Secretariat of the Convention on Biological Diversity.* 2000. www.biodiv.org/doc/legal/cartagena-protocol-en.pdf [Accessed: April 22, 2012].
44. Paarlberg R: **GMO foods and crops: Africa's choice.** *New Biotechnol* 2010. doi:10.1016/j.nbt.2010.07.005. 2010
45. Adenle AA: **Response to issues on GM agriculture in Africa: Are transgenic crops safe?** *BMC Research Notes* 2011, **4:**388. doi:10.1186/1756-0500-4-388
46. Britten N: **Qualitative Research: Qualitative interviews in medical research.** *BMJ* 1995, **311:**251.
47. Thomas DR: *A general inductive approach for qualitative data analysis.* New Zealand: School of Population Health University of Auckland; 2003. http://www.fmhs.auckland.ac.nz/soph/centres/hrmas/_docs/Inductive2003.pdf. Accessed 10 November 2011.
48. FAO: *A cassava industrial revolution in Nigeria.*: Cassava production; 2004. http://www.fao.org/docrep/007/y5548e/y5548e07.htm. Accessed 10 November 2011.
49. Laumann EO, Knoke D: *The Organisational State.* The University of Wisconsin Press, Madison: Social Choice in National Policy Domains; 1987.
50. Sanvido O, Romeis J, Bigler F: **Ecological impacts of genetically modified crops: ten years of field research and commercial cultivation.** *Adv Biochem Eng Biotechnol* 2007, **107:**235–278.
51. Carpenter JE: **Impacts of GM crops on biodiversity.** *Landes Bioscience.* 2011, **2:**1–17.
52. Carman J, Is GM: **Food Safe to Eat?** In *Recoding Nature Critical Perspectives on Genetic Engineering.* Edited by Hindmarsh R, Lawrence G. Sydney: UNSW Press; 82:93–2004.
53. ILSI: **Nutritional and safety assessments of foods and feed nutritionally improved through biotechnology.** *Comput Rev Food Sci Food Safety* 2004, **3:**3–64.
54. Konig A, Cockburn A, Crevel RW, Debruyne E, Graftstroem R, Hammerling U, *et al*: **Assessment of safety of foods derived from genetically modified (GM) crops.** *Food Chem Toxicol* 2004, **42:**1047–1088.
55. EFSA: **Safety and Nutritional Assessment of GM plants and derived food and feed: The role of animal feeding trials. Report of EFSA GMO Panel Working Group on Animal Feeding Trials.** *Food Chem Toxicol* 2008, **46:**S2–S70.
56. Ho M-W: *"Golden Rice"- an exercise in how not to do science." Third World Network Biotechnology and Biosafety Series No. 6 (2000): April 1.* 2010.
57. Chassy B, Egnin M, Gao Y, Glenn K, Kleter GA, Nestel P, *et al*: **Recent Developments in the Safety and Nutritional Assessment of Nutritionally Improved Foods and Feeds.** *Comprehensive Reviews in Food Science and Food Safety* 2008, **7**(1):50–113.
58. Potrykus I: **Lessons from the 'Humanitarian Golden Rice' project: regulation prevents development of public good genetically engineered crop products.** *New Biotechnology* 2010, **27:**466–472.
59. Corpuz GA, Yee-Bee D, Hoffman JL: *No gold in golden rice, says GMO expert. All Voices. 25 August.* 2011. http://www.allvoices.com/contributed-news/10146190-no-gold-in-golden-rice-says-gmo-expert. Accessed 10 November, 2011.
60. Vanderschuren H, Stupak M, Fütterer J, Gruissem W, Zhang P: **Engineering resistance to geminiviruses – review and perspectives.** *Plant Biotechnol J* 2007, **5:**207–220.
61. Bull SE, Ndunguru J, Gruissem W, Beeching JR, Vanderschuren H: **Cassava: constraints to production and transfer of biotechnology to African laboratories.** *Plant Cell Rep* 2011, **30:**779–787.
62. Sawahel W: *"GM cassava has 'super size' roots", SciDev.net, 15 May.*: ; 2006. http://www.scidev.net/en/news/gm-cassava-has-super-size-roots.html> Accessed 11 August 2012.
63. Nassa N: *"Boosting cassava roots the non-GM way", SciDev.Net, 31 May.* 2006. http://www.scidev.net/en/editor-letters/boosting-cassava-roots-the-nongm-way.html> Accessed 11 August 2012.
64. EAM: *Doubts Over Cassava Project. East African Magazine (EAM)11 September.* 2006. http://www.gene.ch/genet/2006/Sep/msg00030.html. Accessed 11 August 2012.
65. GMWATCH: *GM Cassava "Our only hope".* 2009. http://www.gmwatch.org/gm-myths/11156-gm-cassava-qour-only-hopeq. Accessed 10 November 2011.
66. Broach S: *Genetically Modified Cassava. The Yale Globalist. May 10.* 2011. http://tyglobalist.org/front-page/features/genetically-modified-cassava-new-technologies-have-complicated-the-challenge-of-feeding-africa/> Accessed 11 August 2012.
67. ABC. Agricultural Biotechnology Council (ABC): *GMObserver: News about farming and GM in the United Kingdom.* 2006. http://www.abcinformation.org/library/files/Issue%209v3%20FINAL.pdf. Accessed 10 November 2011.
68. FoEA. Friends of the Earth Africa (FoEA): *Genetically Engineered Cassava. A threat to Africa's Food Sovereignty.* 2006. http://www.eraction.org/publications/cassava_fact_sheet.pdf. Accessed 10 November 2011.
69. Burns G: *A plant of last resort: Biocassava a meal ticket for famine-stricken East Africa. 1 October, Chicago Tribune.* 2011. http://articles.chicagotribune.

com/2011-10-01/news/ct-oped-1002-cassava-20110930_1_cassava-genetic-engineering-famine. Accessed 10 November, 2011.

70. Biopact: *Controversy over transgenic cassava project: US accused of pushing GM cassava for biofuels from Africa. September 13.* 2006. http://news.mongabay.com/bioenergy/2006/09/controversy-over-transgenic-cassava.html. Accessed 10 November 2011.
71. Ashby JA: *Evaluating technology with farmers: In A Hand Book of IPRA project.* CIAT, Colombia: CIAT and the Kellog foundation; 1990.
72. Sayre RT: *Biofortification of cassava for Africa: the BioCassava Plus Program. Chapter 11.* London UK: Smith-Gordon; 2011:113–120.
73. Vanderschuren H, Alder A, Zhang P, Gruissem W: **Dose-dependent RNAi-mediated geminivirus resistance in the tropical root crop cassava.** *Plant Mol Biol* 2009, **70:**265–272.
74. Gruissem W: *email of 14 November 2011 sent to Ademola Adenle from Gruissem Wilhelm.* 2011.
75. NU-IAS: *Interviews conducted by Dr. Ademola Adenle at the United Nations University-Institute of Advanced Studies (UNU-IAS) on Genetic Modified (GM) Agriculture in Egypt, June.* 2012.
76. Pray C, Naseem A: **Intellectual Property Rights on Research Tools: Incentives or Barriers to Innovation? Case Studies of Rice Genomics and Plant Transformation Technologies.** *AgBioforum* 2005, **8**(2&3):108–117.
77. WEMA: *Water Effecient Maize for Africa (WEMA). Progress Report. March.* 2011. http://www.aatf-africa.org/userfiles/WEMA-Progress-Report_2008-2011.pdf. Accessed 11 August 2012.
78. AATF: *African Agricultural Technology Foundation (AATF). WEMA Management Response to ESCC Audit report 2010 Key Findings and Recommendations.* 2010. http://www.aatf-africa.org/userfiles/WEMA-Management-Response-ESCC-Audit-2010.pdf. Accessed 11 August, 2012.
79. Mashood E: *The GM Debate-Who Decides? Analysis of decision-making about genetically modified crops in developing countries. Panos Report No 49.* 2004. http://panos.org.uk/wp-content/files/2011/03/gm_debate_reportwZacCL.pdf> Accessed 11 August 2012.
80. Gonzalez C, Johnson N, Qaim M: **Consumer Acceptance of Second-Generation GM Foods: The Case of Biofortified Cassava in the North-east of Brazil.** *J Agric Econ* 2009, **60:**604–624.
81. Wolson RA: **Assessing the Prospects for the Adoption of Biofortified Crops in South Africa.** *AgBioforum* 2007, **10:**184–191.
82. Aerni P, Bernauer T: **Stakeholder Attitudes Toward GMOs in the Philippines, Mexico, and South Africa:The Issue of Public Trust.** *World Dev* 2006, **34:**557–575.
83. Krishna VV, Qaim M: *Consumer attitudes towards GM food and pesticide residues in India Review of Agricultural Economics.* 2008, **30:**233–251.
84. Morris EJ: **A semi-quantitative approach to GMO risk-benefit analysis.** *Transgenic Res* 2011, **20:**1055–1071.
85. Stein AJ, Meenakshi JV, Qaim M, Nestel P, Sachdev HPS, Bhutta ZA: *Analyzing the Health Benefits of Biofortified Staple Crops by Means of the Disability-Adjusted Life Years Approach: a Handbook Focusing on Iron, Zinc and Vitamin A. HarvestPlus Technical Monograph 4.*: Washington, DC and Cali: International Food Policy Research Institute (IFPRI) and International Center for Tropical Agriculture (CIAT); 2005. http://www.ifpri.org/sites/default/files/publications/hptech04.pdf>. Accessed 11 August 2012.
86. Meenakshi JV, Johnson N, Manyong VM, De Groote H, Javelosa J, Yanggen D, *et al*: **How cost-effective is biofortification in combating micronutrient malnutrition? An ex-ante assessment.** *World Dev* 2010, **38:**64–75.
87. Labonté R, Schrecker T: **Globalization and social determinants of health: The role of the global marketplace (part 2 of 3).** *Glob Heal* 2007, **3:**6. doi:10.1186/1744-8603-3-6.

11

Empowering smallholder rice farmers in Tanzania to increase productivity for promoting food security in Eastern and Southern Africa

Costancia Peter Rugumamu

Abstract

Background: Despite low rice yield levels in Tanzania, the country is the leading producer in Eastern and Southern Africa. Given that this business is dominated by smallholder farmers and that the country is endowed with extensive landscapes suitable for rice production coupled with a high domestic and foreign potential market, there is an urgent need to enhance the yield as a key to promote food security from household to national and international levels. This paper seeks to demonstrate the importance of empowering smallholder farmers as a strategy for identifying and applying improved rice husbandry technologies in a holistic and integrated manner by employing a Farmer Field School (FFS) approach.

Methods: The study was conducted in two sample villages, namely Kipera Njiapanda in the Morogoro region and Bwawani Visegese in the Pwani region, where culturally and ecologically rice is grown. The FFS approach was applied with the involvement of smallholders and local authorities to develop innovative ways to increase yield and promote food security. Data and information on current rice farming practices were collected using FFS participants and qualitatively analyzed at each rice production stage.

Results: Findings revealed that farmers' socioeconomic conditions varied, thus reflecting differences in farm size, crop husbandry and associated yield as well as in non-market production along with consumption and/or market exchange. The FFS approach demonstrated that community participatory processes have the capacity to create conducive conditions for empowering smallholders to identify their rice production problems and opportunities related to access to technical and financial support in the realm of seedbed preparation and irrigation, including pest and disease control and crop marketing.

Conclusion: The FFS approach is a robust tool for the identification of problems/challenges as well as pertinent opportunities to increase rice yield and promote food security. The paper calls for support from development partners in managing the rice production systems at all stages in the production cycle given the current potential of both the smallholder farmers and the arable land to increase rice yield, reduce poverty and promote food security at household, national and international levels.

Keywords: Rice yield, Smallholder farmers, Farmer Field School, Technology, Food security

Background

In Tanzania, agriculture is the mainstay of the national economy, with rice being one of the major food and cash crops, as acknowledged by Agritrade [1] for being the largest rice producer accounting for about 80% of total production in Eastern Africa. Approximately 80% of Tanzanians are employed in the agriculture sector, which accounts for half of the national income and three quarters of merchandise exports in the country [2]. As stated by Agritrade [1] and Kafitiriti *et al.* [3], the country ranks second within Eastern, Central and Southern Africa in terms of rice production and consumption after Madagascar. Despite this promising trend, it is further noted that rice production is heavily dependent on rainfall, and under good weather conditions and improved

Correspondence: wrugu@udsm.ac.tz
Department of Zoology and Wildlife Conservation, University of Dar es Salaam, PO Box 35064, Dar es Salaam, Tanzania

husbandry the country is potentially capable of achieving food self-sufficiency as measured by levels of grain production [4].

Rice is grown under three major ecosystems, namely rain-fed lowland, irrigation (or flooded conditions) and upland rice (non-flooded conditions), with each type characterized by a relatively low yet differing production potential depending on varying soil and climatic conditions [5]. Reviews of the major causes of low rice yields have been undertaken by Concern Worldwide [4], the Ministry of Finance and Economic Affairs [6], Kaburire and Ruvuga [7] and Bucheyeki *et al.* [8], among others. These include hostile environmental conditions due to climate change, low level and inadequate technology dissemination channels, crop pests and diseases and low yielding rice varieties as well as gender dynamics.

A number of programmes have been put in place to address some of these challenges [6,9-11]. In 2001, the Government of Tanzania launched the Agriculture Sector Development Strategy (ASDS), followed in 2003 by the Agricultural Sector Development Programme (ASDP), which was implemented at district level under the District Agriculture Development Plan (DADP) focusing on increasing agricultural productivity, profitability and farm incomes [9,10].

It is in this light that the Government of Tanzania's commitment to address the food insecurity issue was reflected in the National Strategy for Growth and Reduction of Poverty phase 2 (NSGRP II) and the Millennium Development Goals (MDGs) [6]. In essence, food security refers to a state of affairs where all people at all times have access to safe and nutritious food to maintain a healthy and active life [12]. In the same vein, working in partnership with other stakeholders, such as the AfricaRice project, Tanzania deemed it mandatory to invigorate rice production in order to promote food security and increase foreign exchange earnings [13]. It is worth noting, therefore, that making progress in achieving food security calls for effective participation of stakeholders in the decisions and actions that affect their lives in the rice production chain.

In this regard, it may be stated that the existing rice production potential for increased yield and promoting food security in Tanzania has not been examined or addressed in a participatory, holistic and integrated manner. On the contrary, it is obvious that various institutions conventionally address issues pertinent to their own missions, thus overlooking the whole set of challenges faced by the farming communities. This logically implies that the identified problems are compartmentalized issues, since they are studied through inadequate understanding of the relationship between specific system components being the basis for guiding input, output and feedback.

It is against this premise that this study was carried out using a Farmer Field School (FFS) approach, as advocated by Stathers *et al.* [14] in order to achieve integrated rice production processes and pest and disease management to enhance yield. Smallholder farmers in Tanzania dominate the agricultural sector, with average farm sizes being between 0.9 and 3.0 hectares and cultivating 5.1 million hectares annually, of which 85% is food crops [2]. By teaming up with the smallholders in the field school situation, the thrust of the study was to improve organization and management skills by empowering the participants through education and innovative communication to make better use of their existing knowledge and resources to handle their own on-farm decisions in the full cropping season. The ultimate goal was to inculcate into the minds of smallholder farmers the best rice production practices, and hence high yield coupled with food security.

The specific objectives of the study were twofold. First, to investigate the current rice production technologies used by smallholders in order to identify attendant problems and opportunities/solutions; and second, to inform development partners of the need to empower smallholders in order to enhance yields in an integrated management approach and to improve household food security and beyond. This paper, therefore, underscores the effectiveness of the FFS approach as a tool for communicating challenges and opportunities faced by smallholder farmers and their development partners in the rice production process on one side and the researcher/facilitator on the other.

Materials and methods

Selection of study area

This field study was conducted in the Morogoro and Pwani regions of Tanzania, located to the west of the Indian Ocean (Figure 1). Four factors influenced the choice of the study areas. First, the area, extending up to about 300 km inland, falls within a suitable agro-ecological zone for rice production. Second, the two regions have a similar sociocultural set-up and spatially they form the hinterland of Dar es Salaam city, a major market for rice. Third, they are known to be the key rice producing regions accounting for 67% of total rice production [15] and form the largest population groups in the country with a significant portion of the local rice market. Finally, there is potential demand for rice grown in these regions due to accessibility by road networks, the Tanzania-Zambia Railway Authority (TAZARA) and the Central Railway line, as shown in Figure 1. As noted above, this means that rice from the two regions could be easily transported to the Southern African region and to most of the countries in the Lake Zone as an export crop [1]. To this end, the two model rice growing villages selected for the study are: Kipera Njiapanda in the Mvomero district, Morogoro

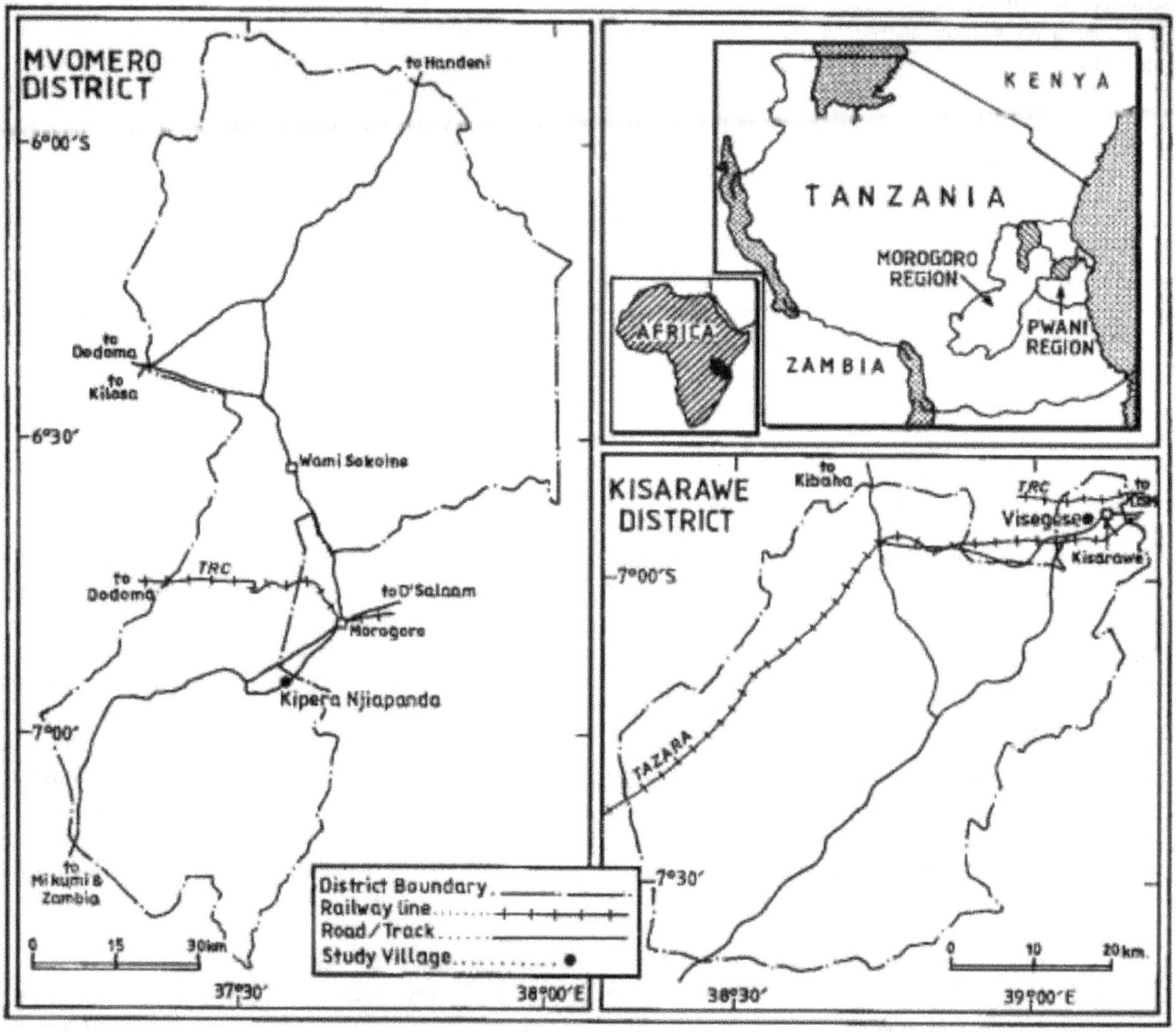

Figure 1 Location of the study areas, Kipera Njiapanda and Bwawani Visegese. Note the road and railway line networks.

region; and Bwawani Visegese in the Kisarawe district, Pwani region (Figure 1).

Data collection

Primary and secondary data sources of integrated rice production processes and management formed the basis of this study. A literature survey was conducted in order to better understand the state-of-the-art of rice farming in the country in general and attendant theories. The field study entailed participatory observation and recording and analysis of real problems and opportunities in all rice production processes and activities, from arable land ownership and selection to consumption and/or processing for the market.

A participatory approach grounded in the FFS was pursued, with 20 smallholder rice growers purposefully chosen in each of the two research villages to participate in the FFS [14]. The selection of participants was based on income levels in order to capture a representative cross-section of the communities within their socioeconomic context. Duwayri *et al.* [16] reported that yields vary among farmers in the same location, with good farmers usually reaping more benefits from improved technologies than mediocre farmers at the same place. The research was thus carried out during the major rice farming activities between August 2012 and June 2013 in order to better understand and verify data and information originating from the location of specific field schools and the secondary sources, as advocated by Gyawali *et al.* [17].

With respect to pest and disease control, the FFS participants were introduced to the management procedures. Insects and other arthropod pests and vectors of disease found on rice plants in farms were captured by employing a sweep net sampling method [18]. Sampling of all pests was conducted by the participants at critical phases of rice growth, particularly tillering, which is a vegetative phase, and booting and milking stages, referred to as the reproductive phase. Sampling of 30 sweeps was carried out at the centre part of the rice plots and also close to the bund (about 1 m) while walking slowly through the rice plot during dew-free periods in the day [18]. Most larvae borers were sampled from plant parts as indicated by damage symptoms. Rice plants showing root-knot symptoms were sampled for investigation of nematodes. Weeds in the farms were collected for identification. Thereafter, all organisms captured were introduced in labeled plastic bags and a preliminary identification by FFS participants guided by the researcher was carried out. The samples were then preserved for detailed identification in the laboratory, according to Southwood and Henderson [18], Hill [19] and Picker *et al.* [20].

Data analysis

Qualitative techniques were employed to analyze the data [21]. Data and information recorded as verbal responses provided by the FFS participants were contextually analyzed to generate more complete, meaningful and useful understanding of the rice production processes.

Results and discussion

Farmland setting and preparation

The rice farm holdings were private property whereby farmers controlled their rice farm holdings. The smallholders who formed the rice FFS predominantly shared a similar socioeconomic and cultural background. The size of farm holdings varied among the participants. This was the result of the predominance of traditional technologies, namely hand tools, humus economy and household labour constraints, coupled with sporadic diseases such as malaria among family members, which contributed to labour shortage and subsequent cultivation of small plots.

The FFS members' ingenuity in Bwawani enabled them to classify their land using physical properties, as follows: 1) flat landscape unit; 2) readily seasonally flooded land; 3) clayey soils; 4) predominance of luxurious grasses belonging in the rice Family Poaceae (also called Gramineae); and 5) land easily reached from the homesteads. The choice of the plot sites in floodplains was hence dependent on indigenous/traditional environmental knowledge. Members upheld that landscapes with the aforementioned attributes were scarce. Therefore, the cultivation of isolated/fragmented plots in Bwawani reflected spatial discontinuity of soils suitable for rice cultivation.

In Kipera, all participants were members of the irrigation scheme. The land under the irrigation agriculture was limited by the small capacity of the scheme, which serves members from four sub-villages, namely Kipera Njiapanda, Kipera Mbuyuni, Kipera Kinyenzo and Kipera Luguno, and therefore supports only about 120 farmers. Land preparation in the scheme included the repairing of canals, land and patch banding, as supported by Katambara *et al.* [11]. The participants also owned plots adjacent to the scheme, and the most popular traditional rice cultivation practice in those valleys was characterized by flooded rainwater.

Given that farmers in both villages predominantly used hand tools, including hand hoes, matchets, pick axes, slashers, wooden sticks and knives, they all expressed the need to access tractors to promote timely sowing with the onset of the rains and flooding of the farms, a precondition of increased yield as articulated by Katambara *et al.* [11]. Farmers in BwawaniVisegese expressed the desire to be enabled to access tractors through loans from the Kisarawe District Council. The farmers felt that it was viable to settle the loan after harvesting and marketing the produce. It was revealed that all available family members were involved in farm preparation,with women spending more hours on the activity than men, which reflected gender inequity in labour input. This viewpoint is supported by Leavens and Anderson [22] who reported that in Tanzania, agriculture comprises a greater part of women's economic activity than men's, whereby 81% of women compared to 73% of men are engaged in agricultural activity, while in rural areas that proportion rises to 98% for women.

Additionally, farmers called for technical support in the realm of nursery preparation, seedling production and improvement in land leveling, paddling and banding as well as canal construction, in order to improve water retention capacity for efficient seed germination and growth. Furthermore, farmers appealed to development partners to assist in the acquisition of technical knowledge on ponding rainwater for irrigation, such as dam construction (charcoal), which would ensure adequate flooding conditions and irrigation throughout the rice growing and development stages, should rain fail them (like it did in the 2011/2012 growing season). During training, it was realized by all members that rice growth and development is heavily constrained by unreliable onset, amount and distribution of rainfall. Katambara *et al.* [11] reported that the demand for water for irrigation in Tanzania outstrips the amount of water available for irrigation and other purposes, and that there is a need to apply appropriate technologies in the farming practices to ensure high yield while minimizing water utilization demand.

Characteristics of rice seed varieties and sowing activity

During the training sessions, the FFS participants established that seed varieties for sowing were locally acquired by recycling the previous year's harvests, purchasing from either the village markets and/or neighbours and from friends and relatives in the same or distant villages. They reported that the rice varieties widely grown in both villages varied and the major characteristic considered for the choice of varieties was taste. The common varieties grown in Bwawani were Machale, Supa, Borakupata, Kahogo, Tule na Bwana and Kibakuli, while in Kipera, Supa, Kahogo, Mbagara and Moshi wa Paa were grown. The varieties were grown on separate portions of farms, and choices could vary in different seasons as they were selected depending mainly on availability.

Concurring with Joshi *et al.* [23], the farmers demonstrated their traditional environmental knowledge of seed genetics as they reported the criteria employed in selecting good quality seeds and varieties for sowing. Their observations were further supported by Joshi *et al.* [23] who noted that smallholder farmers often selected crop varieties that were well-adapted in particular ecosystems. The farmers unanimously acknowledged that varieties of rice which were high yielding, aromatic and marketable coupled with grain heaviness and pest, disease and drought resistant-were the most preferred, but were often inaccessible.

In both study areas, two rice farming systems were broadly identified. In Bwawani, the sowing of maize, carried out immediately at the onset of the November/December rains, precedes that of rice, while in Kipera, tomato cultivation proceeds rice harvesting by capitalizing on the residual soil moisture. These systems were conceived as strategies for food security in the areas. The participants emphasized that sowing early at the onset of rains allowed an adequate amount of water for a desirable period and hence a good harvest. They narrated their experience of a 2- to 3-week drought after sowing which resulted in a low rice yield, a feature that underscores the need for installation of irrigation infrastructure to meet the moisture deficit.

The FFS participants affirmed the practice of two types of sowing rice, namely drill and broadcast sowing. The former type employed uniquely small hand hoes and wooden sticks that lead to desirable patterns and spacing. Establishing nurseries was advocated in order to maximize the utilization of the available arable land for increased yield. They further noted with concern the labour demand at this stage of rice production, appealing for support to access improved rice sowing technology based on the tillering capacity of a particular rice variety.

Management of pest infestation and disease infection

FFS participants shared their knowledge regarding pest attacks and diseases which reduce rice yield. Weed pest control was considered by all members of the FFS to be one of the determinants of high rice yield. The common weeds collected and co-identified by both common and scientific names in both villages included 'Kongwa', *Monochoria vaginalis*,and 'Ndago', *Cyperus iri*. As noted by Euroconsult [24], the facilitator emphasized that two to three weeding operations should be mandatory until crop maturity. As an integrated weed pest control strategy, it was highlighted and noted that an adequate floodwater layer was the most appropriate technology for controlling weeds because few plant species can tolerate flood conditions.

Together with the facilitator, insect pest species were identified as belonging predominantly in the Order Lepidoptera (Table 1). The identified insects were among those reported by Sehgal *et al.* [25] who illustrated that the rice crop can be attacked from the seedling stage through to harvest and storage. Members appreciated the fact that insects attack different parts of the rice plant and that this depends on the type of their mouth parts. It was made clear, for instance, that adult and nymph hemipterans use their piercing and sucking mouth parts, and lepidopteran and Diptera larvae and adult and nymph orthopterans damage plant parts by using their biting and chewing mouth parts. Thus, the insects were conceived to be stem/stalk borers, stem and root feeders and leaf and panicle feeders as well as suckers, extracting liquid known as milk from grains and saps from leaves (Table 1). It was made clear to the members during field investigations that sucking insect vectors spread disease viruses, which turn

Table 1 Insect pests and disease vectors of rice identified in FFS farms in the study areas

Common name	Species	Order/family	Pest/vector
Spotted stalk borer	*Chilo partellus* Swinh.	Lepidoptera/Pyralidae	Pest: larvae feed in the leaf funnel and the dead central leaves form 'dead heart'.
White stem borer	*Maliarpha separatella* Rag.	Lepidoptera/Pyralidae	Pest: larvae bore stem.
Pink stalk borer	*Sesamia calamistis* Hmp.	Lepidoptera/Noctuidae	Pest: larvae bore stem.
Stalk-eyed fly	*Diopsis thoracica* West.	Diptera/Diopsidae	Pest: maggots eat the central shoot and flower head.
Rice stem gall midge	*Orseolia oryzae* W.M.	Diptera/Cecidomyiidae	Pest: maggots feed at base of the growing shoot causing the formation of a tuber gall.
Hispid beetle	*Trichispa sericea* Gn.; *Dactylispa lenta*	Coleoptera/Chrysomelidae	Pest: cause severe defoliation;vector: of rice yellow mottle virus (RYMV).
Small rice grasshopper	*Oxya chinensis* Thnb.	Orthoptera/Acrididae	Pest: adults eat foliage and panicle; nymph eats foliage.
Green stink bug	*Nezara viridula* L.	Hemiptera/Pentatomidae	Pest: punctures and feeds on young seed; vector:spread fungi.

leaves to brown and spotty, a condition frequently observed in their farms.

The FFS groups became aware that damaged leaves result in a low rice yield because the reduced surface area of the leaves lessen their capability to produce food by photosynthesis. These implications are well collaborated by Matteson [26] who reported that plants refill their grains after early attacks, but the rice plant cannot recover from later attacks and the resultant rice yield becomes contaminated by bacteria or fungi injected into the grains by infected insects, and that such harvests are called 'pecky rice'. This knowledge instilled into the participants the sense to undertake rapid control measures.

The group members revealed that rodents were controlled by using baited traps, while birds were scared by throwing stones and sticks as well as erecting human dolls of various shapes and sizes. These measures were undertaken immediately at the onset of the reproductive phase through to ripening and maturity and up to harvesting in 'programmed' rotations. An interesting technology observed in Kipera was the hanging of cassette ribbons across farm plots, which when blown by the wind produced noise that scared birds. The activity was deemed to be labour intensive. It was also observed that water snails occurred rarely, and that these pests cut stems or leaves in young rice plants. It was confirmed by members that control was by physical destruction. According to Sehgal *et al.* [25], this approach could be a palliative measure because in the presence of water females have a high reproductive performance.

During observations of the rice roots, participants noted that some had swelling tips. Further investigation revealed the occurrence of root-knot nematodes, *Meloidegyne graminicola*, as rare pests that cause the swelling of root tips. It was explained by the facilitator that such a condition causes disruption and hypertrophy of the root cortex resulting in stunted plants with chlorotic leaves.

Sporadic armyworm outbreaks were reported in Bwawani, where farmers inform government authorities for intervention. In the event of delayed response, they immediately purchase chemical pesticides from local business persons. However, it was underscored that often such chemicals proved ineffective. Another pest control strategy introduced to members of the FFS was cultivating two or more plant species in the vicinity of rice plots to provide alternative hosts for some pests and vectors, and thus to reduce damage levels. According to Abate *et al.* [27] and Heinrichs and Barrion [28], such diverse traditional cropping systems enhance natural enemy abundance and generally keep pest numbers below economic injury levels (EILs). It was also reported by Matteson [26] that most rice pests are controlled by a complex and rich web of generalist and specialist predators and parasites that live in or on the rice plant, paddy water or soil. Although rare, thefts were cited to occur in both villages when the crop was ripe as well as during harvesting, drying and/or storage in homesteads.

The members and facilitator emphasized the intensification of the existing control practices as well as the utilization of acquired knowledge and skills to develop appropriate technology components in an integrated management approach. The participants expressed confidence and were empowered to rapidly identify pests at different post-embryonic development stages and also the damaging stages as well as specific parts attacked. Such skills were deemed necessary to enable farmers to apply various timely components of Integrated Pest Management (IPM) control methods and to keep the pests and vectors below EILs.

Application of fertilizers

Members of FFSs in both villages were knowledgeable that industrial fertilizer application increases rice yield, yet considered it to be very expensive. They were aware of urea and calcium ammonium nitrate fertilizers and also cow dung. The participants and facilitators thus clearly expressed the inter-linkages of fertilizer-soil-water management to avoid the risk of fertilizers being washed away by seasonal floods.

Rice harvesting and transportation

The harvesting of rice was carried out between May and August depending on the ripening status of the crop, that is, exactly when the crop has attained physiological maturity and there is no further addition of dry matter. Most participants in both villages owned simple tools. Sickles and small knives were used for separating panicles from rice plants to be collected in baskets and/or mats. Wooden sticks were used for thrashing panicles to remove paddy, which was spread on mats for initial sun drying.

Subsequently, the harvested crop, having been carefully packed in baskets, bags (polythene) and even wrapped in pieces of cloth, was immediately transported to the homesteads. In Kipera, hired trucks, bicycles and wheelbarrows were used to transport the crop, while in Bwawani trucks were hired only by farmers residing in Kisarawe town and Dar es Salaam city. Moreover, some farmers carried the harvest on their heads and/or shoulders, which was not a rare phenomenon. Harvesting and transportation were conceived to be one of the tedious activities that calls for extra labour, where the demand is always dictated by the yield.

Rice yield

At their homesteads, FFS farmers spread paddy on large mats to dry in the open air to attain a locally acceptable moisture content level of the crop as tested by experienced farmers. Thereafter, paddy was weighed and packed. It became evident during discussions in both study villages that there were varying production levels and indeed yields. The FFS participants in both villages reported that the yields obtained by the smallholder farmers for the past 3 years using the itemized technologies were varying and fluctuating. In Bwawani Visegese village, the well-to-do farmers had yields varying between 20 and 30 bags of 120 kg each, while the middle group had between 10 and 20 bags and the poor produced between 1 and 5 bags per hectare (Table 2). The estimated average rice yield in the small irrigation scheme in Kipera Njiapanda was 12 bags per hectare. The members were optimistic about gaining increased rice yields once the learnt technologies and innovations could be applied in an integrated manner.

Storage of paddy

FFS members aware of post-harvest crop losses had invented traditional storage technologies to minimize damage and/or losses. This culture was also reported by Rugumamu [29] in the maize storage systems in Tanzania. In both villages, farmers used locally constructed containers and polythene bags which were placed on raised platforms, a method deemed to retain the required level of moisture content of the crop and protect from rodent attack. In this phase, the FFS farmers expressed concern and care that they attached to the harvest in order to minimize post-production losses, which have a direct bearing on food security. They articulated the need for constructing improved warehouses/godowns for storage of paddy as a means of increasing the safe storage period.

Marketing of rice

The production of rice in the two villages was predominantly for subsistence and hence a small proportion was sold to meet other basic needs. In the course of discussions, the members in Kipera observed that economic liberalization was already in place as farmers readily sold their commodity at their weekly local markets. Given the advantageous geographical location and relatively high yield, large quantities of rice and/or paddy were purchased by petty traders who re-sold them to businessmen and women from the Morogoro municipality and metropolitan Dar es Salaam. These are the transactions that form the start of the international rice trade.

Like in other areas in Tanzania [1], it was observed that marketing of the available rice was highly fragmented, with millers and brokers playing a central role in the trading process. Ultimately, as reported by the Tanzania Agricultural Partnership [9], the supply channels are generally long and the produce changes many hands before reaching the final consumer. The FFS members and facilitator, supported by Kaburire and Ruvuga [7], believed that any significant improvement in the incomes of subsistence farmers accruing from improved agricultural production and productivity would have significant backward and forward linkages to the rest of the economy. Such multiplier effects resulting from increased purchasing power of farmers for agricultural inputs, outputs and services from other sectors, and from increased supplies of food and other farm products, would help to lower the cost of living for both rural and urban dwellers.

Table 2 Estimated average rice yield (kg/ha) for 3 years among three socioeconomic groups of smallholder farmers in Bwawani Visegese village

Group of farmers	Rice yield (kg/ha)
Well-to-do	2,400 to 3,600
Middle	1,200 to 2,400
Poor	120 to 600

Conclusions

In order to effectively engage communities in achieving household food security and beyond, it is deduced in this paper that participatory learning processes through FFSs have the capacity to create conducive conditions for empowering smallholder farmers to identify problems and propose opportunities/solutions directed towards increased yield. In a nutshell, the approach has been able to raise community awareness about revitalizing the rice production cycle in a holistic and integrated manner within the improved socioeconomic and ecological production set-up. The knowledge and skills gained and shared among and between the smallholder farmers and their development partners during the FFSs at each stage of the rice production cycle, when integrated, form tools for implementation and dissemination of appropriate technologies required to enhance yield and promote food security.

To this end, it is proposed that the next research agenda will be to explore the implementation of the identified technologies in the study villages in order to form a platform to monitor and evaluate the impact of training to improve rice yields for FFSs. The output will provide demonstration avenues for other rice producing communities of western and southern Tanzania, where the industry is gaining momentum in terms of increased yield to meet both domestic consumption and market demand.

Abbreviations

ASDP: Agricultural Sector Development Programme; ASDS: Agriculture Sector Development Strategy; DADP: District Agriculture Development Plan; EIL: Economic Injury Level; FFS: Farmer Field School; IPM: Integrated Pest Management; MDGs: Millennium Development Goals; NSGRP II: National Strategy for Growth and Reduction of Poverty phase two; TAZARA: Tanzania-Zambia Railway Authority.

Competing interests

The author declares no competing interests.

Author's contribution

CPR is solely responsible for the manuscript and the research work.

Author's information

Academically, CPR holds a B.Sc. (biological sciences and education), M.Sc. (Applied Zoology) and Ph.D. (Applied Zoology). CPR is a Senior Lecturer and specialist in crop pest management. CPR is currently the Dean at the Faculty of Science, Dar es Salaam University College of Education. CPR is a Fellow of the Tanzania Academy of Sciences and Chairperson for the Association of Science Technology and Innovation for Women in Tanzania. CPR has been serving in different boards, including the Board of the National Museum of Tanzania as a Chairperson.

Acknowledgements

This paper benefited from generous support provided by individuals, groups and institutions. The author is grateful to the Gender Centre of the University of Dar es Salaam for facilitating the final development of the research proposal. The author also wishes to pay tribute to the Dar es Salaam University College of Education for availing a four-wheel drive vehicle and a driver for ease of transport in the study villages. Finally, the author wishes to record sincere appreciation to the smallholder farmers, local leaders and extension staff who diligently participated in the FFSs.

References

1. Agritrade: *USDA Review of Policy Constraints on Competitive EAC Rice Production: News and Analysis.* Wageningen: CTA; 2012.
2. United Republic of Tanzania: *(URT) Agricultural Sector Development Programme Performance Report.* Dar es Salaam: Government of Tanzania; 2011.
3. Kafitiriti EM, Dondeyne S, Msomba S, Deckers JA, Raes D: **Variations in agronomic characteristics of irrigated rice varieties: lessons from participatory trials in South Eastern Tanzania.** *Food Agric Environ* 2003, **1**:273–277.
4. Worldwide C: *Responding to the Needs of Marginal Farmers: A Review of Selected District Agricultural Development Plans in Tanzania.* London: Concern Worldwide; 2008.
5. Kato F: **Development of major rice cultivation area in the Kilombero Valley Tanzania.** *Afr Study Monogr* 2007, **36**:3–18.
6. Ministry of Finance and Economic Affairs: *National Strategy for Growth and Reduction of Poverty II (NSGRP II): Policy.* Dar es Salaam: Ministry of Finance and Economic Affairs, United Republic of Tanzania; 2010.
7. Kaburire L, Ruvuga S: **Networking for agricultural innovation. MVIWATA national network of farmers' groups in Tanzania.** *Bull* 2006, **374**:80–86.
8. Bucheyeki TL, Shennkalwa E, Kadadi D, Lobulu J: **Assessment of rice production constraints and farmers preferences in Nzega and Igunga districts.** *J Adv Dev Res* 2011, **2**(1):30–37.
9. Tanzania Agricultural Partnership: *Value Chain Analysis of Rice and Maize in Selected Districts in Tanzania: Report.* Dar es Salaam: Agricultural Council of Tanzania/Tanzania Agricultural Partnership; 2010.
10. Vermeulen S, Cotula L: *Making the Most of Agricultural Investment: A Survey of Business Models that Provide Opportunities for Smallholders: Report.* IIED/FAO/IFAD/SDC: London/Rome/Bern; 2010.
11. Katambara Z, Kahimba FC, Mahoo HF, Mbungu WB, Mhenge F, Reuben P, Maugo M, Nyarubamba A: **Adopting the system of rice intensification (SRI) in Tanzania: A review.** *AgricSci* 2013, **4**:369–375.
12. Food and Agriculture Organization of the United Nations (FAO): *World Food Summit: Food for All.* Rome: FAO; 1996.
13. Lugongo B: **Tanzania government unveils strategy to triple rice output.** *Citizen Newspaper* 2010.
14. Stathers T, Namanda S, Mwanga ROM, Khisa G, Kapinga R: *Manual for Sweet potato Integrated Production and Pest Management Farmer Field Schools in Sub-Saharan Africa.* International Potato Center: Kampala; 2005.
15. Afari-Sefa V, Beed F: *Enhancing Vegetable Value Chains in Rice-Based and Sole Crop Production Systems to Improve Household Income and Consumption in Morogoro. Presented at Africa RISING East and Southern Africa Research Review and Planning Meeting, Arusha, Tanzania, 1 to 5 October 2012.* AVRDC, The World Vegetable Center: Tainan; 2012.
16. Duwayri M, Van Tran D, Nguyen VN: *Reflections on Yield Gaps in Rice Production: How to Narrow the Gaps.* Rome: Food and Agriculture Organization of the United Nations (FAO); 2003.
17. Gyawali S, Sunwar SS, Ubedi M, Tripathi M, Joshi KD, Witcombe JR: **Collaborative breeding with farmers can be effective.** *Field Crop Res* 2007, **101**:88–95.
18. Southwood TRE, Henderson PA: *Ecological Methods.* 3rd edition. Wiley-Blackwell: Hoboken, NJ; 2000.
19. Hill DS: *Pests of Crops in Warmer Climates and their Control.* Berlin: Springer; 2008.
20. Picker M, Griffiths C, Weaving A: *Field Guide to Insects of South Africa.* Cape Town: Struik Publishers; 2004.
21. Marshall C, Rossman GB: *Designing Qualitative Research.* 4th edition. London: Sage Publications; 2006.
22. Leavens MK, Anderson CL: **Gender and agriculture in Tanzania.** *EPAR Brief* 2011, **134**:1–14.
23. Joshi KD, Musa AM, Johansen C, Gyawali S, Harri D, Witcombe JR: **Highly client-oriented breeding using local preferences and selection produces widely adapted rice varieties.** *Field Crop Res* 2007, **100**:107–116.
24. Euroconsult (Eds) (Ed): *Agricultural Compendium for Rural Development in the Tropics and Subtropics.* 3rd edition. Oxford: Elsevier; 1989.
25. Sehgal M, Jeswani MD, Kaira N: **Management of insect, disease and nematode pests of rice-wheat in the indo-gangetic plains.** *J Crop Prod* 2001, **4**(1):167–226.
26. Matteson PC: **Insect pest management in tropical Asian irrigated rice.** *Annu Rev Entomol* 2000, **45**:549–574.
27. Abate T, van Huis A, Ampofo JKO: **Pest management strategies in traditional agriculture: an African perspective.** *Annu Rev Entomol* 2000, **45**:631–659.
28. Heinrichs EA, Barrion T: *Rice-Feeding Insects and Selected Natural Enemies in West Africa: Biology, Ecology, Identification.* International Rice Research Institute: Metro Manila; 2004.
29. Rugumamu CP: **Assessment of post-harvest technologies and gender relations in maize loss reduction in Pangawe village eastern Tanzania.** *Tanz J Sci* 2009, **35**:67–76.

12

Home gardens: a promising approach to enhance household food security and wellbeing

Dilrukshi Hashini Galhena*, Russell Freed and Karim M Maredia

Abstract

With the global population expected to reach over 9 billion by 2050, there is a continuous need to increase food production and buffer stocks. In this scenario, countries around the world, especially developing countries where the pervasiveness of hunger and food scarcity is more acute, are resorting to various counter strategies to meet the growing demand and to avert food insecurity and famine. Over the recent years there has been growing interest to strengthen and intensify local food production in order to mitigate the adverse effect of global food shocks and food price volatilities. Consequently, there is much attention towards home gardens as a strategy to enhance household food security and nutrition. Home gardens are an integral part of local food systems and the agricultural landscape of developing countries all over the world and have endured the test of time.

Through a rigorous literature review, this paper first examines definitions and characteristics of home gardens and then provides a global review of their social, economic, and environmental contributions to communities in various socio-economic contexts. Many of the compositions on home gardens share research and experiences of developing countries in Africa, Asia, and Latin America. These studies recognize positive impacts of home gardens towards addressing food insecurity and malnutrition as well as providing additional benefits such as income and livelihood opportunities for resource-poor families and delivering a number of ecosystem services. However, only a handful of case studies were found on post-crisis settings. While providing a general overview of some of these studies, this review investigates the home garden experiences of post-conflict Sri Lanka, where home gardening has been practiced for centuries. While emphasizing multiple benefits, we also highlight constraints to home garden food production. In conclusion, we emphasize the need for more research and empirical data to appraise the role of home gardens in crisis and post-crisis situations, as well as assessing their economic value and their impacts on food security, nutrition, economic growth, and gender issues.

Keywords: Home gardens, Food security, Livelihood, Post-conflict, Sri Lanka

Introduction

The vast majority of hungry and malnourished people live in developing countries under sub-standard living conditions [1] and over half a billion of the global population suffer from chronic food insecurity[a]. With the global population expected to reach over 9 billion by 2050, there will be a continuous need to increase food production and buffer stocks to meet the growing demand and efficiently cope with volatilities in food production and prices. It has been projected that global food production will need to increase by 70% in order to meet the average daily caloric requirement of the world's population in 2050[b]. Moreover, the need for interventions are stressed as the resources available for food production - including land, water, labor and credit - are becoming scarce and costly. The drive for agricultural innovation is further convoluted by the growing issues of climate change and natural resource degradation.

Multiple strategies are required to address the issue of food production and food security[c]. The choice of feasible approaches hinges on the existing social, political, and economic conditions and resources available to design and implement the intervention. Home gardens are a time-tested local strategy that are widely adopted and practiced in various circumstances by local communities with limited resources and institutional support. It is evident from the literature that home gardens are a part

* Correspondence: galhenad@msu.edu
College of Agriculture and Natural Resources (CANR), Michigan State University, East Lansing, MI 48824, USA

of the agriculture and food production systems in many developing countries and are widely used as a remedy to alleviate hunger and malnutrition in the face of a global food crisis [2].

Globally, home gardens have been documented as an important supplemental source contributing to food and nutritional security and livelihoods. 'Food production on small plots adjacent to human settlements is the oldest and most enduring form of cultivation' [3]. For centuries, home gardens have been an integral component of family farming and local food systems. Home gardening is an ancient and widespread practice all over the world. In the literature, home gardens are classified as mixed, kitchen, backyard, farmyard, compound or homestead garden [4-7].

This paper presents the developing country experiences of home gardens and looks at the specific case of post-conflict Sri Lanka[d]. An extensive literature search was conducted through the review of over 100 publications, reports, and book chapters, covering various aspects of home gardening to develop the theoretical framework. The inherent characteristics of home gardens as well as the contextual attributes, benefits, and constraints captured in the literature are summarized in the following sections.

About home gardens

Home gardens are found in both rural and urban areas in predominantly small-scale subsistence agricultural systems [8]. The very beginning of modern agriculture can be dated back to subsistence production systems that began in small garden plots around the household. These gardens have persistently endured the test of time and continue to play an important role in providing food and income for the family [9]. Since the early studies of home gardens in the 1930s by the Dutch scholars Osche and Terra on mixed gardens in Java, Indonesia [10], there has been extensive contributions to the subject synthesizing definitions, species inventories, functions, structural characteristics, composition, socio-economic, and cultural relevance. Home gardens are defined in multiple ways highlighting various aspects based on the context or emphasis and objectives of the research [11]. Gupta pointed out that the background and gender of the researcher or scientist may also bias their perception on home gardens and may not entirely reflect the opinion of the family involved in home gardening activities [12].

Definition of home gardening

Relying on research and observations on home gardens in developing and developed countries in five continents, Ninez formulated the following definition [13]:

> 'The household garden is a small-scale production system supplying plant and animal consumption and utilitarian items either not obtainable, affordable, or readily available through retail markets, field cultivation, hunting, gathering, fishing, and wage earning. Household gardens tend to be located close to dwelling for security, convenience, and special care. They occupy land marginal to field production and labor marginal to major household economic activities. Featuring ecologically adapted and complementary species, household gardens are marked by low capital input and simple technology.'

Generally, home gardening refers to the cultivation of a small portion of land which may be around the household or within walking distance from the family home [14]. Home gardens can be described as a mixed cropping system that encompasses vegetables, fruits, plantation crops, spices, herbs, ornamental and medicinal plants as well as livestock that can serve as a supplementary source of food and income. Fresco and Westphal [15] specify home gardens as a cropping system composed of soil, crops, weeds, pathogens and insects that converts resource inputs - solar energy, water, nutrients, labor, etc. - into food, feed, fuel, fiber and pharmaceuticals. Kumar and Nair [16], while acknowledging that there is no standard definition for 'a home garden', summarize the shared perception by referring to it as '...an intimate, multi-story combinations of various trees and crops, sometimes in association with domestic animals, around homesteads', and add that home garden cultivation is fully or partially committed for vegetables, fruits, and herbs primarily for domestic consumption.

Adding to this, others have described a home garden as a well-defined, multi-storied and multi-use area near the family dwelling that serves as a small-scale supplementary food production system maintained by the household members, and one that encompasses a diverse array of plant and animal species that mimics the natural eco-system [11,17-19].

Characteristics of a home garden

Michelle and Hanstad [20] list five intrinsic characteristics of home gardens: 1) are located near the residence; 2) contain a high diversity of plants; 3) production is supplemental rather than a main source of family consumption and income; 4) occupy a small area [21]; and 5) are a production system that the poor can easily enter at some level [9].

There is a vast body of literature presenting research and case studies focusing on the role of home gardens as agroforestry or food production systems, or a combination of both. Home gardens are ecologically divided into two categories: tropical and temperate [13]. Much of the literature focuses on home gardens in the tropical areas in Central and South America [22-25]. There is also a substantial interest for home gardens in South and South-East Asia [26-29] and Africa [30-34]. Conversely, only a few documented studies exist on home gardens from temperate regions [35-38] and from developed countries [32,39-42].

Ninez [13] lists general tendencies with respect to home garden food production systems based on 15 type-specific characteristics adopted from Ruthenberg [5] (Table 1), and presents an ethnographical synthesis of home gardens across the globe. Home gardens are commonly established on lands that are marginal or not suitable for field crops or forage cultivation because of their size, topography, or location [11]. The specific size of a home garden varies from household to household and, normally, their average size is less than that of the arable land owned by the household. However, this may not hold true for those families that do not own agricultural land and for the landless. New innovations and techniques have made home gardening possible even for the families that have very little land or no land at all [43]. The home gardens may be delimited by physical demarcations such as live fences or hedges, fences, ditches or boundaries established through mutual understanding. Application of kitchen waste, animal manure, and other organic residues has been a practice amongst home gardeners and this exercise has helped to considerably increase the productivity and fertility of these gardens [11,44,45].

While some similarities exist across the board, each home garden is unique in structure, functionality, composition, and appearance [13,46-48] as they depend on the natural ecology of the location, available family resources such as labor, and the skills, preferences, and enthusiasm of family members [45,49,50]. Home garden cultivation tends to be quite dynamic [17,18]. The decisions related to the selection of crops, procuring inputs, harvesting, management, and so forth are mostly driven by the consumption and income generation needs of the household [27,45]. A study from Indonesia observed that the structure, composition, intensity of cultivation, and diversity of home gardens can be subjected to the socioeconomic status of the household [51]. For instance, as the families became economically stable their cultivation shifted from staples to horticultural crops and some families began to raise livestock. Based on the economics of the household, Niñez [13] differentiated two types of home gardens: 1) subsistence gardens and 2) budget gardens. Access to planting material and social capital are noted as important attributes to species diversity in gardens [52]. Collectively, the ecological potential, economic status, and social elements influence the presence of food and non-food crops and animals in the garden [28,53]. Additionally, Moreno-Black and colleagues [54] identified that limitations resulting from factors such as opportunities for off-farm employment and family structure as well as local customs influence the development and composition of the gardens.

Table 1 The key characteristics of a typical home garden

Characteristic	General practice
Species density	High
Species type	Staples, vegetables, fruits, medicinal plants
Production objective	Home consumption
Labor source	Family (women, elderly, children)
Labor requirements	Part-time
Harvest frequency	Daily, seasonal
Space utilization	Horizontal and vertical
Location	Near dwelling
Cropping pattern	Irregular and row
Technology	Simple hand tools
Input-cost	Low
Distribution	Rural and urban areas
Skills	Gardening and horticultural skills
Assistance	None or minor

Source: Niñez [3].

The home garden frequently uses family labor [18] - women, children, and elders are of particular importance in their management [46,48,55-57] – but, depending on the economic capacity and affordability, households may hire wage laborers to cultivate and maintain the home garden that in turn affect the composition and intensity of home garden activities [22,55,56]. Like any other food production system, home gardens may be vulnerable to harsh environmental conditions such as drought and floods [57,58]. Despite the fact that home gardening activities demand a lesser amount of horticultural and agronomic know-how, crop losses and other negative implications can be reduced when the household members are empowered with better skills and knowledge [59].

Experiences of home gardens from developing countries

Home gardens have been an integral part of local food systems in developing countries around the world. Many studies provide descriptive evidence and analysis of home gardens in developing countries in Asia, Africa, and Latin America and pinpoint their numerous benefits to communities and families. They encapsulate perpetual small-scaled subsistence agricultural systems established by the households to obtain and supplement the food requirements of the family. Home gardens are mainly intended to grow and produce food items for family consumption, but they can be diversified to produce outputs that have multiple uses including indigenous medicine and home remedies for certain illnesses, kindling and alternative fuel source, manure, building material, and animal feed. Chris Landon-Lane [60] provides an overview of the benefits of home gardens (Table 2) and describes home gardens as a 'place for innovation' with the potential to improve the livelihood of peri-urban and rural communities. In-depth exploration of past and more recent compositions on home gardens worldwide not only affirms Landon-Lane's insight but also recognize additional advantages. We broadly categorized benefits of home gardening into three components: (1) social; (2) economic; and (3) environmental benefits. These benefits are presented and explained through the vast experiences on home gardens from developing nations around the world.

Social benefits

Enhancing food and nutritional security

Reviews of studies from various countries reveal that the degree and combination socio-cultural impacts on societies engaged in home gardening vary across the board. Multiple social benefits of home gardens include enhancing food and nutritional security in many socio-economic and political situations, improving family health and human capacity, empowering women, promoting social justice and equity, and preserving indigenous knowledge and culture [20].

The most fundamental social benefit of home gardens stems from their direct contributions to household food security by increasing availability, accessibility, and utilization of food products[e]. Home gardens are maintained for easy access to fresh plant and animal food sources in both rural and urban locales. Food items from home gardens add substantially to the family energy and nutritive requirements on a continuous basis. A pioneering research study on home gardens conducted by Ochse and Terra in the early 1930s [10] states that home gardens led to 18% of the caloric and 14% of the protein consumption by households in Kutowinangun, Indonesia. Subsequent studies on the Javanese home gardens point out a direct link between successful home gardens and households' nutritional status [61], and observe an increase in households' food consumption with intensification of home food production [62]. Javanese experiences illustrating the potential of home gardens to add to households' food supply and nutrition [13,21], as well as their eminence as multi-storied agro-ecosystem [55,63] in the tropics, heightened the global attention towards home gardens.

Table 2 The key benefits of home gardening

- Improved food security
- Increased availability of food and better nutrition through food diversity
- Income and enhanced rural employment through additional or off-season production
- Decreased risk through diversification;
- Environmental benefits from recycling water and waste nutrients, controlling shade, dust and erosion, and maintaining or increasing local biodiversity

Source: Landon-Lane [60].

Foods from home gardens varied from horticultural crops to roots to palm and animal products; further plants from the gardens are also used as spices, herbs, medicines, and fodder for the animals [53,64-67]. Although home gardens are not generally reputed as a staple crop production base, Thaman [65] documented that Pacific Islanders obtained their main staple root crops from home gardens. Similar reports were found from Nepal [19], Yucatan Peninsula [66], Bangladesh [67], Peru [68], Ghana [69], and Zimbabwe [56]. Resource-poor families often depended more on home gardens for their food staples and secondary staples than those endowed with a fair amount of assets and resources such as land and capital [51,70]. For poor and marginalized families unable to afford expensive animal products to fulfill their nutritional needs, home gardens offer a cheap source of nutritive foods [71]. Through gardening, households can have better access to a diversity of plant and animal food items that lead to an overall increase in dietary intake and boost the bioavailability and absorption of essential nutrients [72].

As stated by Marsh [9], home gardens provide easy day-to-day access to an assortment of fresh and nutritious foods for the household and accordingly those homes obtained more than 50% of the vegetables, fruits, tubers, and yams from their garden. Supporting this premise, different studies conclude that, while adding to the caloric quantity, home gardens supplement staple-based diet with a significant portion of proteins [48], vitamins [16], and minerals [73], leading to an enriched and balanced diet [74,75] particularly for growing children and mothers [76]. Additionally, plants from the gardens - especially spices and herbs - are used as flavor enhancers, teas, and condiments [77]. Recently, countries like Bangladesh have been successful in increasing the availability and consumption of vitamin A-rich food items through national home gardening programs [72].

Furthermore, the integration of livestock and poultry activities into home gardening reinforces food and nutritional security for the families as milk, eggs, and meat from home-raised animals provided the main and, in many instances, the only source of animal protein [66]. In some places, home gardeners are also engaged in mushroom cultivation and beekeeping [75] and even small fresh water fish ponds are incorporated into the garden space adding to the share of proteins and other nutrients available for the family [27].

Evidence from around the world suggests that home gardens can be a versatile option to address food insecurity in various challenging situations, and thus they have attracted sponsorship by numerous government and non-governmental organizations. Consequently, home garden production has significantly increased in the country and has been instrumental in reducing 'hidden hunger' and disease cause by micronutrient deficiency. In an attempt to assess the dynamics of home garden evolution in Java and Sulawesi in Indonesia, Wiersum [51] notes that home gardens make available a small but continuous flow of subsistence food products for the household. Also, home gardens provide the main source of staple food for people in heavily degraded and densely populated areas with limited croplands [44,78].

Home gardens can ensure food to underprivileged and resources-poor households as they can be established and maintained within a small patch of land or with no land using a few inputs [20,43]. A study of home gardens in Cuba reveals that they were used as a strategy to increase resilience and ensure food security in the face of economic crisis and political isolation [79]. To mitigate recurring food shortage and malnutrition, Cuban households obtained basic staple foods (rice and beans) through rations, but the households relied on their home gardens to obtain additional produce to diversify the family diet [80]. Ensuring a reliable and convenient source of food, fiber, and fuel for the family, they are viewed as a robust food system in circumstances where population pressures and numerous resource limitations persist [81,82]. In the Peruvian capital of Lima, home gardening has led to nutritional benefits to families living in slum areas by increasing the availability of carbohydrates as well as nutrient-rich vegetables and fruits that are not economically accessible for poor slum dwellers [83].

The Global Hunger Index specified that the lack of political stability has escalated hunger and poverty in countries affected by conflicts [84]. Similarly, environmental disaster can also have devastating impacts on communities and disable food production systems [85]. Even though there are only a few published narratives, home gardens have been proposed as an option for food and nutritional security in disaster, conflict, and other post-crisis situations[f] [9,45,86]. Home gardens based on enset and coffee are an integrated farming system that not only provide subsistence and complementary food products for Ethiopian families, especially during famines, but also provide the primary means of employment for the household [78].

Tajikistan became independent from the Soviet Union in 1991 but was plagued by a civil war soon after. Rowe [7] showed that, during the post-soviet era, Tajik families tormented by civil war, agricultural downfall, and drought heavily depended on their gardens for food. This trend still continues, and home gardens continue to significantly supplement household food security and sustenance. In recent years, several countries transitioning towards peace and stability and those that are recovering from natural disaster have been adopting policies that support home gardening to reduce the prevalence and severity of hunger and malnutrition [45,87].

Bandarin et. al. [88] point out that, in a post-conflict setting, assistance and reconciliation mechanisms work best and result in environmental, social and economic benefits when there is a cultural or traditional linkage between the target population and the intervention. Hence, home garden projects offer a realistic solution as in most countries home gardening is a regular day-to-day activity amongst the household, especially for women. In addition, home gardens when properly managed provide a four-in-one solution to the food and nutrition problem by increasing household food availability, enabling greater physical, economic and social access, providing an array of nutrients, and protecting and buffering the household against food shortages.

Improving health

Plants are an important source of medicine for humans and livestock and are used as biological pesticides to protect crops from diseases and pest infestations. Herbs and medicinal plants are grown in home gardens all over the world, and in developing countries nearly 80% of the people use them to treat various illnesses, diseases, and also to improve their health conditions [89]. A generous portion of the plants found in home gardens have some medicinal value and they can be used to treat many common health problems in a cost-effective manner. For instance, Perera and Rajapksa, in their assessment of Kandyan gardens in Sri Lanka [90], note that out of the 125 plant species found about 30% were exclusively used for medicinal uses and about 12% for medical and other purposes. Medicinal plants were documented to be an important plant group second only to high-value species in Sri Lanka [90] and in Bangladesh [91]. Home gardens in Bukoba district of Tanzania contained plant species grown entirely for medicinal purposes [92]. Around 70% of the plant species identified in forests and gardens in the Yucatan had a medicinal use [93], and in traditional Mayan home gardens nine species of the 77 useful plants found were exclusively used for medicinal purposes and 26 species had mixed uses as medicines, food, spices, and ornamentals [94].

Food insecurity and economic hardships force people to consume less and to settle for food that is of low nutritional quality. Adverse health effects due to inadequate intake of basic macronutrients are further compounded by deficiencies in micronutrients such as vitamins and minerals. More than 35% of the fatalities worldwide are caused by factors attributed to nutritional deficits [95]. Amongst them, vitamin A deficiency is a major health issue in many low-income countries and pose serious health problems, particularly for pregnant women and their babies and growing children. Reports indicate more than 7 million women suffering from complications due to vitamin A insufficiency [96] and cause 6 to 8% of the deaths amongst children under the age of 5 years in Africa and Asia [95]. In some countries where this problem is acute, homestead food production programs have been launched to assist and address vitamin A deficiency and to improve the quality of diet by facilitating a year-round production of vegetables and fruits [85].

The global incidence of anemia is primarily attributed to iron deficiency. Iron insufficiency elevates the risk of mortality during pregnancy by 20% [97]. Moreover, estimates suggest that nearly one-third of the global population live in countries with high zinc deficiency [98]. Micronutrient deficiency can raise the vulnerability to other infectious diseases and the risks of mortality due to illnesses such as diarrhea, pneumonia, malaria, and measles [99]. It may also lead to poor physical and cognitive development and impairment of motor skills in young children as well as other short-term and long-term health effects. Furthermore, a vicarious cycle of undesirable socio-economic effects can be triggered as peoples' ability to actively engage in physical and economic activities are hindered by illness, disability, and reduced life expectancy. In different contexts, home gardening initiatives have been proposed and implemented as potential strategies to address health issues resulting from malnutrition [100,101]. Although the opportunity is real, minimal efforts have been made to identify and maximize the gross benefits of home gardening for better health.

Uplifting the status of women

In many cultures, women play an important role in food production but at times their worth is somewhat undermined. They are also active in home gardening, though their involvement in the home garden tends to be determined by socio-cultural norms [20]. In most scenarios women's contribution to household food production is immense, but this does not imply that home gardening is predominantly a female activity. Women's participation and responsibilities in home gardening varies across cultures, including land preparation, planting, weeding, harvesting, and marketing [54,102,103]. In fact, in some cultures, women are the sole caretakers of household gardens [35,72] while, in others, they play more or less a supportive role [11]. Howard's 2006 analysis [104] of 13 home gardens case studies in South America revealed that women are the main managers of home gardens across the region. Home gardening activities are vital and fit well with their day-to-day domestic activities and employment patterns along with their cultural and aesthetic values. On the other hand, in the Indonesian context women take part during planting and harvesting [49] and, in Sri Lanka, they provide labor during peak times [55]. Regardless, particularly for women and disadvantaged groups, home gardening is an avenue for social and economic enrichment.

Home gardens stimulate social change and development. Amongst the Achuar Indian community in the upper Amazon, a woman's ability to maintain a lush home garden not only demonstrates her agronomic competency but also her status in society [105]. Similarly, for the Saraguro women of the Andes, a plentiful garden help elevates a woman's social eminence and demonstrates her commitment to the family's wellbeing [25]. Based on a study conducted in Senegal by Brun and colleagues [106], evaluating the food and nutritional impact of home gardening, it was found that, although the gardens did not make a major contribution to food consumption and nutrition, they were instrumental in improving the women's income and social status as well as their awareness of evolving food habits in urban areas. For some women, sales of garden products are often the only sources of income or livelihood [9]. In Tajikistan, where many of the men were killed during the civil war or have migrated to Russia and other countries for employment, women, elders and children have been providing invaluable family labor and resources to local agricultural economy. Kitchen gardens, as they are referred to by Rowe [7], are very important especially for women-headed families in terms of meeting their everyday food consumption needs and generating income. In fact, nearly half of the food consumed at home and one-third of the food sold in the market came from these garden lots. Other studies have shown that, in situations where women are leading home gardens, there has been improvement to household nutrition, especially child nutrition [72,107].

Through home gardening women have developed proficiency related to plants and garden practices that helps them become better home and environment managers. Their labor is indispensable to maintain the garden and to help keep production cost low. As home managers, women have useful knowledge of numerous domestic needs. By their involvement in the production process, they are able to meet family needs more easily and economically [44]. Home gardeners in Peru indicated that women gardeners are inclined to produce food primarily for family consumption while men gardeners typically focus on high value crops for marketing [83]. While home gardens provide a respectable path for women to contribute to household subsistence, eminence, and character, they hold a greater socio-cultural and spiritual importance for women [104]. Furthermore, they are a key source of gardening knowledge and information [105].

Research suggests that, in some societies and cultures, a woman's role in family decision making is rather limited; however, many accounts confirm that when it comes to home gardening women tend to have more autonomy and decision making capabilities [108]. Moreno-Black and colleagues [54] conducted a study of 49 women's home gardens in Northeastern Thailand where the rural women constantly indicated that they were the key decision makers and carried out most of the activities of the home gardens.

Preserving indigenous knowledge and building integrated societies

Home gardens consist of a variety of components and species that represent social and traditional aspects of different societies. This rich indigenous culture and communal knowledge base is expressed through home gardening by the selection of plants and animal species as well as the farming practices used by the local community [16,70]. Home gardens serve as a valuable repository for preserving and transferring indigenous crops and livestock species, production knowledge and the skills from one generation to another [109-111].

Interactions in and around the home garden create and reinforce social status and ties between the household and the community. Home gardeners habitually exchange or gift planting materials, vegetables, fruits, leaves, herbals and medicinal plants for social, cultural, and religious purposes [109,110]. Such interactions are essential for social integration and building social capital. The social dimension of home gardening is yet not fully explored.

Economic benefits

The economic benefits of home gardens go beyond food and nutritional security and subsistence, especially for resource-poor families. Bibliographic evidence suggests that home gardens contribute to income generation, improved livelihoods, and household economic welfare as well as promoting entrepreneurship and rural development [111,112]. Through the review of a number of case studies, Mitchell and Hanstad [20] assert that home gardens can contribute to household economic well-being in several ways: garden products can be sold to earn additional income [17,48,83]; gardening activities can be developed into a small cottage industry; and earnings from the sale of home garden products and the savings from consuming home-grown food products can lead to more disposable income that can be used for other domestic purposes. Studies from Nepal, Cambodia, and Papua New Guinea report that the income generated from the sale of home garden fruits, vegetables, and livestock products allowed households to use the proceeds to purchase additional food items as well as for savings, education, and other services [85,113]. Families in mountain areas of Vietnam were able to generate more than 22% of their cash income through home-gardening activities [111].

Home gardens are widely promoted in many countries as a mechanism to avert poverty and as a source of income for subsistence families in developing countries. Although home gardens are viewed as subsistence-low production systems, they can be structured to be more efficient commercial enterprises by growing high-value crops and animal husbandry [43]. A number of research studies have focused on evaluating the potential or real economic contribution to the household and local economy as well as social development [114]. A study from Southeastern Nigeria reported that tree crops and livestock produced in home gardens accounted for more than 60% of household income [115]. In many cases the sale of produce from home gardens improves the financial status of the family providing additional income, while contributing social and cultural amelioration [116]. The fact that home production is less cost-intensive and requires fewer inputs and investment is extremely important for resource-poor families that have limited access to production inputs. Yet it has been assessed that moderately rigorous crop and livestock production in home gardens can generate as much revenue per unit area as field crop production [9,62]. Where land constraints exist, innovative tools have been used to make efficient use of limited space [43]. Also, livestock housed in gardens diversify risk due to crop losses and provide a cash buffer and asset to the household [117].

Environmental benefits

Home gardens provide multiple environmental and ecological benefits. They serve as the primary unit that initiates and utilizes ecologically friendly approaches for food production while conserving biodiversity and natural resources. Home gardens are usually diverse and contain a rich composition of plant and animal species. Hence they make interesting cases for ethno-botanical studies [110,118].

Gardens are complex and may resemble ecological agricultural production systems that sponsor biodiversity conservation. The rich diversity and composition of species and the dense distribution of faunal and floral strata denote extraordinary features of home garden ecology [20,46]. Buchmann's 2009 assessment [79] of 25 home gardens in Central Cuba noted 182 plant species. Other reports from around the world also identify a significant concentration of plants used as vegetables, fruits, herbs, medicines, yams, and spices [78,114]. Home gardens also contain a wide spectrum of plant species, some of which are landraces, rare or threatened species, and specific cultivars selected for a set of desirable traits [119]. Thus they become ideal sites for *in situ* conservation of biodiversity and genetic material [111,120].

Home gardens also provide a number of ecosystem services such as habitats for animals and other beneficial organisms, nutrient recycling, reduced soil erosion, and enhanced pollination [121]. The high density of plants within the home garden provides the ideal environment and refuge for wildlife species such as birds, small mammals, reptiles, and insects [122]. Calvet-Mir et. al. [112] highlight a number of ecosystems services provided by home gardens such as production of quality food, maintenance of landraces, cultural services, pest control, and pollination. They conclude that the most important ecosystems services provided by home gardens differ from large-scale and commercial agriculture.

Nutrient cycling is another ecological benefit of home gardens [120,123]. The abundance of plant and animal litter and continuous recycling of organic soil matter contributes to a highly efficient nutrient cycling system. Another potential benefit of home gardens is the reduction of soil erosion and land conservation [70,124]. The attraction of honey bees provides added benefits including improved pollination and increased fruit dispersal [64].

Individuals of the household, animals, and plants all maintain a symbiotic relationship within the home gardens. For instance, the plants and animals provide food and other benefits for the family and the family in turn takes care of the home gardens. Plant materials are used as fodder for the animals and animal manure is incorporated into the compost to fertilize plants, hence reducing the need for chemical fertilizer [20]. Livestock and poultry manure can add a significant amount of organic soil matter, nitrogen, potassium, and phosphorus into the soil. The integration of livestock activities into home gardening can expedite nutrient cycling in ecosystem and help retain moisture [125].

Home gardens in Sri Lanka

The economy of Sri Lanka is founded on agriculture. More than 35% of the 20 million people of Sri Lanka are engaged directly or indirectly in the agrarian sectors. Home gardening has been a long-standing practice among the rural and urban households in Sri Lanka for centuries [55]. Despite the traditional basis for home gardening, over the recent years national policies have focused on promoting home gardening in the country. Currently, two national initiatives are underway and are receiving notable patronage from the government for the initiation of a countrywide food production drive to establish one million home gardens across the island[g]. The programs highlight the key role home gardens play in the face of food insecurity, economic downturn and malnutrition by providing a diversified source of food and a way of generating income.

In spite of the growing interest in home gardening, literature discussing home gardens in Sri Lanka is rather limited. The bulk of the available excerpts almost exclusively focuses on Kandyan Gardens, also known as Kandyan Forest Gardens (KFG). KFG are a common traditional agroforestry system found in the wet central hills in Sri Lanka. They encompass a mixed cropping system, which includes a diverse collection of economically valuable perennial and semi-perennial crops situated around the household [126] along with animal species that were raised to suit the necessities of the family, the environment, and the recommendations by scientists and extension workers [127].

One of the earliest studies on KFG in Sri Lanka was published by McConnell and Dharmapala [126]. From a survey of 30 KFGs established through the use of a farming systems approach they conclude that, although in the short-run KFG were not as productive and profitable as the commercial farming systems, they lead to multiple benefits over time. Jacob and Alles [55] differentiate KFGs from other mixed forest-gardening systems found in South Asia and South-east Asia with respect to the diversity of plants grown. They also stated that these garden systems improved the well-being of people who nurtured them through the provision of various food products and timber, livelihood opportunities, and sustainability of the production system. Furthermoe, they emphasized the need to inspect existing agronomic practices and to design feasible models that can improve the productivity per unit of land.

Perera and Rajapaksa [90] characterize various components of KFGs based on ownership, structure, species composition, livestock composition, and management practices. Their baseline survey of 50 randomly selected gardens in the Kandy District showed that the various species in the KFG had numerous uses including food, cash, timber, fuel wood, construction material, green manure, fodder, medicines, shade, and beautification. KFG also have significant implications to the region in terms of *in situ* germplasm conservation [128], watershed management [90], preservation of habitat and other ecological contributions [129].

A book by Hochegger [130] offers a comprehensive overview of the ecological, economic, and cultural relevance of KFG in Sri Lanka investigating six locations in the central hills. The Green Movement in Sri Lanka pioneered by Kumarathunga [131] has published a guide on environmentally friendly agriculture with key emphasis on home gardens. This publication has been initially written in the Sinhala language and is in the process of being translated to the Tamil language. The guide provides step-by-step instructions to home gardeners and farmers on environmentally friendly cultivation and management practices and strategies to boost the efficiency of small agricultural production systems.

Ranasinghe [43] has developed a detailed manual drawing on the ideas of family business gardens and low/no-space agriculture. This manual was shaped primarily to attract urban households to develop their home crop production into a small agribusiness. In addition this publication attempted to reach a wider audience that includes professionals, non-professionals, entrepreneurs, as well as policy makers concerned with issues related to food and nutritional security in the face of limited resources such as land. It highlights that, through improved management, home garden cultivations can be transformed into agricultural ventures through the systematic adoption of economical and eco-friendly technologies and interventions.

Constraints and opportunities

While there are multiple benefits of home gardening for developing countries, the literature also reveals the key constraints to the productivity and sustainability of home gardens and makes recommendations for improving the home gardens and making them a viable and sustainable enterprise. Hoogerbrugge and Fresco [11] and Mitchell and Hanstad [20] provide a review of key constraints to home gardening. Among several constraints, they identified the access to suitable and sufficient land to establish a home garden along with lack of ownership and usage rights of some form as the most important limiting factors. The other constraints include access to capital or credit, access to water, seeds and planting materials, weak extension and advisory services, access to labor, and access to markets. The cultural acceptance of home gardening is also an important constraint. Table 3 summarizes the most common constraints to home gardening specified in literature by Hoogerbrugge and Fresco and others.

Conclusions

Overall, the literature review supports the inclusion and promotion of home gardens as an eco-friendly sustainable agricultural practice to improve food security and enhance economic growth.

The structure, functions, and contributions of home gardens vary in geographic regions. The literature shows that home gardens fulfill social, cultural and economic needs, while providing a number of ecosystem services. While these benefits are broadly distinguished here for better illustration, these benefits are not mutually exclusive. In the real world, there is substantial overlap and dependence between the various beneficial elements resulting in a bundle of advantages making home gardening initiatives even more attractive.

In the wake of a global food crisis and the soaring food prices, there has been increased emphasis on enhancing and building local food systems. In this context, there is renewed attention to food production and livelihood enhancement through home gardens. However, more empirical evidence on the value and importance of home gardens in conflict and post-conflict situations needs to be researched and documented. There is also a need for research on the cost-benefit analysis of home gardening to determine the economic value and to derive viable models that hold the most promise in diverse circumstances. The areas of nutrition, access to new technologies, extension

Table 3 Selected publications presenting constraints to home gardening

Constraint	References
Limited access to agricultural inputs such as seeds, planting material, tools, and capital	Brownrigg [21]; Niñez [83]; Talukder *et al.* [72]; Fernandes and Nair [46]
Shortage of land and lack of land tenure security	Hoogerbrugge and Fresco [11]; Fernandes and Nair [46]; Jose and Shanmugaratnam [132]; Howard [104]
Inadequate access to water	Thaman [133]; Niñez [83]; tho Seeth *et al.* [35]
Damage due to insect pests, diseases, animals, and theft	Thaman [133]; Brownrigg [21]; Niñez [83]; Pandey *et al.* [103]
Poor environmental conditions	Howard [104]; Marsh [9]
Lack of knowledge, information, and advisory services	Thaman [133]; Hoogerbrugge and Fresco [11]
Shortage of family or hired labor	Howard [104]; Fernandes and Nair [46]
Poor soil fertility and soil erosion	Howard [104]; Thaman [133]
Limited access to quality livestock breeds	Brownrigg [21]; Mitchell and Hanstad [20]
Limited marketing opportunities	Howard [104]
Excessive post-harvest losses	Okafor and Fernandes [30]
Inadequate research and development on home gardens	Evers, et al. [134]
Social and cultural Barriers	Marsh [9]; Ninez [83]; Miura *et al.* [135], Talukder *et al.* [72]
Lack of information on nutritional benefits of home gardening	Miura *et al.* [135]; Talukder *et al.* [72]

and advisory services, economic and non-economic benefits, women empowerment, and long-term sustainability of home gardens specifically in post-conflict situations need further research.

Recognizing the value and potential of home gardens for enhancing food security and livelihoods, numerous initiates have been launched by governmental, non-governmental, and international organizations in many developing countries that are providing support and building local capacity to enhance the productivity and also for scaling up home garden activities. In this light, a number of resource materials, manuals, and guides have been developed through various home garden-related projects that can be used to improve and promote home gardening programs to enhance food security [43,60,87].

Endnotes

[a]Food insecurity occurs in three forms: chronic food security is the most severe category where a person is unable to consume the minimum amount of food needed for healthy life over a long period usually due to poverty or lack of productive recourses to generate income to purchase food [136]. Other types include transitional (short-term) food insecurity, which is further subdivided into temporary (limited time period due to shocks) and seasonal or cyclical (trend) food insecurity.

[b]The Food and Agriculture Organization [137] reported an average consumption per person of 3,130 kcal per day by the year 2050 based on their baseline projections. Alexandratos [138] estimated a slightly lower average daily caloric availability per person of 3,047 kcal per day by the year 2050.

[c]Since its inception in the early 1970s, the concept of food security has undergone many revisions and has held multiple connotations in research and in policy arenas. The two widely adopted conceptualizations are defined by: 1) the Food and Agriculture Organization - 'food security is a situation that exists when all people, at all times, have physical, social and economic access to sufficient, safe and nutritious food that meets their dietary needs and food preferences for an active and healthy life' [139]; and 2) by the United States Department of Agriculture - 'food security for a household means access by all members at all times to enough food for an active, healthy life. Food security includes at a minimum: the ready availability of nutritionally adequate and safe foods; and an assured ability to acquire acceptable foods in socially acceptable ways' [140].

[d]For nearly 30 years, the country underwent erratic disorder and military action that caused substantial destitutions to the political and socio-economic structure in the country as well as the infrastructure. The civil conflict was brought to an end in 2009 by the Sri Lankan government forces.

[e]As noted earlier there are numerous definitions to food security; nevertheless, these definitions highlight three broad dimensions: food availability, accessibility, and adequacy/utilization [139-141] (FAO, 2003). Food availability refers to the supply of food made available through domestic production, net imports, food reserves, donations, etc. Accessibility is ensured when an individual is able to obtain food without any physical, social, or economic barriers. Food adequacy/utilization is achieved through various biological and non-biological processes that ensure sufficient energy and nutrient intake.

[f]One general connotation for crisis is defined by Gasser and colleagues [142]. Crisis situations are identified as unique and complex in nature and are due to various factors [141]. At times, crisis is unforeseen and inevitable - as in the case of natural disasters - while others may be more protracted and influenced by economic, social, and political changes - such as civil or armed conflict. Irrespective of the origin, crisis adversely affects society by depriving affected groups of their rudimentary and ancillary needs and services including food, shelter, income, health care, security, and infrastructure.

[g]The two major national home gardening programs in Sri Lanka are "Api Wavamu, Rata Nagamu" (Let us cultivate to uplift the nation) and "Divinaguma" (Livelihood upliftment). In addition, a number of other regional and village level gardening programs are coordinated and managed by international and non-governmental organizations.

Abbreviation

KFG: Kandyan forest gardens.

Competing interest

The authors declare that they have no competing interests.

Authors' contributions

DHG initiated this review, conducted the bibliographic analysis, and developed the content of this manuscript. RF and KMM contributed to the enhancement of the framework and writing. All authors reviewed and approved the manuscript.

Acknowledgement

The authors would like to thank the many reviewers for their feedback and suggestions for the early drafts of this manuscript.

References

1. Food and Agriculture Organization of the United Nations and World Food Programme: *The State of Food Insecurity in the World - Addressing Food Insecurity in Protracted Crises.* Rome, Italy: Food and Agriculture Organization of the United Nations; 2010.
2. Johnson-Welch C, Alemu B, Msaki TP, Sengendo M, Kigutha H, Wolff A: *Improving Household Food Security: Institutions, Gender and Integrated Approaches.* Davis CA, USA: Paper prepared for the Broadening Access and Strengthening Input Market Systems (BASIS) Collaborative Research Support Project (CRSP); 2000.
3. Niñez VK: **Household gardens: theoretical and policy considerations.** *Agr Syst* 1987, **23**:167–186.

4. Terra GJA: **Farm systems in southeast Asia.** *Neth J Agric Sci* 1958, **6**:157–182.
5. Ruthenberg H: *Farming Systems in the Tropics.* 3rd edition. Oxford, UK: Clarendon Press; 1980.
6. Puri S, Nair PKR: **Agroforestry research for development in India: 25 years of experiences of a national program.** *Agrofor Sys* 2004, **61**:437–452.
7. Rowe WC: **"Kitchen gardens" in Tajikistan: the economic and cultural importance of small-scale private property in a post-soviet society.** *Hum Ecol* 2009, **37**(6):691–703.
8. Nair PKR: *An Introduction to Agroforestry.* Dordrecht, The Netherlands: Kluwer Academic Publishers; 1993.
9. Marsh R: **Building on traditional gardening to improve household food security.** *Food Nutr Agr* 1998, **22**:4–14.
10. Ochse JJ, Terra GJA: *Research on the Economic and Agricultural Condition and Food Consumption in Koetowinangoen. Landbouw.* The Netherlands: Publisher unknown; 1934:10.
11. Hoogerbrugge I, Fresco LO: *Homegarden Systems: Agricultural Characteristics and Challenges.* London, UK: International Institute for Environment and Development; 1993. Gatekeeper Series No. 39.
12. Gupta AK: **Scientists' views of farmers' practices in India: barriers to effective interaction.** In *Farmer First. Farmer Innovation and Agricultural Research.* Edited by Chambers R, Pacey A, Thrupp LA. London, UK: Intermediate Technology Publications; 1989.
13. Niñez VK: *Household Gardens: Theoretical Considerations on an Old Survival Strategy.* Peru, Lima: International Potato Center; 1984.
14. Odebode OS: **Assessment of home gardening as a potential source of household income in Akinyele Local Government Area of Oyo State.** *Nig J Horticulture Sci* 2006, **2**:47–55.
15. Fresco LO, Westphal E: **A hierarchical classification of farm systems.** *Exp Agric* 1988, **24**:399–419.
16. Kumar BM, Nair PKR: **The enigma of tropical homegardens.** *Agrofor Syst* 2004, **61**:35–152.
17. Eyzaguirre PB, Linares OF: **Intorduction.** In *Homegardens and Agrobiodiversity.* Edited by Eyzaguirre PB, Linares OF. Washington DC, USA: Smithsonian Books; 2010:1–28.
18. Sthapit BR, Rana RB, Hue NN, Rijal DR: **The diversity of taro and sponge gourds in traditional home gardens in Nepal and Vietnam.** In *Home Gardens and Agrobiodiversity.* Edited by Eyzaguirre PB, Linares OF. Washington DC, USA: Smithsonian Books; 2004:234–254.
19. Krishna GC: *Home Gardening as a Household Nutrient Garden.* Pokhara, Nepal: Paper presented at Home Gardens in Nepal: Proceeding of a workshop on "Enhancing the contribution of home garden to on-farm management of plant genetic resources and to improve the livelihoods of Nepalese farmers: Lessons learned and policy implications" (2004); 2006.
20. Mitchell R, Hanstad T: *Small Homegarden Plots and Sustainable Livelihoods for the Poor.* Rome, Italy: LSP Working Paper 11; 2004.
21. Brownrigg L: *Home Gardening in International Development: What the Literature Shows.* Washington DC, USA: The League for International Food Education; 1985.
22. Mendez VE, Lok R, Somarriba E: **Interdisciplinary analysis of homegardens in Nicaragua: micro-zonation, plant use and socioeconomic importance.** *Agrofor Syst* 2001, **51**:85–96.
23. Leiva JM, Azurdia C, Ovando W, López E, Ayala H: *Contribution of Home Gardens to in situ Conservation in Traditional Farming Systems—Guatemalan Component.* Witzenhausen, Germany: Paper presented at Proceedings of the Second International Home Gardens Workshop; 2001.
24. Aguilar-Stoen M, Moe SR, Camargo-Ricalde SL: **Home gardens sustain crop diversity and improve farm resilience in Candelaria Loxicha, Oaxaca, Mexico.** *Hum Ecol* 2009, **37**:55–77.
25. Finerman R, Sackett R: **Using home gardens to decipher health and healing in the Andes.** *Med Anthropol Q* 2003, **17**(4):459–482.
26. Cai CT, Luo LS, Nan YZ: **Energy and economic flow in home gardens in subtropical Yunnan, SW China: a case study on Sanjia village.** *Int J Sustain Dev World Ecol* 2004, **11**:199–204.
27. Ali AMS: **Home gardens in smallholder farming systems: Examples from Bangladesh.** *Hum Ecol* 2005, **33**:245–270.
28. Abdoellah OS, Hadikusumah HY, Takeuchi K, Okubo S, Parikesit: **Commercialization of homegardens in an Indonesian village: vegetation composition and functional changes.** *Agrofor Syst* 2006, **68**:1–13.
29. Sunwar S, Thornstrom CG, Subedi A, Bystrom M: **Home gardens in western Nepal: opportunities and challenges for on-farm management of agrobiodiversity.** *Biodivers Conserv* 2006, **15**:4211–4238.
30. Okafor JC, Fernandes ECM: **Compound farms of South Eastern Nigeria: a predominant agroforestry homegarden system with crops and small livestock.** *Agrofor Syst* 1987, **5**:153–168.
31. Fernandes ECM, Oktingati A, Maghembe J: **The Chagga homegardens: a multistoried agroforestry cropping system on Mt. Kilimanjaro, Northern Tanzania.** *Agrofor Syst* 1984, **2**:73–86.
32. Faber M, Venter SL, Benade AS: **Increased vitamin A intake in children aged 2–5 years through targeted home-gardens in a rural South African community.** *Public Health Nutr* 2002, **5**(1):11–16.
33. Drescher AW, Hagmann J, Chuma E: **Homegardens – a neglected potential for food security and sustainable land management in the communal lands of Zimbabwe.** *Der Tropenlandwirt – Beiträge zur tropischen Landwirtschaft und Veterinärmedizin (Journal of Agriculture in the Tropics and Subtropics)* 1999, **100**:163–180.
34. Thompson JL, Gebauer J, Hammer K, Buerkert A: **The structure of urban and peri-urban gardens in Khartoum, Sudan.** *Genet Resour Crop Evol* 2010, **57**:487–500.
35. tho Seeth H, Chachnov S, Surinov A: **Russian poverty: muddling through economic transition with garden plots.** *World Dev* 1998, **26**(9):1611–1623.
36. Vogl CR, Vogl-Lukasser B: **Tradition, dynamics and sustainability of plant species composition and management in homegardens on organic and non-organic small scale farms in Alpine Eastern Tyrol.** *Austria. Biol Agric Hortic* 2003, **21**:149–166.
37. Bassullu C, Tolunay A: **General characteristics of traditional homegarden involving animal practices in rural areas of Isparta Region of Turkey.** *J Anim Vet Adv* 2010, **9**:455–465.
38. Reyes-García V, Aceituno L, Vila S, Calvet-Mir L, Garnatje T, Jesch A, Lastra JJ, Parada M, Rigat M, Vallès J, Pardo-De Santayana M: **Home gardens in three mountain regions of the Iberian peninsula: description, motivation for gardening, and gross financial benefits.** *J Sustain Agric* 2012, **36**:249–270.
39. Cleveland DA, Orum TV, Ferguson N: **Economic value of home vegetable gardens in an urban desert environment.** *Hort Sci* 1985, **20**(4):694–696.
40. Agelet A, Bonet MÀ, Vallès J: **Home gardens and their role as a main source of medicinal plants in mountain regions of Catalonia (Iberian Peninsula).** *Econ Bot* 2000, **54**:295–309.
41. Calvet-Mir L, Calvet-Mir M, Vaqué-Nuñez L, Reyes-García V: **Landraces in situ conservation: A case study in high-mountain home gardens in VallFosca, Catalan Pyrenees, Iberian Peninsula.** *Econ Bot* 2011, **65**:146–157.
42. Bleasdale T, Crouch C, Harlan SL: **Community gardening in disadvantaged neighborhoods in Phoenix, Arizona: Aligning programs with perceptions.** *Journal of Agric, Food Syst Community Dev* 2010-2011, **3**(1):99–114.
43. Ranasinghe TT: *Manual of Low/No-Space Agriculture cum-Family Business Gardens.* AN Leusden, The Netherlands: RUAF Foundation; 2009.
44. Soemarwoto O, Conway GR: **The Javanese homegarden.** *J Farming Syst Res Ext* 1991, **2**(3):95–118.
45. Galhena DH, Mikunthan G, Maredia KM: **Home Gardens for Enhancing Food Security in Sri Lanka.** *Farming Matters* 2012, **28**(2):12.
46. Fernandes ECM, Nair PKR: **An evaluation of the structure and function of tropical homegardens.** *Agr Syst* 1986, **21**:279–310.
47. Nair PKR: *Agroforestry Systems in the Tropics.* Dordrecht, The Netherlands: Kluwer Academic Publishers; 1989.
48. Torquebiau E: **Are tropical agroforestry gardens sustainable?** *Agric Ecosyst Environ* 1992, **41**:189–207.
49. Christanty L, Abdoellah OL, Marten GG, Iskandar J: **Traditional agroforestry in West Java: the Pekaranagan (home garden) and Kebun-Talun (annual-perennial rotation) cropping systems.** In *Traditional Agriculture in South East Asia.* Edited by Marten GG. Boulder CO, USA: Westview Press; 1986.
50. Asfaw Z: *Home Gardens in Ethiopia: Some Observations and Generalizations.* Witzenhausen, Germany: Paper presented at Proceedings of the Second International Home Gardens Workshop; 2002.
51. Wiersum KF: **Diversity and change in homegarden cultivation in Indonesia.** In *Tropical Homegardens: A Time-Tested Example of Sustainable Agroforestry. Vol 3.* Edited by Nair BMKPKR. Dordrecht, The Netherlands: Springer Science; 2006.
52. Perrault-Archambault M, Coomes OT: **Distribution of agrobiodiversity in home gardens along the Corrientes River, Peruvian Amazon.** *Econ Bot* 2008, **62**(2):109–126.
53. Peyre A, Guidal A, Wiersum KF, Bongers F: **Dynamics of homegarden structure and functions in Kerala, India.** *Agrofor Syst* 2006, **66**:101–115.
54. Moreno-Black G, Somansang P, Thamathawan S: **Cultivating continuity and creating change: woman's home garden practices in northeastern Thailand.** *Agr Hum Val* 1996, **13**(3):3–11.

55. Jacob VJ, Alles WS: **The Kandyan gardens of Sri Lanka.** *Agrofor Sys* 1987, **5**:123–137.
56. Maroyi A: **Traditional home gardens and rural livelihoods in Nhema, Zimbabwe: A sustainable agroforestry system.** *Int J Sustain Dev World Ecol* 2009, **16**(1):1–8.
57. Turner BL, Hanham RQ, Portararo AV: **Population pressure and agricultural intensity.** *Assoc Am Geogr* 1977, **67**:384–396.
58. Meertens HCC, Fresco LO, Stoop WA: **Farming systems dynamics: Impact of increasing population density and the availability of land resources on changes in agricultural systems. The case of Sukumaland, Tanzania.** *Agric Ecosys and Environ* 1996, **56**:203–215.
59. Turner BL, Brush SB: **The nature of farming systems and views of their changes.** In *Comparative Farming Systems.* Edited by Turner BL, Brush SB. New York, USA: Guilford Press; 1987.
60. Landon-Lane C: *Livelihoods Grow in Gardens - Diversifying Rural Income Through Home Garden*, Volume 2. Rome, Italy: Food and Agriculture Organization of the United Nations; 2011.
61. Ochse JJ: **Horticulture and its importance in diet research.** *Landbouw* 1937, **13**:202–225.
62. Danoesastro H: *The Role of Homegardens as a Source of Additional Daily Income.* Bandung, Indonesia: Paper presented at the Seminar on the Ecology of Homegardens III; 1980.
63. Soemarwoto O, Soemarwoto I, Karyono Soekartadireja EM, Ramlan A: **The Javanese home garden as an integrated agroecosystem.** *Food Nutr Bull* 1985, **7**(3):44–47.
64. Michon G, Mary F: **Conversion of traditional village gardens and new economic strategies of rural households in the area of Bogor, Indonesia.** *Agrofor Syst* 1994, **25**:31–58.
65. Thaman RR: **Urban food gardening in the Pacific Islands: a basis for food security in rapidly urbanising small-island states.** *Habitat International* 1995, **19**(2):209–224.
66. de la Cerda HEC, Mukul RRG: **Homegarden production and productivity in a Mayan community of Yucatan.** *Hum Ecol* 2008, **36**:423–433.
67. Alam M: **Tropical homegardens in Bangladesh: characteristics and sustainability.** In *Sustainable Agriculture Reviews 6: Alternative Farming Systems, Biotechnology, Drought Stress and Ecological Fertilisation.* Edited by Lichtfouse E. Dordrecht, The Netherlands: Springer Science; 2011.
68. Coomes OT, Ban N: **Cultivated plant species diversity in home gardens of an Amazonian peasant village in northeastern Peru.** *Econ Bot* 2004, **58**(3):420–434.
69. Bennett-Lartey SO, Ayernor GS, Markwei CM, Asante IK, Abbiw DK, Boateng SK, Anchirinah VM, Ekpe P: *Contribution of Home Gardens to in situ Conservation of Plant Genetic Resources in Farming Systems in Ghana.* Witzenhausen, Germany: Paper presented at Proceedings of the Second International Home Gardens Workshop; 2002.
70. Soemarwoto O: **Homegardens: a traditional agroforestry system with a promising future.** In *Agroforestry: A Decade of Development.* Edited by Steppler H, Nair PKR. Nairobi, Kenya: International Council for Research in Agroforestry; 1987.
71. Ali MY, Ahmed MM, Islam MB: *Homestead Vegetable Gardening: Meeting the Need of Year Round Vegetable Requirement of Farm Family.* Dhaka, Bangladesh: Paper presented at the National workshop on Multiple cropping; 2008.
72. Talukder A, Kiess L, Huq N, de Pee S, Darnton-Hill I, Bloem MW: **Increasing the production and consumption of vitamin A–rich fruits and vegetables: lessons learned in taking the Bangladesh homestead gardening programme to a national scale.** *Food Nutr Bull* 2000, **21**(2):165–172.
73. Asfaw Z, Woldu Z: **Crop associations of home gardens in Welayta and Gurage in southern Ethiopia.** *Ethiopian J Sci* 1997, **20**:73–90.
74. Abdoellah OS, Gunawan B, Hadikusumah HY: *Home Gardens in the Upper Citarum Watershed, West Java: A Challenge for in situ Conservation of Plant Genetic Resources.* Witzenhausen, Germany: Paper presented at Proceedings of the Second International Home Gardens Workshop; 2001.
75. Pulami RP, Poudel D: *Home Garden's Contribution to Livelihoods of Nepalese Farmers.* Pokhara, Nepal: Paper presented at Home Gardens in Nepal: Proceeding of a workshop on Enhancing the contribution of home garden to on-farm management of plant genetic resources and to improve the livelihoods of Nepalese farmers: Lessons learned and policy implications (2004); 2006.
76. Alvarez-Buylla Roces ME, Lazos Chavero E, García-Barrios JR: **Homegardens of a humid tropical region in Southeast Mexico: an example of an agroforestry cropping system in a recently established community.** *Agrofor Syst* 1989, **8**:133–156.
77. WinklerPrins AMGA: **House-lot gardens in Santarém, Para, Brazil: linking rural with urban.** *Urban Ecosyst* 2002, **6**(1&2):43–65.
78. Abebe T, Wiersum K, Bongers F, Sterck F: **Diversity and dynamics in homegardens of southern Ethiopia.** In *Tropical Homegardens: A Time-tested Example of Sustainable Agroforestry.* Edited by Nair BMKPKR. Dordrecht, The Netherlands: Springer Science; 2006.
79. Buchmann C: **Cuban home gardens and their role in social-ecological resilience.** *Hum Ecol* 2009, **37**:705–721.
80. Wezel A, Bender S: **Plant species diversity of homegardens of Cuba and its significance for household food supply.** *Agrofor Syst* 2003, **57**:39–49.
81. Salam MA, Babu KS, Mohana KN: **Home garden agriculture in Kerala revisited.** *Food Nutr Bull* 1994, **16**(3):220–223.
82. Holden ST, Hvoslef H, Simajuntak R: **Transmigration settlements in Seberida, Sumatra. Deterioration of farming system in a rain forest environment.** *Agr Syst* 1996, **49**:237–258.
83. Niñez VK: **Working at half-potential: constructive analysis of homegarden programme in the Lima slums with suggestions for an alternative approach.** *Food Nutr Bull* 1985, **7**(3):6–13.
84. von Grebmer K, Torero M, Olofinbiyi T, Fritschel T, Wiesmann D, Yohannes Y, Schofield L, von Oppeln C: *The 2011 Global Hunger Index.* Bonn, Germany; Washington DC, USA; Dublin, Ireland: International Food Policy Research Institute, Concern Worldwide, and Welthungerhilfe; 2011.
85. Iannotti L, Cunningham K, Ruel M: *Improving Diet Quality and Micronutrient Nutrition: Homestead Food Production in Bangladesh.* Washington DC, USA: International Food Policy Research Institute; 2009. Discussion Paper 00928.
86. Wanasundera L: *Rural Women in Sri Lanka's Post-Conflict Rural Economy.* Colombo, Sri Lanka: International Labor Office; 2006. RAP Publication 13.
87. Helen Keller International - Asia-Pacific: **Homestead food production model contributes to improved household food security, nutrition and female empowerment – experience from scaling-up programs in Asia (Bangladesh, Cambodia, Nepal and Philippines).** *Nutr Bull* 2010, **8**(1):1–8.
88. Bandarin F, Hosagrahar J, Albernaz FS: **Why development needs culture.** *J Cult Herit Manag Sustain Dev* 2011, **1**(1):15–25.
89. Rao MR, Rajeswara Rao BR: **Medicinal plants in tropical homegardens.** In *Tropical Homegardens: A Time-tested Example of Sustainable Agroforestry.* Edited by Kumar BM, Nair PKR. Dordrecht, The Netherlands: Springer Science; 2006.
90. Perera AH, Rajapakse RMN: **A baseline study of Kandyan forest gardens of Sri Lanka: structure, composition and utilization.** *For Ecol Manage* 1991, **45**:269–280.
91. Millat-e-Mustafa M, Teklehaimanot Z, Haruni AKO: **Traditional uses of perennial homestead garden plants in Bangladesh.** *Forests Trees Livelihoods* 2002, **12**:235–256.
92. Rugalema GH, Okting'ati A, Johnsen FH: **The homegarden agroforestry system of Bukoba district, Northwestern Tanzania. 1. Farming system analysis.** *Agrofor Syst* 1994, **26**:53–64.
93. Rico-Gray V, Chemas A, Mandujano S: **Use of tropical deciduous forest species by the Yucatecan Maya.** *Agrofor Sys* 1991, **14**:149–161.
94. De Clerck FAJ, Negreros-Castillo P: **Plant species of traditional Mayan homegardens of Mexico as analogs for multistrata agroforests.** *Agrofor Sys* 2000, **48**:303–317.
95. World Health Organization: *Global Health Risks: Mortality and Burden of Disease Attributable to Selected Major Risks.* Geneva, Switzerland: World Health Organization; 2009.
96. West KP Jr, Darnton-Hill I: **Vitamin A deficiency.** In *Nutrition and Health in Developing Countries.* Edited by Semba RD, Bloem MW. Totowa NJ, USA: Humana Press; 2008.
97. Stoltzfus R, Mullany L, Black R: **Iron deficiency anaemia.** In *Comparative Quantification of Health Risks: Global and Regional Burden of Disease Attributable to Selected Major Risk Factors.* Edited by Ezzati M, Lopez A, Rodgers A, Murray C. Geneva, Switzerland: World Health Organization; 2004.
98. de Benoist B, Darnton-Hill I, Davidsson L, Fontaine O, Hotz C: **Conclusions of the joint WHO/UNICEF/IAEA/IZiNCG interagency meeting on zinc status indicators.** *Food Nutr Bull* 2007, **28**(3):S480–S79.
99. Black R, Allen L, Bhutta Z, Caulfield L, de Onis M, Ezzati M, Mathers C, Rivera J: **Maternal and child undernutrition: global and regional exposures and health consequences.** *Lancet* 2008, **371**(9608):243–260.
100. Talukder A, Sapkota G, Shrestha S, de Pee S, Bloem MW: *Homestead Food Production Program in Central and Far-Western Nepal Increases Food and Nutrition Security: An Overview of Program Achievements.* Pokhara, Nepal: Paper presented at Home Gardens in Nepal: Proceeding of a workshop on Enhancing the contribution of home garden to on-farm management of

plant genetic resources and to improve the livelihoods of Nepalese farmers: Lessons learned and policy implications (2004); 2006.

101. Bloem M, de Pee S, Darnton-Hill I: **New issues in developing effective approaches for the prevention and control of vitamin A deficiency.** *Food Nutr Bull* 1998, **19**(2):137–148.
102. Keys E, Kaqchikel G: *Women, Children, and Multiple Roles of Gardens among the Maya of Highland Guatemala. Yearbook, Conference of Latin Americanist Geographers.* Austin TX, USA: Paper presented at Conference of Latin Americanist Geographers; 1999.
103. Pandey CB, Rai RB, Singh L, Singh AK: **Homegardens of Andaman and Nicobar, India.** *Agr Syst* 2007, **92:**1–22.
104. Howard PL: **Gender and social dynamics in swidden and homegardens in Latin America.** In *Tropical Homegardens: A Time-Tested Example of Sustainable Agroforestry.* Edited by Nair BMKPKR. Heidelberg, The Netherlands: Springer Science; 2006.
105. Descola P: *In the Society of Nature: A Native Ecology in Amazonia.* New York, USA: Cambridge University Press; 1994.
106. Brun T, Reynaud J, Chevaussus-Agnes S: **Food and nutrition impact of one home garden project in Senegal.** *Ecol Food Nutr* 1989, **23:**91–108.
107. Kumar SK: *Role of the Household Economy in Child Nutrition at Low Incomes: A Case Study in Kerala.* Ithaca NY, USA: Department of Agricultural Economics, Cornell University; 1978. Occasional Paper No. 95.
108. Adhikari A, Singh D, Suwal R, Shrestha PK, Gautam R: *The Role of Gender in the Home Garden Management and Benefit-Sharing from Home Gardens in Different Production System of Nepal.* Pokhara, Nepal: Paper presented at Home Gardens in Nepal: Proceeding of a workshop on Enhancing the contribution of home garden to on-farm management of plant genetic resources and to improve the livelihoods of Nepalese farmers: Lessons learned and policy implications (2004); 2006.
109. Soemarwoto O: **The Javanese rural ecosystem.** In *An Introduction to Human Ecology Research on Agricultural Systems in Southeast Asia.* Edited by Rambo AT, Sajise PE. Los Banos, The Philippines: University of Philippines; 1984.
110. Blanckaert I, Swennen RL, Paredes Flores M, Rosas López R, Lira SR: **Floristic composition, plant uses and management practices in homegardens of San Rafael Coxcatlán. Valle de Tehuacán-Cuicatlán, México.** *J Arid Environ* 2004, **57:**39–62.
111. Trinh LN, Watson JW, Hue NN, De NN, Minh NV, Chu P, Sthapit BR, Eyzaguirre PB: **Agrobiodiversity conservation and development in Vietnamese home gardens.** *Agric Ecosyst Environ* 2003, **97:**317–344.
112. Calvet-Mir L, Gómez-Bagetthun E, Reyes-García V: **Beyond food production: Home gardens" ecosystem services. A case study in Vall Fosca, Catalan Pyrenees, northeastern Spain.** *Ecol Econ* 2012, **74:**153–160.
113. Vasey DE: **Household gardens and their niche in Port Moresby, Papua New Guinea.** *Food Nutr Bull* 1985, **7**(3):37–43.
114. Kehlenbeck K, Maass BL: **Crop diversity and classification of homegardens in Central Sulawesi, Indonesia.** *Agrofor Syst* 2004, **63**:53–62.
115. Okigbo B: **Home Gardens in Tropical Africa.** In *Tropical Home Gardens.* Edited by Landauer K, Brazil M. Tokyo, Japan: United Nations University Press; 1990.
116. Wilson RT: *Livestock Production Systems.* London, UK: Macmillan; 1995.
117. Devendra C, Thomas D: **Smallholder farming systems in Asia.** *Agr Syst* 2002, **71:**17–25.
118. Albuquerque UP, Andrade LHC, Caballero J: **Structure and floristics of homegardens in Northeastern Brazil.** *J Arid Environ* 2005, **62:**491–506.
119. Watson JW, Eyzaguirre PB: *Homegardens and in-situ Conservation of Plant Genetic Resources in Farming Systems.* Witzenhausen, Germany: Paper presented at 2nd International Homegarden Workshop; 2002.
120. Gajaseni J, Gajaseni N: **Ecological rationalities of the traditional homegarden system in the Chao Phraya Basin, Thailand.** *Agrofor Syst* 1999, **46:**3–23.
121. Pushpakumara DKNG, Wijesekara A, Hunter DG: **Kandyan homegardens: A promising land management system in Sri Lanka.** In *Sustainable use of Biological Diversity in Socio-ecological Production Landscapes. Background to the 'Satoyama Initiative for the Benefit of Biodiversity and Human Well-being'.* Edited by Bélair C, Ichikawa K, Wong BYL, Mulongoy KJ. Montreal, Canada: The Secretariat of the Convention on Biological Diversity; 2010.
122. Christanty L: **Home Gardens in Tropical Asia, with Special Reference to Indonesia.** In *Tropical Home Gardens.* Edited by Landauer K, Brazil M. Tokyo, Japan: United Nations University Press; 1990.
123. Seneviratne G, Kuruppuarachchi KAJM, Somaratne S, Seneviratne KACN: **Nutrient cycling and safety-net mechanism in the tropical home gardens.** *Int J Agric Res* 2010, **5**(7):529–542.
124. Terra GJA: **Mixed-garden horticulture in Java.** *Malayan Journal of Top Geogr* 1954, **3:**33–43.
125. Powell JM, Williams TO: *Livestock, Nutrient Cycling and Sustainable Agriculture in the West African Sahel.* London, UK: International Institute for Environment and Development; 1993. Gatekeeper Series No. 37.
126. McConnell DJ, Dharmapala KAE: *The Economic Structure of Kandyan Forest Garden Farms.* Peradeniya, Sri Lanka: UNDP/SF/FAO Diversification Project; 1973.
127. Pushkaran K: *Home Gardens in Kerala as an Efficient Agroecosystem for Conservation and Sustainable Management of Biodiversity.* Witzenhausen, Germany: Paper presented at Proceedings of the Second International Home Gardens Workshop: Contributions of home gardens to in situ conservation of plant genetic resources in farming systems; 2002.
128. Perera AH: *Role of Kandyan Forest Gardens in Germplasm Conservation.* Peradeniya, Sri Lanka: Paper presented at the regional workshop on conservation methodology in South Asia; 1988.
129. Senanayake FR: **Analog forestry as conservation tool.** *Tiger Paper* 1987, **15:**25–28.
130. Hochegger K: *Farming like the forest: traditional home garden systems in Sri Lanka.* Weikersheim, Germany: Margraf Verlag; 1998.
131. Kumarathunga N: *Mihikatha Surakina Parisara Govikama.* Nugegoda, Sri Lanka: Govikam Asiriya 2; 2008.
132. Jose D, Shanmugaratnam N: **Traditional homegardens of Kerala: a sustainable human ecosystem.** *Agrofor Sys* 1993, **24:**203–213.
133. Thaman RR: **"Urban root crop production in South West Pacific".** In *Regional meeting on the production of root crops. Technical paper - South Pacific Commission;* 1977:73–82.
134. Evers G, Keleta E, Kirway T: *A farming system study in the lowland wet zone of Sri Lanka, Agalawatta Division, Kalutara District.* Wageningen, The Netherlands: International Course for development oriented Research in Agriculture; 1985.
135. Miura S, Osamu K, Susumu W: **Home gardening in urban poor communities of the Philippines.** *Int J Food Sci Nutr* 2003, **54**(1):77–88.
136. Food and Agriculture Organization: *An Introduction to the Basic Concepts of Food Security.* Rome, Italy: Food Security Information for Action, Practical Guides; 2008.
137. Food and Agriculture Organization: *World Agriculture: Towards 2030/2050.* Rome, Italy: Interim report; 2006.
138. Alexandratos N: *World Food And Agriculture to 2030/2050: Highlights and Views From mid-2009.* Rome, Italy: In Proceedings from FAO Expert Meeting on How to Feed the World in 2050; 2009.
139. Food and Agriculture Organization: *An Introduction to the Basic Concepts of Food Security. Food Security Information for Action, Practical Guides.* Rome, Italy: EC - FAO Food Security Programme; 2008.
140. United States Department of Agriculture: *Definitions of Food Security.* http://www.ers.usda.gov/topics/food-nutrition-assistance/food-security-in-the-us/measurement.aspx (accessed January 2013).
141. Pinstrup-Andersen P: **Food security: definition and measurement.** *Food Secur* 2009, **1**(1):5–7.
142. Gasser M, Salzano C, Meglio RD, Lazarte-Hoyle A: *Post-crisis Situations: Problems and Needs. Local Economic Development In Post-Crisis Situations: Operational Guide.* International Labour Organization: Geneva, Switzerland; 2004.

13

Supporting food security in the 21st century through resource-conserving increases in agricultural production

Norman Uphoff[1,2]

Abstract

The Green Revolution was accomplished under a set of demographic, economic, climatic and other conditions in the 20th century that have been changing and will surely be different and more difficult in the decades ahead. The suitability and sustainability of any given agricultural technology depends on factors like resource availability and productivity, energy costs, and environmental constraints. The achievements of Green Revolution technologies in the 1960s and 1970s came at a critical time of impending food shortages, and the world's people would be worse off without them. However, the rate of yield improvement for cereal production has been slowing since the mid-1980s.

Looking ahead at the foreseeable circumstances under which 21st century agricultural producers must try to assure food security, there will be need for technologies that are less dependent on resources that are becoming relatively scarcer, like arable land and water, or becoming relatively more costly, like energy and petrochemical-based inputs. This paper considers agroecologically-based innovations that reduce farmers' dependence on external inputs, relying more on endogenous processes and existing potentials in plants and soil systems. Such resource-conserving production represents a different approach to meeting food security goals.

While these innovations are not yet fully understood and are still being researched, there are good agronomic reasons to account for their effectiveness, and scientific validations are accumulating. Enough successes have been recorded from making changes in the management of plants, soil, water and nutrients that more attention from researchers, policy-makers and practitioners is warranted, especially given the need to adapt to, and to mitigate the effects of, climate change. The same agroecological concepts and management methods that are enhancing factor productivity in rice production are giving similar results with other crops such as wheat, finger millet, sugarcane, mustard, and tef.

Genetic potentials are the starting point for any and all agricultural production, and current efforts to improve food security and nutrition through plant breeding efforts should continue. However, future research and production strategies could beneficially seek to capitalize on biological processes and potentials existing within crops and in their supporting soil systems, rather than focusing so predominantly on making modifications in genetic factors. Scientific advances in the domains of microbiology, soil ecology and epigenetics could foreseeably assist farmers in meeting production and income goals with resource-economizing methods. It remains to be seen to what extent agroecologically-informed methods can help farmers meet expected agricultural production requirements to ensure global food security, but this direction deserves more attention and support.

Keywords: Agroecology, Food security, Green revolution, Soil biota, System of rice intensification

Correspondence: ntul@cornell.edu
[1]SRI International Network and Resources Center (SRI-Rice), Cornell University, Ithaca, NY 14853, USA
[2]Cornell International Institute for Food, Agriculture and Development (CIIFAD), Cornell University, Ithaca, NY 14853, USA

Introduction

An overriding challenge for agricultural research and practice in the 21st century will be to produce more food and fiber with less investment of our land, labor, capital and (particularly) water resources. This large order must be met in the context of less favorable weather conditions for much if not most of our planet, as weather shifts and variability are becoming ever-more evident constraints for agricultural scientists, policy-makers, and practitioners.

Dr. Norman Borlaug and his generation of agricultural scientists gave leadership in the 20th century to meeting the challenge faced then of 'feeding a hungry planet'. Fortunately, they achieved some quick and remarkable technological advances under the banner of the Green Revolution. It does not detract from their accomplishment of raising cereal yields substantially to note, however, that these gains were made mostly in Asia, and that the opportunities were not equally distributed across all sizes of landholding. A more crucial consideration at present is that these improvements have been slackening over the past 25 years.

Without the increases in production that resulted from the Green Revolution, millions of people would have suffered from sickness and hunger, and many would have died if there had not been significant increases in food availability in the latter part of the past century. This accomplishment is evident from a 50% decline in the relative real price of food over a four decade period.

But this favorable trend has been reversed, with food prices again rising in recent years. Moreover, we have to deal now with some very stark constraints foreseeable in this century. Many millions of agricultural producers, particularly in Africa, need to raise their production very substantially with the resources that they have at hand if life-saving levels of food security are to be achieved and maintained, to keep hunger, poverty and other woes from becoming more widespread.

Reckoning with a changing world

The world situation 50 years ago with rapid population growth presented an ominous, even dire, threat to the well-being of all people, rich and poor alike. In the 21st century, this pressure has fortunately abated as population growth rates have fallen dramatically over the last three decades. Still, the momentum of even decelerating population growth means that we must reckon with 9 to 10 billion people as the lowest peak population that can be expected in this century, and some of the poorest countries still have very rapid population growth.

It is good news that the 'population bomb' now appears unlikely to explode. A world population which stabilizes at less than 10 billion, and then possibly declines, is only half as large as what demographers were predicting when I began working on development issues over four decades ago. Deceleration of population growth, to be sure, brings some of its own challenges, as there may not be enough productive younger countrymen and women to support elderly populations. However, such problems are less difficult to cope with than the multiple 'bad news' trends that we need to grapple with now.

In this century, we face a threat to our welfare and even to our existence that is even more inexorable and potentially devastating than population growth - *climate change*. It is neither alarmist nor paranoid to point out that the continuation of our human species depends on remaining within the ranges of temperature, precipitation and other climate parameters within which our and other species have evolved.

While humans are unique in their ability to expand the range of environments in which they can live, drawing on our intelligence, culture and technology, the means for extension and adaptation of human habitats do not come free, and the ranges that are physiologically as well as economically tolerable are fairly narrow. There are significant economic and environmental costs as well as some social and cultural costs involved.

Agriculture as a sector is uniquely sensitive to, and dependent on, weather factors. As we think about our countries' agricultural sectors in the future, we need to figure out how we can cultivate our crops, raise our livestock, and fish our water bodies with more resource-economizing methods, no longer using water, metals, chemicals and fuels as profligately as in the past. Furthermore, how can we devise ways to protect our soil systems, watersheds, forests, hydrological cycles and agroecosystems - indeed, how can we restore and replenish them?

Most of the concepts, principles, laws and even methodologies in our scientific domain have been developed within the realms of chemistry, physics and engineering, where closed-system thinking is realistic, indeed necessary. Biology, on the other hand, operates with the nonlinear logic of open systems, where small inputs can produce large outputs, but also where huge inputs can give no output at all. While biology has inputs and outputs parallel to those that drive chemical and physical processes and that get channeled by engineering, its *throughputs* function quite differently.

When trying to understand, maintain and improve open-system phenomena, we should guard against mindsets and assumptions that were derived from, and are appropriate to, closed systems. While engineering concepts can be applied to agriculture, this does not mean that agriculture should be seen or managed simply as an industrial enterprise. Although agricultural practice has many industrial characteristics and opportunities, it remains essentially biological. This becomes even more evident as we gain a better understanding of some micro-biological dimensions of agriculture as discussed below.

When doing more of the same won't do

The changing circumstances under which agriculture will be practiced in the 21st century are likely to make less effective and less economic the agricultural technologies that were successful in the preceding century, even with some modifications and evolutions. The fore-seeable world presents us with rather different conditions and constraints than we had to deal with in the past.

As explained by Hayami and Ruttan [1], past changes in agricultural technology have been driven by changing factor proportions. These inexorably reshape and redirect our future technological paths, reflecting the relative amounts and productivities of land, labor and other resources. The added impacts now of climate change will make the need for different, even novel technological solutions even more certain.

Agriculture in the 21st century will have to deal with major alterations in the physical and institutional landscape:

- *Arable land per capita* is declining. While different definitions of land quality can lead to different numbers, the combination of continuing population growth and land degradation is surely lowering this ratio. By some calculations, by the middle of the 21st century there will be about one-third as much cultivable area per person as 100 years earlier, falling from about 0.25 ha to 0.08 ha per person. *Land-extensive* strategies will thus become less tenable as the ratios of population-to-land continue to change substantially [2].
- *Supply of water,* absolutely essential for agricultural production, is becoming less reliable as well as less available for farmers. This is a result of competing uses, of declines in water quality and of shifting weather patterns, so economizing on water is becoming an agricultural and social imperative; getting 'more crop per drop'.
- *Energy costs* in the 21st century will almost certainly be higher than in the 20th century, making energy-intensive production more costly. This applies to the mechanization of operations as well as to our present dependence on petroleum-based agrochemical inputs. Also, patterns of agricultural production and consumption that involve long-distance movement of commodities will become less economic as transport costs rise. This will make local production economically more competitive with distant agricultural operations that were previously considered to have a comparative advantage. Moreover, the increasing use of grains to manufacture biofuels in response to the rising prices for energy contributes to increased prices of food grains, which adversely affects particularly the food-insecure.
- *Environmental concerns* are affecting the acceptability of various agricultural practices in many places. Negative externalities such as adverse effects on water, soil and air quality and biodiversity losses are becoming less tolerated than in the past, given larger and denser human populations and more urban political influence.
- *Pest and disease problems* are likely to increase, partly in response to climate changes, but also with pest resistance to agrochemical controls as a factor. Despite a 14-fold increase in the amount of pesticides used in the US after World War II, the percentage of crops lost to pests increased in this period from 6% to 13% [3]. The theory of 'trophobiosis' may help account for this phenomenon of chemical-induced vulnerability of crops to pest and disease losses [4].
- *Climate changes* will make all of these other trends more challenging for agricultural producers. Altered temperature and rainfall patterns will not be felt equally because there will be winners and losers from change; but especially the poorer countries, and the more marginalized farmers within them, appear likely to bear the greatest brunt of new weather regimes and of the greater uncertainty and variability that these bring. 'Extreme events' are likely to become both more frequent and more extreme: droughts, storms, flooding, heat waves, cold periods, disadvantaging both producers and consumers at the same time.
- To make matters worse, *government capabilities* to deal with these many problems appear less coherent, less stable, and less statesmanlike, with less commitment and less ability to achieve effective functioning of their economies, societies and administrations. Our legislative and governing bodies are increasingly beset by narrow partisanship and jockeying for short-term personal or factional advantages. Also, we are seeing in many parts of the world an attenuation of cultural and political inhibitions against large-scale use of violence - that genie which governments previously kept bottled up with at least some success.
- Governments' *fiscal capacities* are almost everywhere stretched to the breaking point. Accumulating debts and voter disaffection are reducing public-sector willingness and ability to invest in (or subsidize) inputs and improvements in

agriculture, even though the central importance of food has not diminished. Farm subsidies and export restrictions in many countries, maintained as a result of governments' political weakness rather than strength, continue to exert upward pressures on food grain prices that affect the food-insecure.

- Furthermore, the forces of *globalization*, while they may open up many opportunities, seem also to contribute to socio-economic instability and inequity that make national governance ever more difficult. Simply the greater scale and complexity of modern societies and economies appears to be making our institutions and practices of governance less effective.
- *Slowing gains in agricultural productivity* cast an ominous shadow over this whole scene. The rates of growth in cereal production during the 1970s and 1980s were truly remarkable, although we should notice that the rate of yield increase was even higher in the 1960s before Green Revolution technologies began taking effect. World grain production per capita, according to FAO and USDA statistics, peaked in 1984, with the rate of grain production increases falling ever since. The world's total grain production has been stagnant or declining since the mid-1990s [5-7].
- The *global food crisis* in 2007 to 2008 with its huge price jumps was prompted by multiple factors; but an underlying cause was the tightening of world food supplies relative to demand. It is true that with available resources and improved technology, more food could have been produced in recent years if there had been higher market prices for food. However, if food had been more expensive, this would have increased the extent of hunger and poverty in the world, both of which are still unacceptably pervasive.

All these considerations make it imperative to find ways to raise our agricultural productivity, to make our available resources, land, labor, water and capital more productive. We need to be enhancing food supply particularly on the basis of *productivity gains* rather than trying to evoke more food production through the incentive of higher prices, as this will increase food insecurity for millions of households.

There are ongoing efforts to improve food security through genetic improvements that focus on yield, resistance to various biotic and abiotic stresses, and improved nutritional value through biofortification. While some of these gains may be some years away, they are worth pursuing. But in the present and near future, we need to be achieving greater productivity of the key factors of production: land, labor, and especially water, as well as capital.

The more resources which are require for meeting people's food needs, the fewer resources are available for meeting other needs, for investment in human resources, for housing and infrastructure, for expanding research and development in other fields. Resources expended in food production cannot be used to improve output and well-being in other sectors. This makes the quest to find ways to produce more food with fewer resources important for the whole world as we head into an uncertain, fairly ominous future.

Capitalizing on existing biological opportunities

Making improvements in our global ability to meet people's needs for sustenance will not solve all of the many problems that face our governments and societies around the world. But if we cannot produce and provide sufficient food to everyone - if we tolerate large pockets of hunger and misery, and have large domains of desperation with loss of hope for the future - most of the other problems that loom on the global agenda will remain beyond solution.

The limited and, in many respects, shrinking resources that are available for agricultural production dictate that we *intensify* our production practices rather than continue with the *land-extensive* strategies that expanded in the 20th century, relying heavily on mechanization and economies of scale. In the 21st century, these strategies for food production, being energy-intensive and chemical-dependent, will become less attractive and indeed less economically viable as factor proportions and relative prices change.

The word 'intensification' when applied to agricultural production has usually referred to an intensification of *input use* [8]. The logic of such intensification - and the cost-benefit analysis that guides it - has been to compare how much more of what inputs will yield how much more in resulting outputs? If the value of the resulting outputs from production is sufficiently greater than the cost of the inputs needed to produce them, this represents a net gain. This strategy has been bolstered by plant-breeding efforts which have developed genotypes that are more responsive to additional inputs of water, fertilizer, and agrochemicals. There has been an underlying assumption that we always need to increase inputs to raise production, with the expectation that a greater increase in outputs will result to justify the expenditure.

An alternative kind of intensification is the intensification of *management* involving both more knowledge and skill - to achieve more output from whatever inputs are used. Both approaches, of course, confront the limitations of diminishing returns. Management here refers to making modifications in, and manipulations of, the growing environment (E), aiming to produce better phenotypes (P) from the genotypes (G) in use. This is represented by the symbolic equation well-known to plant breeders:

$$P = f\,(G \times E)$$

Current agricultural development efforts focus particularly on improving G to achieve maximum increases from the resources invested. A management-oriented strategy, conversely, starts with the best available G, focusing on how to improve those factors that constitute the E for their crops' growing environments both above and below ground. The logic of cost-benefit analysis is turned on its head when we try to calculate a ratio between *reduced* inputs and *greater* output. This may be why agroecological alternatives are widely viewed with skepticism.

Raising food production, especially for the food-insecure

An agroecological innovation that has demonstrated remarkable gains in food output is the System of Rice Intensification (SRI) developed in Madagascar [9-11]. This methodology for raising the productivity of all the resources devoted to rice production (land, labor, seeds, water and capital) departs from the intensification strategy that relies on new varieties with genetic improvement and then on greater external inputs.

Instead, the productivity gains from SRI management derive from changes made in the way that the rice plants, soil, water and nutrients are managed; their timing, spacing, and so on. The phenotypical changes resulting from modifications in management are easily seen (Figures 1 and 2) and have been documented and explained by many researchers, for example [12-16]. Over 300 published articles on SRI are listed now on the SRI-Rice website at http://sri.ciifad.cornell.edu/research/index.html#journals. Results vary considerably because the changes achieved derive more from biological processes, which are vulnerable as well as potent, than from genetic blueprints or material inputs. The preponderance of evidence is certainly positive.

The yield effects of intensifying crop management with SRI methods have been a matter of some controversy among rice scientists, for example [17-20]. But governments in China, India, Indonesia, Cambodia and Vietnam, where two-thirds of the world's rice is produced, have become satisfied from their own evaluations that factor productivity is greatly enhanced by SRI changes in rice crop management. Two of the world's most eminent rice scientists, Dr. MS Swaminathan and Prof. Yuan Long-ping, have satisfied themselves from their own evaluations that SRI methods lead to more productive rice plants and have recommended SRI use in their respective countries, India and China [21].

In 2007 the Ministry of Agriculture and Rural Development in Vietnam, after three years of evaluations, concluded and declared that SRI represented a 'technical advance'. At that time, fewer than 10,000 Vietnamese farmers were using SRI methods. Four years later, in 2011, the Ministry announced that over 1 million Vietnamese farmers were using most or all of the recommended SRI practices - with higher yields, less water, lower production costs, and improvements in environmental quality [22]. See news report at http://qdnd.vn/qdndsite/en-US/75/72/182/156/189/164012/Default.aspx.

Figure 1 Phenotypical effects of SRI (System of Rice intensification) management: (a) Nepali farmer on left holds a rice plant grown with SRI methods from a single seed; (b) on right, a Cuban farmer holds two rice plants of the same variety (VN2084) and same age (52 days after sowing). Both Cuban plants were started in the same nursery, but the plant on right was removed at 9 days and was transplanted into a field where SRI growing conditions were maintained. A video showing, week by week, the contrasting and divergent growth of SRI and conventional rice plants on this same Cuban farm in the following year is posted at: http://www.youtube.com/watch?v=1zOMi9Ibao4&feature=relmfu. Pictures courtesy of Rajendra Uprety, District Agricultural Development Office, Biratnagar, Nepal; and Dr. Rena Perez, then food security advisor, Ministry of Sugar, Havana, Cuba.

Figure 2 Comparisons of growth response of different varieties to SRI (System of Rice Intensification) management: trial plots at Al-Mishkhab Rice Research Station, Najaf, Iraq, evaluating differences in varietal response to SRI methods. On the left in each pair of trials are SRI plants, being compared with rice plants of same variety grown with current best management practices on right. Picture courtesy of Dr. Khidhir Hameed, MRRS.

The management changes introduced with SRI include: use of younger seedlings if the rice crop is established by transplanting (note: direct-seeding is also beginning to be used with the other SRI practices); greatly reduced plant populations by planting single seedlings in widely-spaced hills; no continuous flooding of paddy fields; controlling weeds with a mechanical weeder that aerates the soil; plus enhancement of the soil's organic matter as much as possible.

When these methods are used together as recommended, yield increases have ranged from 20% to 100% or sometimes even more, with large reductions in seed requirements (by 80 to 90%), in irrigation water (by a quarter to a half), in chemical fertilizer (by 50% or even 100%), and in applications of agrochemicals (varying, according to how much pest and disease resistance is engendered by SRI methods). Often labor requirements are also reduced once the methods have been learned and mastered. Unfortunately, SRI was initially regarded and dismissed by some as too labor-intensive for farmer acceptance; however, many farmers in China and India consider *reduction* in labor requirements to be one of SRI's main attractions, particularly for women. Table 1 gives a summary overview of SRI results across 13 countries, as assessed by a variety of institutions - international agricultural research centers, universities, government agencies, NGOs, private firms (Nippon Koei and Syngenta/Bangladesh), and a donor agency (GTZ).

It is understandable that SRI results have been received with some incredulity. However, the bases for these increases in productivity are becoming better understood. The most evident factor is the greater *root growth* induced by SRI methods (Figure 1b). Less visible but now being documented are the methods' impacts on *soil biota* that provide benefits to rice crops [42-45]. Similar benefits are now being seen when SRI methods have been adapted to other crops such as wheat and sugarcane [46,47], finger millet, mustard, tef, and some legumes and vegetables. This adds credibility to the remarkable results that have been reported for rice.

For some years, skeptics rejected SRI because they considered some of its reported results to be impossible. The top yields reported from Madagascar were characterized as being above the biological maximum achievable with rice [18]. However, in the 2011 kharif season in the Indian state of Bihar, five first-time SRI farmers in one village (using hybrid varieties) matched or exceeded the previous world-record yield for paddy of 19 tons/ha. According to Bihar Department of Agriculture data, one achieved a yield of 22.4 tons/ha, almost ten times the average paddy yield for the state. The dry weight was an unprecedented 20.16 tons/ha [48]. These measurements, made with standard methods and with hundreds of observers looking on, have been officially accepted.

The controversy over SRI maximum yields has unfortunately deflected attention from the large differences in *average* yields, which should have been the central focus of discussion. SRI methods have often enabled poor farmers to double, triple or even quadruple their yields, not just individually but on a village level, without having to purchase new varieties or agrochemical inputs [49-52]. Such resource-limited farmers started at very low levels of production, it is true. But for them, to go from 1 ton per hectare to 4 tons, or from 2 tons to 8 tons, without added costs of production, makes a huge difference in their food security and well-being. For the 2011 kharif season, the Bihar Department of Agriculture in India has calculated an average SRI yield of 8.08 tons/ha on 335,000 hectares across all 38 districts of the state. This was more than three times the state-wide average of 2.5 tons/ha.

It has been argued that the 'best management practices' (BMP) recommended by scientists can give better results than reported from SRI practices [53]. However, the evidence and analysis presented to support this claim are seriously flawed [54]. More importantly, the paddy yields reported above from Bihar state show the argument to be simply wrong; SRI methods can surpass what best conventional practices can achieve by yielding more productive phenotypes. The five Bihar farmers who got SRI yields matching or exceeding the previous world record got less than half as much yield from their hybrid varieties when these were grown on the same farms, with the same soil and same climate, using conventional crop management methods.

Table 1 Review of SRI management impacts on yield, water saving, costs of production, and farmer income per ha in 13 countries

Country	Farmers (N) or Trials/Area	Conventional yield (t/ha)	SRI yield (t/ha)	Yield increase (%)	Water saving (%)	Impact on cost per ha (%)	Impact on income per ha (%)	Coverage of evaluation and agency doing study [source of data]
AFGHANISTAN	42[d]	5.6	9.3	55%	NM	NM	NM	Aga Khan Foundation program in Baghlan district [23]
BANGLADESH	1,073[d]	5.44	6.86	26%	NM	−7%	+59%	2-year study by NGOs and Syngenta under IRRI program auspices [24]
CAMBODIA	500[c]	1.63	2.29	41%	Rainfed	−56%	+74%	5-province study done for GTZ [25]
Experienced farmers	120[d]	1.34	2.75	105%	Rainfed	−47%	+98%	Study of all 3-year SRI users by CEDAC [26]
CHINA	82[c]	6.6	9.37	42%	44%	−7.4%[f]	+64%	Village study by China Agricultural University in Sichuan province [27]
Sichuan province	Total area 2004 to 2010 : 301,967 ha	7.7	9.5	23%	25.6%	NR	Additional income: US$320 million	Provincial Department of Agriculture data, 2004 to 2010 [28]
INDIA	108[d]	4.12[g]	5.47[g]	32%[g]	Rainfed	−35%	+67%	IWMI-India programme study in W. Bengal [29]
Andhra Pradesh state	1,525[d]	6.31	8.73	34%	40%	NM	NM	Evaluations by AP state university ANGRAU [30]
INDONESIA	12,133[d]	4.27	7.61	78%	40%	−20%	>100%	On-farm trials managed by Nippon Koei TA team, 2002 to 2006 [31]
KENYA	On-station trials	6.2	7.6	26%	28.2%	NM	NM	Mwea irrigation scheme trials, 3 replications [32]
Additional trials	"	8.66	14.85	70%	24%	NM	NM	Mwea irrigation scheme trials, 4 replications [33]
MALI	53[d]	5.5	9.1	60%	10%	+15%	+108%	Timbuktu region under Africare program [34]
MYANMAR	612[d]	2.1	4.4	110%	Rainfed	+0.2%	8.7 times more	FFS results in Kachin and Shan States, 3 years [35]
NEPAL	412[d]	3.3	6.1	82%	43%	−2.2%[e]	+163%	District agricultural extension program, Morang [36]
Far Western region	890[d]	4.01	7.58	88%	>60%	+32%	+164%	EU-FAO Food Facility Programme [37,38]
PANAMA	46[d]	3.44	4.75	38%	71-86%	NM	NM	10 communities, evaluation by NGO [39]
SRI LANKA	120[c]	3.84	5.52	44%	24%	−12%	+104%	IWMI study in 2 districts [40]
VIETNAM	1,274[d]	5.58	6.79	22%	33%	−30%	+36%	MARD FFS results from 13 districts [41]
Total N and Averages	18,870[a]	4.77	7.12	50%	37.5%	−16%	94%[a,b]	

NM: Not measured, NR: Not reported, FFS: Farmer Field Schools.
[a]Not including Sichuan/China data; [b]Not including Myanmar data.
[c] Based on random samples; [d]Results from all farmers using SRI in area; no sampling involved.
[e]Labor-saving hand weeders were not yet available in district to reduce labor inputs and costs.
[f]Extension personnel were promoting the purchase of modern seeds and fertilizer simultaneously with SRI methods.
[g]50% increase in the village with normal rainfall, 12% in the drought-stricken village.

The contention that BMPs are superior to SRI disregards the fact that many millions of small and marginal farmers do not have access to, or cannot afford, the new seeds and other inputs required for what are considered as BMP. For hunger and poverty to be reduced, we need methods that substantially improve farmers' productivity with the resources and means that they have at hand. A variety-based, input-dependent strategy of agricultural development has inherent limitations for reducing world hunger and food insecurity in a significant way.

Re-biologizing agriculture

Experience with SRI methods has directed our attention to the symbiotic relationships between plants and soil microbes, which parallel the growing realization that humans and other members of the animal kingdom are themselves thoroughly interdependent with, and for that matter, dependent on microorganisms, referred to collectively for humans as the human *microbiome* [55]. The microbiomes of plants are no less important. Appreciating this fact re-focuses our thinking from engineering and chemical solutions to capitalizing more upon biological and ecological processes and potentials.

It has long been known that the root zones (rhizospheres) of plants are domains of intense and mutually beneficial interaction between plant roots and soil organisms [56,57]. Research has shown that certain soil organisms in fact inhabit the leaves, sheaths and stalks of rice plants, that is, their phyllosphere; and controlled trials have demonstrated that this 'infection' of above-ground plant tissues by soil bacteria is associated with the plants having larger canopies and root systems, higher levels of chlorophyll and more photosynthesis in their leaves, greater water-use efficiency, and higher yield [15,58].

Moreover, the presence of soil rhizobia in the tissues of leaves and sheaths of rice plants has been seen to up-regulate certain genes that produce proteins supportive of photosynthesis and other functions in the canopy; and these same microorganisms, in the roots of rice plants, enhance the expression of genes in tissues there that produce proteins conferring greater protection against pathogens [59]. Plant growth and performance are thus assisted by the presence and influence of soil microbes.

Signaling between roots and shoots via phytohormones has been shown to contribute to higher yield with reduced fertilizer inputs and to healthier plants that resist pathogens [60]. Even more surprising may be the results of replicated trials in which rice plants grown from seeds inoculated with a certain fungus (*Fusarium culmorum*) have been found to have five times more root growth in their early days after germination, and their root hairs emerge two days sooner. The fungus, when residing in seeds, enhances the plants' ensuing growth and robustness [61].

The contributions that symbiotic endophytes, microbes living mutualistically within plant tissues and cells, can make to crop performance and health are becoming more evident [62]. Learning how to utilize and benefit from positive microbial influences could possibly enhance crop productivity in parallel with the improvements that are expected to be made in human health as a result of contemporary research on the human microbiome. Research on the microbiomes of crop plants should be similarly productive.

Focusing research on ecological rather than only on species productivity

These findings suggest that we consider making some shifts in the predominant focus for agricultural research. Present strategies are based on ever-increasing use of *inputs* in conjunction with making improvements in *genotypes*, with a species focus. There is reason to explore more how improving crops' environments above and below ground through changes in plant, soil, water and nutrient *management* can elicit more beneficial *expression* of crops' genetic potentials, recognizing that these are affected by ecological factors.

The focus should be not just on producing more grain, but also on growing plants that can better resist the biotic and abiotic stresses that are likely to become more common as a result of impending climate changes. The impetus for this reorientation is coming not so much from research stations as from farmers' fields, where producers are contending directly with the factors that contribute to food security or insecurity (Figures 3 and 4).

Previously, agricultural researchers have focused separately on individual species of plants or animals in isolation from other species. Finding ways to capitalize upon symbiotic relationships among various species – considering sets of species together in ecological perspective, such as with the 'push-pull' crop management system developed in East Africa to deal with *striga* infestation – can create multiple benefits at low cost [63].

Soil scientists have been evaluating plants' performance often under *axenic* soil conditions, where all life in the soil has been eliminated by fumigation or sterilization. If the effects of soil organisms are studied, possibly just one or two species would be added back to look at their effect under what are called *gnotobiotic* conditions in otherwise axenic soil. This latter word is curious because 'axenic' is constructed from Greek semantic roots which mean 'without strangers' (*a* + *xenos*). This formulation implies that soil organisms are to be considered as strangers or foreigners within their own natural habitat [64].

Future agricultural research, in light of recent findings, will do well to examine the performance of crops and livestock *within their ecological contexts*. Species improvement should be attempted in the context of *interactions* with other species rather than in isolation. This is one way in which we can 're-biologize' agriculture, having been guided for the past half century more by the disciplines of chemistry and engineering in league with

Figure 3 Effects on root and tiller growth with modified rice management: stump of a rice plant with 223 tillers, grown with SRI methods from a single seed, presented to the author in October 2009 by farmers in a farmer field school at Panda'an, East Java, Indonesia, operated by the Sampoerna Foundation as a corporate social responsibility project. This was the farmers' best plant from the previous season, but it showed them what a single seed could produce with optimum growing conditions. Picture by author.

Figure 4 Resistance to biotic and abiotic stresses based on alternative crop management: two adjacent rice paddy fields in Crawuk village, Ngawi district, East Java, Indonesia, after they were both hit by a brown planthopper (BPH) attack and then by a tropical storm in June 2011. Paddy field on left, planted with an improved rice variety (Ciherang) and using inorganic fertilizer and agrochemical protection, gave almost no yield because of BPH burn and lodging. From the field on the right, 1,000 m^2 in area, planted with an aromatic unimproved variety (Sinantur) and having organic SRI management, 800 kg was harvested, a yield of 8 tons/ha. Picture given to the author by Ms. Miyatty Jannah, the farmer who managed the field on right.

the science of genetics. The disciplines of microbiology, soil ecology, and epigenetics will, foreseeably, become more prominent within the agricultural sciences, focusing in particular on understanding better how genetic potentials are expressed (epigenetics), appreciating that it is phenotypes and not genotypes that we consume as food.

Some implications for plant breeding

SRI practitioners have found that the greatest yields obtained with their methods have come from what are called high-yielding varieties (HYVs) or hybrids. So, agroecological approaches do not negate the basis for genetic research. The world-record yield reported above was with a hybrid variety with application of some inorganic fertilizer, although there was primary reliance on organic fertilization, that is, on integrated nutrient management [48]. It is also true, however, that many local or indigenous varieties also respond well to SRI management, with some giving yields over 10 tons/ha. (The world average is about 4 tons/ha.) Since the market price for local varieties is often higher because of consumer taste and other preferences, growing 'unimproved' cultivars with SRI practices can be more profitable than using HYVs.

The kind of plant breeding that Norman Borlaug and his colleagues undertook has been very successful and deserves our appreciation. But it may not be as suitable a strategy for the emerging conditions that will be faced in the decades ahead. Both breeding and management are essential for success, combining the effects of G and E; that is clear. But since the 1990s there have been diminishing returns from breeding strategies, whereas over the past decade we have been seeing from SRI experience that very great impacts on yield can be achieved by making changes in crop management, as apparently this has beneficial impacts on crops' relationships with their microbial symbionts [43,45].

One can find very little in the Green Revolution literature that examines the contributions of roots and soil biota to plants' health and growth. These are the factors that SRI experience and its extrapolation to other crops are pointing to as very potent. For decades, plant breeders have sought through genetic improvements to raise the Harvest Index (HI), that is, the proportion of plant biomass that is found in the edible portion. But the focus has all been on above-ground plant organs. The biomass in plant roots has not even been considered when calculating HI.

The abundant growth of roots has been largely ignored or even deprecated by plant breeders, much as profuse tillering has been disapproved. The prolific growth of roots and shoots was thought to be wasteful and in competition with - rather than supportive of - grain production. (I have been told, twice, by a senior agronomist in the US that 'roots are a waste', because they lower the Harvest Index). However, crops in this current century will need to be more naturally resistant to pest and disease assaults and to be buffered from drought, from rain and wind damage, and from temperature stresses. Root development and more abundant, more diverse soil biota are likely to become more appreciated for their promotion of crop resilience and for their enhancement of productivity in the face of climate-change pressures.

Future research opportunities and modes of operation

There is widespread agreement on the need for 'another Green Revolution', but the operational shape and direction of this effort remains to be determined. Agricultural scientists have begun to investigate and evaluate SRI methods only in the last few years, and these methods are rapidly spreading to other crops with similar effects. SRI ideas and methods were developed inductively by a Jesuit priest [65] based on observations and field experimentation, so there are surely many opportunities for scientists to make improvements and to determine limitations in this methodology.

The symbiotic relationships between crops and beneficial microorganisms appear to be a particularly fruitful area for research [43,61]. Studying rhizosphere biology can illuminate changes in underground dynamics that occur when plants are grown using SRI methods. For example, spatial and temporal variations in redox gradients can lead to qualitative and quantitative changes in microbial populations and activity [42,66]. Recent analysis has underscored the importance of optimizing redox potential which has its basis in soil biology as much as soil chemistry and soil physics [67].

Research at the Indian Agricultural Research Institute has shown that SRI methods of cultivation not only increased water productivity (with a 27% saving of irrigation water in these trials), but they also enhanced soil microbial activity, micronutrient uptake, and the N content in both soil and grain, compared to conventional transplanting methods [45,68]. Biofertilizers and microbial inoculants when used with SRI management may further contribute to the prolific growth of roots and facilitate nutrient mobilization and uptake in rice, with improvements in grain quality. Much research remains to be done.

Along with a refocusing of research in microbiological directions, it should be noted that SRI experience encourages agricultural scientists to work more closely with practitioners, both farmers and NGO personnel, who presently have more knowledge of this innovation than anyone else. Advances in the future are likely to come from observations, experience and thinking from many sources, not just from the formal agricultural disciplines.

Conclusions

There is no reason to think that SRI and other agroecological concepts and practices will replace all other kinds of agriculture; agricultural change does not proceed so discontinuously. Norman Borlaug reasonably insisted that agricultural development be *science-based*. This means that innovations should be examined and evaluated objectively and systematically, although this does not mean that all innovations must originate from the formal scientific community, or that knowledge and data coming from sources other than credentialed scientists should be excluded from consideration.

Competing interests

The author declares that he has no competing interests, having no financial stake in this subject, although he has been personally and professionally involved with trying to make improvements in food security internationally for more than 40 years. He has been directly involved in encouraging the evaluation and, where warranted, dissemination of SRI along with other agroecological innovations for the past 15 years.

Authors' information

Norman Uphoff has been on the faculty of Cornell University since 1970 as a professor of government and (since 1990) of international agriculture. From 1990 to 2005, he served as director of the Cornell International Institute for Food, Agriculture and Development (CIIFAD), and he is currently director of the Cornell Institute for Public Affairs (CIPA), while also senior advisor to the SRI International Network and Resources Center (SRI-Rice) based in CIIFAD.

Acknowledgements

This paper updates a presentation made at a forum organized by the Norman Borlaug Foundation at the 2010 BioVision Alexandria International Conference, held at the Alexandria Library in Egypt in April 2010. Inputs from Dr. Radha Prasanna, senior scientist in the Indian Agricultural Research Institute, New Delhi, are gratefully acknowledged.

References

1. Hayami Y, Ruttan VW: *Agricultural Development: An International Perspective.* Baltimore: Johns Hopkins University Press; 1985.
2. UN Environmental Program: *The Environmental Food Crisis: environment's Role in Averting Future Food Crises.* Nairobi; 2005. http://www.grida.no/publications/rr/food-crisis/.
3. Pimentel D: *Techniques for Reducing Pesticides: Environmental and Economic Benefits.* Chichester, UK: John Wiley; 1997.
4. Chaboussou F: *Healthy Crops: A New Agricultural Revolution.* Charnley, UK: Jon Anderson Press; 2004.
5. Pingali PL, Hossain M, Gerpacio RV: *Asian Rice Bowls: The Returning Crisis?* Wallingford, UK: CAB International; 1997.
6. Dyson T: **World food trends and prospects to 2025.** *Proc Natl Acad Sci (US)* 1999, **9**:5929–5936.
7. Uphoff N: **Agricultural futures: What lies beyond 'modern agriculture'?** *Trop Agric Assoc Newsletter* 2007, **27**:13–19. http://www.bioscienceresource.org/cms/documents/TAA%20%20Bunting%20Lecture.pdf.
8. Reichardt W, Dobermann A, George T: **Intensification of rice production systems: opportunities and limits.** In *Rice in the Global Food System.* Edited by Dowling NG, Greenfield SM, Fischer KS. Los Baños: International Rice Research Institute; 1998:127–144.

9. Uphoff N: **Opportunities to achieve resource-conserving increases in agricultural productivity: learning from the System of Rice Intensification (SRI).** In *New Life Sciences: Future Prospects – Proceedings of BioVision Alexandria 2010*. Edited by Serageldin I, Masood E, El-Faham M, El-Wakil M. Alexandria: Biblioteca Alexandrina; 2011. http://www.bibalex.org/BVA2012/attachments/staticpages/file/Publications/BVA2010_Web.pdf.
10. Stoop WA, Uphoff N, Kassam A: **A review of agricultural research issues raised by the system of rice intensification (SRI) from Madagascar: opportunities for improving farming systems for resource-poor farmers.** *Agric Syst* 2002, **71**:249–274.
11. Uphoff N, Kassam A: **Case study: System of Rice Intensification.** In *Agricultural Technologies for Developing Countries: Final Report, Annex 3*. Brussels: European Parliament; 2009.
12. Hameed KA, Mosa AKJ, Jaber FA: **Irrigation water reduction using System of Rice Intensification (SRI) compared with conventional methods in Iraq.** *Paddy Water Envir* 2011, **9**:121–127.
13. Lin XQ, Zhu DF, Chen HZ, Cheng SH, Uphoff N: **Effect of plant density and nitrogen fertilizer rates on grain yield and nitrogen uptake of hybrid rice (*Oryza sativa* L.).** *J Agric Biotech & Sust Dev* 2009, **1**:44–53.
14. Mishra A, Salokhe VM: **Seedling characteristics and early growth of transplanted rice under different water regimes.** *Exper Agric* 2008, **44**:1–19.
15. Thakur A, Uphoff N, Antony E: **An assessment of physiological effects of system of rice intensification (SRI) practices compared to recommended rice cultivation practices in India.** *Exper Agric* 2010, **46**:77–98.
16. Zhao LM, Wu LH, Li YS, Lu XH, Zhu DF, Uphoff N: **Influence of the System of Rice Intensification on rice yield and nitrogen and water use efficiency with different N application rates.** *Exper Agric* 2009, **45**:275–286.
17. Dobermann A: **A critical assessment of the system of rice intensification (SRI).** *Agric Syst* 2004, **79**:261–281.
18. Sheehy JL, Peng SB, Dobermann A, Mitchell PL, Ferrer A, Yang JC, Zou YB, Zhong XH, Huang JL: **Fantastic yields in the system of rice intensification: fact or fallacy?** *Field Crops Res* 2004, **88**:1–8.
19. Sinclair TR, Cassman KG: **Agronomic UFOs.** *Field Crops Res* 2004, **88**:9–10.
20. Stoop WA, Kassam AH: **The SRI controversy: a response.** *Field Crops Res* 2005, **91**:357–360.
21. Uphoff N: **Comment to 'The System of Rice Intensification: time for an empirical turn' [NJAS: Wageningen Journal of Life Sciences 57 (2011) 217–224].** *NJAS: Wageningen J Life Sci* 2012, **59**:53–60.
22. Castillo GE, Minh NL, Pfeifer K: **Oxfam America: learning from the System of Rice Intensification in Northern Vietnam.** In *Scaling Up in Agriculture, Rural Development and Nutrition*. Edited by. Washington DC: International Food Policy Research Institute; 2012. Brief No. 15.
23. Thomas V, Ramzi MA: **SRI contributions to rice production dealing with water management constraints in northeastern Afghanistan.** *Paddy Water Envir* 2011, **9**:101–109.
24. Husain AM, Chowhan G, Barua P, Razib Uddin AFM, Ziaur Rahman ABM: *Final Evaluation Report on Verification and Refinement of the System of Rice Intensification (SRI) Project in Selected Areas of Bangladesh (SP 36 02)*. Dhaka, Bangladesh: Report to IRRI/Bangladesh Programme from BRAC, POSD, SAFE and Syngenta/Bangladesh Ltd; 2004.
25. Anthofer J: *The potential of the System of Rice Intensification (SRI) for poverty reduction in Cambodia*. Berlin: Phnom Penh: Summary of report prepared for GTZ; 2004. http://www.tropentag.de/2004/abstracts/full/399.pdf.
26. Tech C: *Ecological System of Rice Intensification (SRI) Impact Assessment (2001–2003)*. Phnom Penh: Report of the Cambodian Center for Study and Development in Agriculture (CEDAC); 2004.
27. Xu XL, Li XY, Li H: ***Socioeconomic impact analysis of SRI in China.*** *China Rural Economics (in Chinese)*; 2006. English version available: http://scholar.google.com/scholar_url?hl=en&q=http://ciifad.cornell.edu/SRI/countries/china/cnciadeng.pdf&sa=X&scisig=AAGBfm1lE2WtonKhHMxx_5aPAnSAuB5QXA&oi=scholarr.
28. Zheng JG, Zhou L, Chi ZZ, Jiang XL: *Agricultural water savings possible through SRI for future water management in Sichuan, China*. Chengdu: paper for Crop Research Institute, Sichuan Academy of Agricultural Sciences; 2011. http://www.google.com/#hl=en&sclient=psy-ab&q=zheng+Zhou+Chi+Jiang+Agricultural+Water+Savings+SRI&oq=zheng+Zhou+Chi+Jiang+Agricultural+Water+Savings+SRI&gs_l=hp.3...126.23924.0.24415.83.66.16.1.2.0.302.7794.25j35j5j1.66.0.les%3B..1.0...1c.CDwSwJjZcg4&pbx=1&bav=on.2,or.r_gc.r_pw.r_qf.&fp=7e9041a74f73c7f&biw=1179&bih=600. Data from Sichuan Provincial Department of Agriculture.
29. Sinha SK, Talati J: **Productivity impacts of the system of rice intensification (SRI): A case study in West Bengal, India.** *Agric Water Manage* 2007, **87**:55–60.
30. Satyanaryana A, Thiyagarajan TM, Uphoff N: **Opportunities for water saving with higher yield from the system of rice intensification.** *Irrig Sci* 2007, **25**:99–115.
31. Sato S, Uphoff N: **A review of on-farm evaluations of system of rice intensification methods in Eastern Indonesia.** *CAB Reviews: Perspectives in Agriculture, Veterinary Science, Nutrition and Natural Resources* 2007, **2**:54.
32. Ndiri JA, Mati BM, Home PG, Odongo B, Uphoff N: **Comparison of water savings of paddy rice under System of Rice Intensification (SRI) growing rice in Mwea, Kenya.** *Intl J Curr Res Rev* 2012, **4**:63–73.
33. Nyamai M, Mati BM, Home PG, Odongo B, Wanjugo R, Thuranira EG: **Improving crop productivity and water use efficiency in basin rice cultivation in Kenya through SRI.** *Agric Engin Intl: CIGR Journal* 2012, **14**:1–9.
34. Styger E, Attaher MA, Guindo H, Ibrahim H, Diaty M, Abba I, Traore M: **Application of system of rice intensification practices in the arid environment of the Timbuktu region in Mali.** *PaddyWater Envir* 2011, **9**:137–144.
35. Kabir H, Uphoff N: **Results of disseminating the System of Rice Intensification with Farmer Field School methods in northern Myanmar.** *Exp Agric* 2007, **43**:463–476.
36. Uprety R: *Economic analysis of System of Rice Intensification (SRI) methods in Morang district of Nepal, main season, 2005*. Biratnagar, Nepal: Report from the District Agricultural Development Office, Ministry of Agriculture; 2006.
37. FAO-EU: *FAO-EU Food Facility Programme: after One and a Half Years in Nepal*. Lalitpur, Nepal: FAO; 2010.
38. Khadka RB: *Project Completion Report of EU Food Facility Project*. Dhangadhi, Kailali, Nepal: FAYA-Nepal; 2011.
39. Turmel MS, Espinosa J, Franco L, Pérez C, Hernández H, González E, Fernández G, Rojas C, Sánchez D, Fernández N, Barrios M, Whalen JK, Turner BL: **On-farm evaluation of a low-input rice production system in Panama.** *Paddy Water Envir* 2011, **9**:155–161.
40. Namara R, Bossio D, Weligamage P, Herath I: **The practice and effects of the system of rice intensification (SRI) in Sri Lanka.** *Qtly J Intl Agric* 2008, **47**:5–23.
41. Dung NT, Minh LN: *System of Rice Intensification - Advancing Small Farmers in Mekong Region*. Oxfam America: Report prepared by the Plant Protection Department (PPD)/Ministry of Agriculture and Rural Development (MARD) and East Asian Regional Office; 2008.
42. Uphoff N, Anas I, Rupela OP, Thakur AK, Thiyagarajan TM: **Learning about positive plant-microbial interactions from the System of Rice Intensification (SRI).** *Aspects Appl Biol* 2009, **98**:29–54.
43. Anas I, Rupela OP, Thiyagarajan TM, Uphoff N: **A review of studies on SRI effects on beneficial organisms in rice soil rhizospheres.** *Paddy Water Envir* 2011, **9**:53–64.
44. Lin XQ, Zhu DF, Lin XJ: **Effects of water management and organic fertilization with SRI crop management practices on hybrid rice performance and rhizosphere dynamics.** *Paddy Water Envir* 2011, **9**:33–39.
45. Priyanka S, Singh YV, Prasanna R, Shivay YS: ***Influence of methods of crop establishment and rice varieties on nutrient uptake and soil properties.*** New Delhi: Research report for Indian Agricultural Research Institute; under review.
46. Prasad A: *Going against the grain: The system of rice intensification is now being adapted to wheat – with similar good results. Outlook Business*, New Delhi; 2008.
47. ICRISAT-WWF: *Sustainable Sugarcane Initiative: improving Sugarcane Cultivation in India – Training Manual*. Patancheru, Hyderabad, India: ICRISAT-WWF Dialog Project on Food, Water and Environment, Intl Crop Research Inst for Semi-Arid Tropics; 2008.
48. Diwakar MC, Kumar A, Verma A, Uphoff N: ***Report on the world-record SRI yield in kharif season 2011. Nalanda district, Bihar state, India.*** *Agric Today*, New Delhi; 2012:53–56.
49. Uphoff N: **Agroecological implications of the System of Rice Intensification (SRI) in Madagascar.** *Envir Dev & Sust* 1999, **1**:297–313.
50. Cook G, O'Connor T: *Rice aplenty in Aceh. Caritas*: Spring; 2009. 10–11. www.caritas-europe.org/module/FileLib/RiceaplentyinAceh.pdf.
51. Singh MP: **Organic rice cultivation transforming lives of Damoh farmers.** *The Hindu*, Nov. 28, 2011. http://www.thehindu.com/sci-tech/agriculture/

article2668286.ece.

52. Lyman J, Lyman J, Rasmei S, Than YK, Chantea L: *Rice production in the Family Food Production project.* Phnom Penh: Report of LDS Charities; 2007. Posted on SRI website: http://sri.ciifad.cornell.edu/countries/cambodia/camldsrpt07.pdf.
53. McDonald AJ, Hobbs PR, Riha SJ: **Does the System of Rice Intensification outperform conventional best management? A synopsis of the empirical record.** *Field Crops Res* 2006, **96:**31–36.
54. Uphoff N, Stoop WA, Kassam AH: **A critical assessment of a desk study comparing crop production systems: the example of the 'system of rice intensification' vs 'best management practice'.** *Field Crops Res* 2007, **108:**109–114.
55. Turnbaugh PJ, Ley RE, Mahady M, Fraser-Liggett CM, Knight R, Gordon JI: **The Human Microbiome Project.** *Nature* 2007, **449:**804–810.
56. Pinton R, Varanini Z, Nannipieri P: *The Rhizosphere: Biochemistry and Organic Substances at the Soil-Plant Interface.* 2nd edition. Boca Raton, FL: CRC Press; 2007.
57. Mukerji KG, Manoharachary C, Singh J: *Microbial Activity in the Rhizosphere.* Berlin: Springer; 2006.
58. Chi F, Shen SH, Chang HP, Jing YX, Yanni YG, Dazzo FB: **Ascending migration of endophytic rhizobia, from roots to leaves, inside rice plants and assessment of benefits to rice growth physiology.** *Appl Environ Microbiol* 2005, **71:**7271–7278.
59. Chi F, Yang PF, Han F, Jing YX, Shen SH: **Proteomic analysis of rice seedlings infected by *Sinorhizobium meliloti* 1021.** *Proteomics* 2010, **10:**1861–1874.
60. Mattoo AK, Abdel-Baki A: **Crop genetic responses to management: evidence of root-shoot communication.** In *Biological Approaches to Sustainable Soil Systems.* Edited by Uphoff N, Ball A, Fernandes ECM, Herren H, Husson O, Laing M, Palm CA, Pretty JN, Sanchez PA, Sanginga N, Thies JE. Boca Raton, FL: CRC Press; 2006:221–230.
61. Rodriguez RJ, Freeman DC, McArthur ED, Kim YO, Redman RS: **Symbiotic regulation of plant growth, development and reproduction.** *Communic & Integr Biol* 2009, **2:**1–3.
62. Uphoff N, Chi F, Dazzo FB, Rodriguez RJ: **Soil fertility as a contingent rather than inherent characteristic: Considering the contributions of crop-symbiotic soil biota.** In *Principles of Sustainable Soil Management in Agroecosystems.* Edited by Lal R, Stewart B. Boca Raton, FL: Taylor & Francis; 2012. in press.
63. Khan Z, Hassanali A, Pickett J: **Managing polycropping to enhance soil system productivity: a case study from Africa.** In *Biological Approaches to Sustainable Soil Systems.* Edited by Uphoff N, Ball A, Fernandes ECM, Herren H, Husson O, Laing M, Palm CA, Pretty JN, Sanchez PA, Sanginga N, Thies JE. Boca Raton, FL: CRC Press; 2006:575–586.
64. Uphoff N, Ball A, Fernandes ECM, Herren H, Husson O, Laing M, Palm CA, Pretty JN, Sanchez PA, Sanginga N, Thies JE: *Biological Approaches to Sustainable Soil Systems.* Boca Raton FL: CRC Press; 2006.
65. Laulanié H: **Le système de riziculture intensive Malgache.** *Tropicultura* 1993, **11:**110–114. reprinted in 2011, 29:3, 183–187, as English translation.
66. Mishra A, Uphoff N: **Morphological and physiological responses of rice roots and shoots to varying water regimes and soil microbial densities.** *Arch Agron Soil Sci* 2012, **58:**1–27.
67. Husson O: **Redox potential (Eh) and pH as drivers of plant/soil/microorganism systems: A transdisciplinary overview pointing to integrative opportunities in agronomy.** *Plant Soil*, http://rd.springer.com/article/10.1007/s11104-012-1429-7.
68. Priyanka S, Singh YV, Prasanna R, Bhatia A, Shivay YS: **Pattern of methane emission and water productivity under different methods of rice crop establishment.** *Paddy Water Envir* 2011, **9:**1–12.

14

The role for scientists in tackling food insecurity and climate change

John R Beddington[1], Mohammed Asaduzzaman[2], Megan E Clark[3], Adrian Fernández Bremauntz[4], Marion D Guillou[5], Molly M Jahn[6], Erda Lin[7], Tekalign Mamo[8], Christine Negra[9*], Carlos A Nobre[10], Robert J Scholes[11], Rita Sharma[12], Nguyen Van Bo[13] and Judi Wakhungu[14]

Abstract

To adapt to climate change and ensure food security, major interventions are required to transform current patterns and practices of food production, distribution and consumption. The scientific community has an essential role to play in informing concurrent, strategic investments to establish climate-resilient agricultural production systems, minimize greenhouse gas emissions, make efficient use of resources, develop low-waste supply chains, ensure adequate nutrition, encourage healthy eating choices and develop a global knowledge system for sustainability. This paper outlines scientific contributions that will be essential to the seven policy recommendations for achieving food security in the context of climate change put forward by the Commission on Sustainable Agriculture and Climate Change. These include improved understanding of agriculture's vulnerability to climate change, food price dynamics, food waste and consumption patterns and monitoring technologies as well as multidisciplinary investigation of regionally appropriate responses to climate change and food security challenges.

Challenges to food security

The world faces multiple challenges to food security including undernutrition and overconsumption, rising food prices, population growth, rapid diet transitions, threats to agricultural production, inefficient production practices and supply chains, and declining investment in food system research. In addition to causing widespread human suffering, food insecurity contributes to degradation and depletion of natural resources, migration to urban areas and across borders, and political and economic instability.

Food insecurity afflicts communities throughout the world, wherever poverty inhibits purchasing power and prevents assured access to food supplies. Global food prices have risen dramatically in the last few years and are forecast to rise further and become more volatile [1,2], disrupting assumptions that stable or declining food prices and assured supplies can be taken for granted. The estimated number of hungry people in the world rose from 800 million to over 1 billion following the 2007/08 food price spike [3][a]. It is estimated that an additional 44 million people have since fallen into extreme poverty due to the rise in food prices since June 2010 [4]. Globally, diets are shifting towards higher consumption of calories, fats and animal products [5]. A growing number of low-income and middle-income countries are facing a double burden of malnutrition: a persistence of both macronutrient and micronutrient undernutrition, notably among children, along with a quick rise in the number of overweight and obese people, and diet-related chronic diseases [6,7].

The food system faces additional pressure as the global population grows to around 9 billion by 2050 [8]. This dramatic increase in global population will be accompanied by major shifts in the regional distribution of our planet's inhabitants. From 2010 to 2050, the population in Asia is estimated to grow from 4.2 billion to 5.1 billion and Africa's population to grow from 1 billion to 2.2 billion [8]. From 1950 to 2050, the population ratio for developing countries to developed countries is projected to shift from 2:1 to 6:1 [8]. As the world population has grown, the land available per capita has shrunk from 13.5 ha/person in 1950 to 3.2 ha/person in 2005, and is projected to diminish to 1.5 ha/person in 2050 [9].

* Correspondence: christine@ideapaths.org
[9]Commission on Sustainable Agriculture and Climate Change, New York, NY, USA

Agriculture continues to be the economic mainstay of most low-income countries, employing the majority of the population in these countries. The importance of agricultural research and development for food security and poverty reduction has been recognized[b], yet recent decades have seen declining investment from both donor partners and low-income country governments [1,6]. In low-income countries with agriculture-based economies, domestic public support to agriculture is, on average, about 4% of the gross domestic product, and official development assistance provides the balance [6,10]. At the beginning of the 21st century, only 6% of total spending on agriculture research and development in low-income countries came from private companies [1].

In the coming decades, climate change and extreme weather events will exacerbate the fragility of food production systems and the natural resource base, especially in places affected by soil degradation, water stress or desertification [11]. While the overall effect on agriculture will vary among geographic regions, it will be harder for farmers to plan and manage production and prevent crop losses from storms or pests wherever planting seasons and weather patterns change. Already heightened by globalization, a warming climate is likely to increase the incidence and geographic spread of human, animal and plant diseases [1,12,13].

While no legally binding agreement was reached at the United Nations Framework Convention on Climate Change (UNFCCC) 15th Conference of the Parties in 2009, the Copenhagen Accord provided a commitment to hold the increase in global temperatures below 2°C. Over 70 countries submitted emissions reduction targets under the Copenhagen Accord, and more formal UNFCCC agreement on the 2°C limit was achieved at the 16th Conference of the Parties in Cancun in 2010. The 17th Conference of the Parties produced the Durban Platform for Enhanced Action, which commits parties to reach a legal framework for reducing global emissions by 2015 [14]. Despite the agreed 2°C target, greenhouse gas emissions are still rising. Even a 2°C rise is predicted to be problematic owing to increased floods and storms, a shortage of water resources, impacts on food production at low latitudes, greater depth of seasonal permafrost thaw and other changes. Yet greater change in global temperature would be disastrous. A 4°C change in average global temperature (estimated by the Intergovernmental Panel on Climate Change's A1F1 scenario for the end of this century [15]) is predicted to bring about a much greater temperature increase in the Arctic, a substantial impact on major crops in all regions, around 1 billion additional people experiencing water scarcity by 2080, extensive coastal flooding as sea levels rise and other negative impacts [16] (Figure 1).

Agriculture consumes 70% of total global 'blue water' withdrawals from available rivers and aquifers, and will increasingly compete for water with pressures from industry, domestic use and the need to maintain environmental flows [12]. Current farming practices, including land clearing and inefficient use of fertilizers and organic residues, make agriculture a significant contributor to greenhouse gas emissions [15,17]. From the farm gate to consumers, refrigeration and other supply-chain activities are an additional major source of greenhouse gas emissions. As global demand for food, fodder and bioenergy crops grows, many agricultural systems are depleting soil fertility, biodiversity and water resources. In many regions there are large gaps between potential and actual crop yields [18]. Every year, an estimated 12 million hectares of agricultural land, which could potentially produce 20 million tonnes of grain, are lost to land degradation, adding to the billions of hectares that are already degraded [19,20]. Estimates indicate that one-third of food produced for human consumption is lost or wasted across the global food system [21].

This paper reviews the critical contributions required from the scientific community in order to foster integrated, decisive policy action for addressing the interconnected challenges of food insecurity and climate change.

Features of a food-secure future

As a global community, we need to make concurrent, strategic investments to establish climate-resilient agricultural production systems, make efficient use of resources, develop low-waste supply chains, ensure adequate nutrition and encourage healthy eating choices [22]. This will require significant capacity for coordinated action in policy-making, private business and civil society and will not be possible without scientific and technological innovation.

New technology and practices

Farmers will need to produce significantly more food on less land, with less water, using less energy, fertilizer and pesticide without further encroaching on sensitive ecosystems [23]. Intensification of food production must be accompanied by concerted action to reduce greenhouse gas emissions from agriculture to avoid further acceleration of climate change and to avert threats to the long-term viability of global agriculture. Climate-smart agriculture must enhance and secure the livelihoods of rural farmers.

There is a large potential for reducing net food system emissions through efficiency measures in production as well as through demand management such as reduction of loss and waste in supply chains and changing food preferences [12,24]. For example, in Kerala, India, anaerobic digestion technology has been used to recycle domestic organic waste, including food waste, to produce

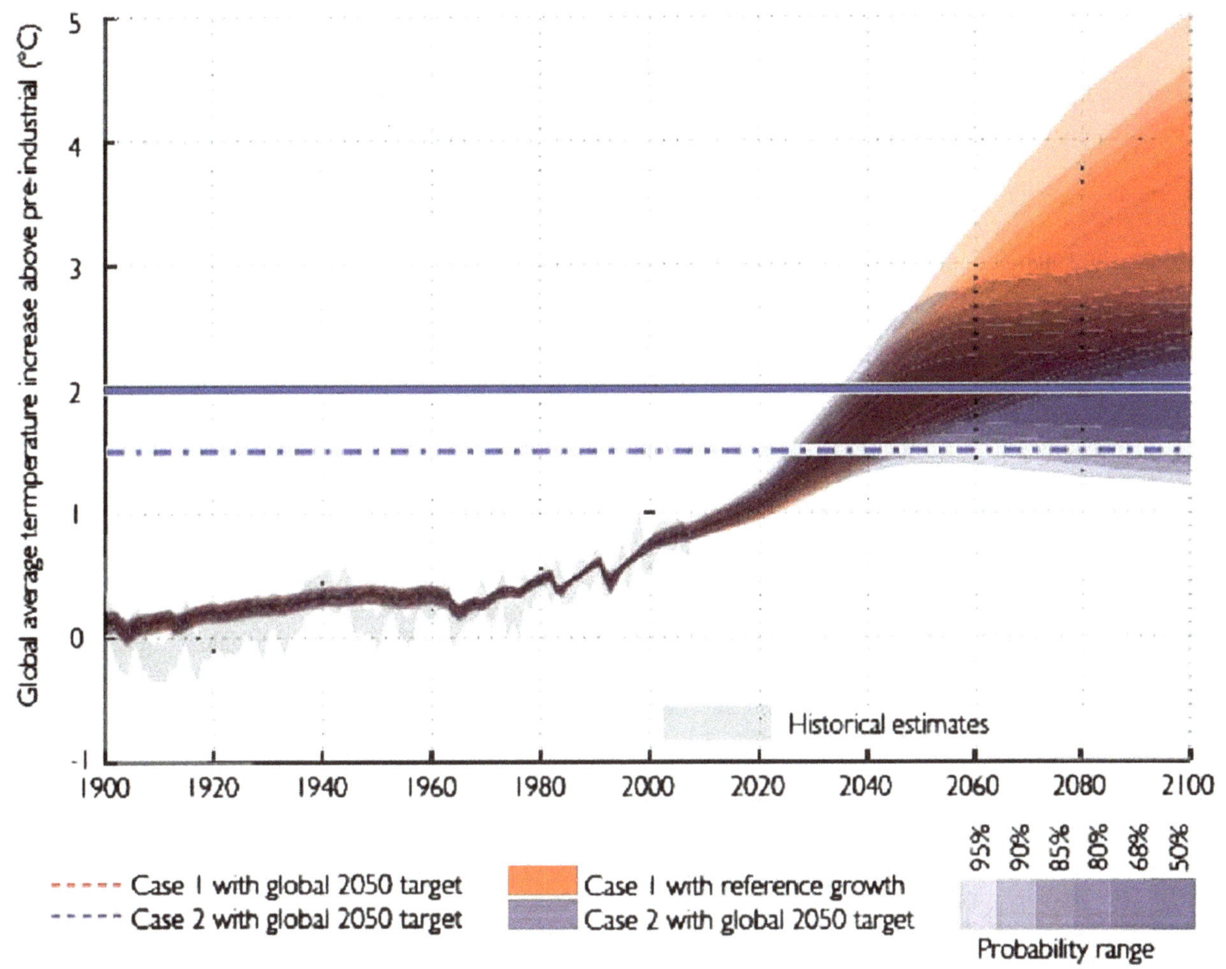

Figure 1 Effects of the Copenhagen Accord on global average temperature through the 21 stcentury [25].

biogas for cooking and electricity [26]. There is also meaningful potential for greenhouse gas sinks associated with a number of agricultural practices. Some of these practices, such as improved land management, have co-benefits for both the reliability of food production and the quality of the environment [27]. For example, intensive rice production techniques have come into use across several regions of Viet Nam and are associated with higher yields as well as reduced use of nitrogen fertilizers and lower nitrous oxide emissions [28]. Market demand for organic and eco-certified products, consumer expectations for social and environmental corporate responsibility and longer-term concerns about sources of supply have contributed to greater attention to sustainability by some agribusinesses [29].

Integrated global knowledge systems

The multiple threats to food security are interconnected, and multi-scale and robust knowledge systems are critical given our increasingly globalized food system [1]. Adaptive management and governance to improve nutritional security, economic prosperity and environmental outcomes will require a much better global system for integrating information about agriculture, ecosystem services, markets and human populations in real time. (The need for shared information in order to address global problems was recognized at the World Summit for Sustainable Development, leading to the formation of the Group on Earth Observations and the development of the Global Earth Observation System of Systems [30]).

Existing and future investments in information and knowledge must be structured to identify limits, inform tradeoffs and deliver practical guidance for a sustainable future, not simply to maximize single components of the food system. Mechanisms should include remote-sensing and ground-based monitoring systems and spatially explicit support systems that integrate biophysical and socioeconomic information. Such an information system will give us a richer understanding of the dynamic systems we depend on and will enable us to renew and broaden our efforts to secure a more sustainable and healthy food system for our own and future generations. It will also enable measurement of progress toward target indicators such as those identified in Bangladesh's 5-year Country Investment Plan for improved tenure of land and water resources, access to financing, private-sector involvement and empowerment of women [31].

A safe operating space

As a global community, we need to navigate toward a safe operating space (see Figure 2) that provides adequate food and nutrition for everyone without crossing critical environmental thresholds. At present we operate outside that safe space, as witnessed by the enormous number of people who are undernourished. If current trends in population growth, diets, crop yields and climate change continue, the world will still be outside this safe operating space in 2050. The situation then will be unsustainable and there will be very little room to maneuver.

Plotting a course towards a safe operating space will require innovative technologies, institutions and policies, and will severely test our social, technological and agricultural ingenuity. There are various changes we can make to either enlarge the safe space or move ourselves into the safe space. First, the global demand for food will increase with population growth but the amount of food per person that needs to be produced can be brought down by eliminating waste in supply chains, ensuring more equitable access to food and moving to more resource-efficient (and healthier) vegetable-rich diets. Secondly, given its large land base, global agriculture represents a major opportunity for mitigating climate change and helping to moderate its overall negative effect on agricultural productivity [27] through a wide range of regionally suitable practices that increase the efficiency of carbon and nitrogen management (for example, livestock feeding regimes that reduce methane emissions) or sequester carbon (for example, agroforestry) [32]. Finally, agricultural innovation, including better management of soil, water and other resources and careful matching of crops to environments, can help adapt food systems to climate change. For example, under China's Plan for the Construction of Protective Cultivation Projects, 1.6 million severely degraded hectares of grassland have been rehabilitated [33] with improvements in soil structure and enhanced carbon storage [34].

Seven areas for action by the scientific community

The transition to a global food system that satisfies human needs, reduces its carbon footprint, adapts to climate change and is in balance with planetary resources requires concrete and coordinated actions, implemented at scale, simultaneously and with urgency. In February 2011 the Commission on Sustainable Agriculture and Climate Change was convened to identify critical leverage points and practical policy actions to be undertaken by key stakeholders and institutions in pursuit of food security in the context of climate change. Drawing on a review of recent major assessment reports, expert consultation and their own knowledge, the Commissioners proposed seven areas for policy action to achieve food security in the face of climate change [22]. For each of these seven recommended policy actions, we now identify relevant contributions needed from the scientific community.

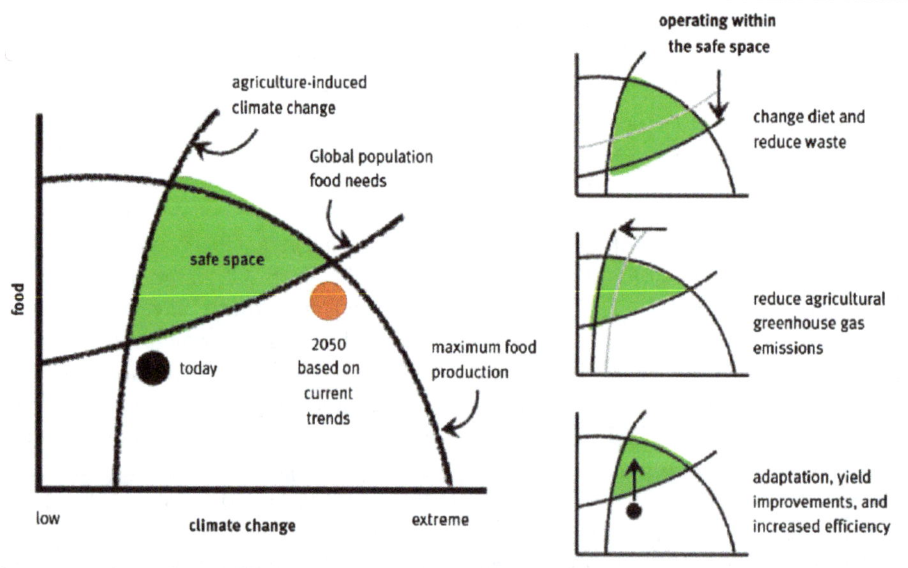

Figure 2 A safe operating space for interconnected food and climate systems [22].

1. Integrate food security and sustainable agriculture into global and national policies

As a first step to inclusion of agriculture in the mainstream of international climate change policy, negotiators should establish a work program on mitigation and adaptation in agriculture under the UNFCCC. Similarly, country representatives to global policy processes should integrate sustainable, climate-friendly agriculture into 'early action' climate finance schemes. To enable coherent dialogue and policy action related to climate change, agriculture, crisis response and food security, at global, regional and national levels, governments and global donors should develop common platforms at global, regional and national levels.

The global food system is managed through a complex mix of public and private-sector action, across local to global scales. Collectively, the policy choices within national governments, United Nations bodies, global treaties and conventions, regional economic communities, political forums (for example, G8, G20) and standard-setting bodies shape the way food is produced, distributed and consumed. The scientific evidence base is an essential foundation for public policies and programs as well as for systems of market and industry governance and of civil society influence and agenda setting.

Global climate change policy is a critical arena for solidifying international support for sustainable agriculture development programs that adapt to and mitigate against climate change. National climate change action plans can also usefully integrate the agriculture sector in country-specific ways. Without a global commitment to reducing greenhouse gas emissions from all sectors, including agriculture, no amount of agricultural adaptation will be sufficient under the destabilized climate of the future [12].

The scientific community can support evidence-based policy-making by quantifying vulnerability of agriculture to climate change and forecasting outcomes under a broad range of potential mechanisms for agricultural adaptation and mitigation. By working across disciplinary boundaries, researchers can develop a pragmatic, multi-disciplinary understanding of what it means to reduce poverty and food insecurity within the context of the planet's boundaries. Scientists can help to mobilize increased investment by detailing how multiple benefits can be achieved through sustainable farming practices and by clarifying geographic and sectoral potential for greenhouse gas mitigation.

2. Significantly raise the level of global investment in sustainable agriculture and food systems in the next decade

Donor governments should implement and strengthen the G8 L'Aquila commitments to sustainable agriculture and food security and enable UNFCCC Fast Start funding, major development banks and other global finance mechanisms to prioritize sustainable agriculture programs that improve infrastructure and rehabilitate land. To reflect the significance of sustainable agriculture in economic growth, poverty reduction and long-term environmental sustainability, governments should increase national research and development budgets, build integrated scientific capacity and support revitalized extension services, technology transfer and communities of practice to increase knowledge of best practices and access to innovation.

By demonstrating the outcomes of alternative farming practices in different regions, farming systems and landscapes and by clarifying the conditions under which local agricultural production systems integrate innovative technologies or approaches, researchers can help to effectively direct investments in agriculture [35]. For example, in the Cerrado region of Brazil, public-sector investment in agricultural research combined with producer innovation has been credited with dramatic gains in productivity and livelihoods despite low natural soil fertility [36].

3. Sustainably intensify agricultural production while reducing greenhouse gas emissions and other negative environmental impacts of agriculture

To enable more productive and resilient livelihoods and ecosystems, with emphasis on closing yield gaps and improving nutrition, multi-benefit farming systems should be developed and rewarded. This includes introducing strategies for minimizing ecosystem degradation and rehabilitating degraded environments, with emphasis on community-designed programs. To empower marginalized food producers and increase crop productivity, improvements are needed in land and water rights, access to markets, finance and insurance, and local capacity [37]. Subsidies that provide incentives for farmers to deplete water supplies or destroy native ecosystems should be modified [1]. To prevent further loss of forests, wetlands and grasslands, the economic incentives for sustainable intensification of agriculture should be coupled with stronger governance of land tenure and land zoning [38].

There is great variety in the pattern of agricultural productivity and land use in different regions. For example, cereal yields in Asia in 2001 were 240% higher than they were in 1961 with minimal change in land use (that is, increased production per unit land area), while in the same period in sub-Saharan Africa land use increased by 80% with only moderate increase in cereal yields [39]. Strategic investments can make an important difference. The agricultural potential in Africa is substantial and existing technologies can be used to create the necessary transformations in increasing productivity.

Through international, regional, national and local collaborations, researchers have a critical role to play in defining the practical meaning of sustainable intensification and elucidating forms of low-emissions agriculture that support long-term productivity and resilience (that is, decoupling increase in yield from emissions). There is a wide array of opportunities to investigate the suitability of sustainable agricultural practices (for example, diversified rotations, agro-ecological processes, improved nutrient and water-use efficiency, agroforestry, minimum tillage) in different regions and farming systems. To boost productivity while reducing greenhouse gas emissions, greater global coordination on research and implementation is needed [1]. Some promising areas include improved breeding and input for crops, livestock and aquatic organisms, diversification of agricultural systems (for example, agroforestry), soil management to sequester carbon and resource-efficient practices for crop production. To promote public trust and inform debate on new advances, scientists must become adept at articulating the benefits and dangers of new technologies in an open and transparent way.

4. Develop specific programs and policies to assist populations and sectors that are most vulnerable to climate changes and food insecurity

To provide rapid relief when extreme weather events affect communities, funds that respond to climate shocks should be developed (for example, index-linked funds) [40]. To moderate excessive food price fluctuations by promoting open and responsive trade systems, country information on production forecasts and stocks should be shared, and early warning systems should be established [41]. Safety nets and other programs to help vulnerable populations become food secure can include cash and in-kind transfers, employment guarantee schemes and education. Humanitarian responses to vulnerable populations threatened by food crises should be rapidly delivered through robust emergency food reserves. Global donor programs, policies and activities should be harmonized, paying particular attention to systematically integrating climate change risk management, adaptation and mitigation co-benefits, and improved local nutritional outcomes [38].

Key areas for multidisciplinary research include clarifying how index-linked funds can best reduce impacts on climate-affected populations (that is, increased hunger and poverty, lost productivity), investigating the criteria and optimal design for effective food reserves and understanding the drivers of food crises to improve targeting of fiscal responses. Research initiatives may be directed toward local-level strategies for risk management, preparedness, institutional capacity-building and household and community food systems.

5. Reshape food access and consumption patterns to ensure basic nutritional needs are met and to foster healthy and sustainable eating patterns worldwide

Chronic undernutrition and hunger should be addressed by harmonizing development policy and coordinating regional programs to improve livelihoods and access to services among food-insecure rural and urban communities. Positive changes in the variety and quantity of diets should be promoted through innovative education campaigns and through economic incentives that align the marketing practices of retailers and processors with public health and environmental goals [12]. A coherent set of evidence-based sustainability metrics and standards should be developed to monitor and evaluate food security, nutrition, health, agricultural productivity and efficiency, resource use and environmental impacts, and food system costs and benefits.

The research community can deliver better knowledge about the variety of food combinations that can deliver a nutritionally appropriate and environmentally low-impact diet. To improve overall food supply, scientists should investigate opportunities to improve agricultural productivity and resilience to climate change through effective deployment of existing and new technologies for producing, processing and distributing food. Research is needed to understand the impact and cost-effectiveness of a range of interventions on dietary behavior among different socioeconomic groups [42]. The toolbox for promoting sustainable diets includes economic interventions (for example, taxation of specific food types), retailers' purchasing guidelines (for example, to restrict consumer choices), public education campaigns (for example, advertising and programs in schools and workplaces) and labeling [12].

6. Reduce loss and waste in food systems, targeting infrastructure, farming practices, processing, distribution and household habits

In all sustainable agriculture development programs, research and investment components focusing on reducing waste, from production to consumption, by improving harvest and postharvest management and food storage and transport should be included. Integrated policies and programs should be developed to reduce waste in food supply chains (for example, economic innovation to enable low-income producers to store food during periods of excess supply). Dialog and working partnerships across food supply chains (producers, processors, retailers, consumers, regulators and researchers) can help to ensure that interventions to reduce waste are effective and efficient (for example, redirecting food waste to other purposes), and do not create perverse incentives.

Research and innovation will be needed to improve understanding of the causes of food loss and waste and support experimentation with reduction strategies [21]. This should include development of effective technological advancements in production, harvesting, and postharvest handling systems, drawing on expertise across plant biology, engineering, agricultural economics, food processing, nutrition, food safety and environmental conservation. Agencies and organizations that fund food systems research should prioritize work on optimizing yield, nutritional quality and postharvest life as well as characterizing the sociological dimensions of food consumption in different cultural and economic settings, including home food management, which is important for designing effective education campaigns [43]. There is a range of opportunities for reducing consumer and food service sector waste in middle-income and high-income countries using public campaigns, advertising, taxes, regulation, purchasing guidelines and improved labeling [1,12]. Raising awareness of food waste and promoting the use of efficiency strategies among food businesses, retailers and consumers will probably need to be targeted at specific economic and cultural characteristics [21].

7. Create comprehensive, shared, integrated information systems that encompass human and ecological dimensions

Increased, sustained investment in regular monitoring, on the ground and by public-domain remote-sensing networks, is essential to track changes in land use, food production, climate, the environment, human health and well-being worldwide. Spatially explicit data and decision-support systems that integrate biophysical and socioeconomic information and that enable policy-makers to navigate tradeoffs among agricultural intensification, nutritional security and environmental consequences should be developed, validated and implemented. To address food price volatility, improved transparency and access to information in global food markets as well as investment in interlinked information systems are needed [44].

The threats posed by climate change to food supplies and livelihoods are likely to be spatially variable. We will need to identify global hotspots where the threats are greatest and to develop specific, practical interventions to boost resilience in these areas. We also need a more robust understanding of our dynamic and increasingly globalized food system if we are to make headway on moderating food price volatility and increasing overall efficiency of the food system. From 1961 to 2003, world food trade increased from 1,500 Gkcal/day to >7,000 Gkcal/day [24]. There is growing integration of global supply chains and the emergence of large economies like Brazil, China and India as major sources of both demand and supply of agricultural products. In many low-income countries, rural and urban areas are ever more interconnected [38] although imperfect connectivity between global and domestic markets inhibits price transmission across global, national and local markets [45].

Scientists are integral to the development of a global system of repeated observations of ecological and human systems with key roles in advancing technical capabilities for monitoring and streamlining remote-sensing data to user communities. Working with governments, researchers should engage stakeholders to design and create novel frameworks that assimilate existing information assets (for example, farmer knowledge, spatial data) and incorporate them into decision-making pathways. Multidisciplinary research effort is needed to characterize the interactive drivers of food price spikes and the effectiveness of possible interventions.

Research activity is needed in a diverse set of areas to improve understanding of agriculture's vulnerability to climate change, food price dynamics, food waste and consumption patterns and monitoring technologies as well as multidisciplinary investigation of regionally appropriate responses to climate change and food security challenges. Making these changes, although technically feasible, requires urgent, collective and substantially increased action internationally, nationally and locally.

Conclusions

The growing threat of global climate change greatly amplifies the urgent need for food systems to shift to better meet human needs and align with planetary resources. This will demand major interventions, at local to global scales, to transform current patterns of food production, distribution and consumption. Investment, innovation and a deliberate effort to empower the world's most vulnerable populations will be required to construct a global food system that adapts to climate change and ensures food security while minimizing greenhouse gas emissions and sustaining our natural resource base. Greatly expanded investments in sustainable agriculture, including improving the supporting infrastructure and restoring ecosystems, are an essential component of long-term economic development. The sooner these investments are made, the greater the benefits will be.

The scientific community has an essential role to play in meeting the global challenge of moving the world into a safe operating space in which agriculture can meet global food needs while reducing its greenhouse gas emissions. Given the already intolerable conditions for many livelihoods and ecosystems, and the time lag between research and development and widespread application, we need to take urgent action.

Endnotes

[a]Note that future Food and Agriculture Organization estimates may be revised downward due to a review of the Organization's estimation methodology.

[b]For example, the 2003 Maputo Declaration on Agriculture and Food Security by African governments committed 10% of national budgets to agriculture.

Abbreviations
UNFCCC: United Nations Framework Convention on Climate Change.

Competing interests
The authors declare that they have no competing interests.

Acknowledgements
The work of the Commission on Sustainable Agriculture and Climate Change was funded by the CGIAR Research Program on Climate Change, Agriculture and Food Security (CCAFS) and the Global Donor Program for Rural Development (GDPRD). We express gratitude to researchers at CSIRO, INRA and the University of Minnesota whose input helped to shape the ideas presented in this manuscript.

Author details
[1]Government Office of Science, London, UK. [2]Bangladesh Institute of Development Studies, Dhaka, Bangladesh. [3]Commonwealth Scientific and Industrial Research Organisation, Campbell, Australia. [4]Universidad Autónoma Metropolitana, Mexico City, Mexico. [5]INRA, French National Institute for Agricultural Research, Paris, France. [6]University of Wisconsin, Madison, WI, USA. [7]Chinese Academy of Agricultural Sciences, Beijing, China. [8]Ministry of Agriculture, Addis Ababa, Ethiopia. [9]Commission on Sustainable Agriculture and Climate Change, New York, NY, USA. [10]Ministry of Science, Technology and Innovation, Brasília, DF, Brazil. [11]Council for Scientific and Industrial Research, Pretoria, South Africa. [12]National Advisory Council, Prime Minister's Office, New Delhi, India. [13]Vietnam Academy of Agricultural Science, Thanh Tri, Ha Noi, Viet Nam. [14]African Center for Technology Studies, Nairobi, Kenya.

Authors' contributions
All authors read and approved the final manuscript. **Through their service (in their personal capacity) on the Commission on Sustainable Agriculture and Climate Change, JB, MA, MC, AFB, MG, MJ, LE, TM, CN, RS, RS, NVB and JW developed the underlying concepts and recommendations and reviewed the manuscript.** MC and CN contributed to supporting studies on food price volatility. MG contributed to supporting studies on changing diet patterns. CN reviewed background literature and drafted and edited the manuscript.

References

1. International Assessment of Agricultural Knowledge, Science and Technology for Development: In *Agriculture at a Crossroads: A Synthesis of the Global and Sub-global IAASTD Reports*. Edited by McIntyre BD, Herren HR, Wakhungu J, Watson RT. Washington, DC: Island Press; 2009.
2. Nelson GC, Rosegrant MW, Palazzo A, Gray I, Ingersoll C, Robertson R, Tokgoz S, Zhu T, Sulser TB, Ringler C, Msangi S, You L: *Climate Change: Impact on Agriculture and Costs of Adaptation and Food Security, Farming, and Climate Change to 2050*. Washington, DC: International Food Policy Research Institute; 2011.
3. Food and Agriculture Organization: *The State of Agricultural Commodity Markets. High Food Prices and the Food Crisis – Experiences and Lessons Learned*. Rome: Food and Agriculture Organization of the United Nations; 2009.
4. **Food Price Watch.** 2011, http://siteresources.worldbank.org/INTPOVERTY/Resources/335642-1210859591030/FPW_April2011.pdf.
5. Foresight: *Tackling obesities: future choices. Project Report*. London: Government Office for Science, Department for Business, Innovation and Skills; 2007.
6. World Bank: *World Development Report, 2008: Development and Agriculture*. Washington, DC: The World Bank; 2008.
7. **Obesity and Overweight. Fact Sheet No. 311.** www.who.int/mediacentre/factsheets/fs311/en/.
8. United Nations Population Division: World Population Prospects: *World Population Prospects: The 2011 Revision*. New York: United Nations Department of Economic and Social Affairs United Nations Population Division; 2011. http://esa.un.org/wpp/unpp/panel_population.htm.
9. United Nations Population Fund: *State of World Population 2007: Unleashing the Potential of Urban Growth*. New York: United Nations Population Fund; 2007.
10. Lipper L, Mann W, Meybeck A, Sessa R: *'Climate-smart' Agriculture: Policies, Practices and Financing for Food Security, Adaptation and Mitigation*. Rome: United Nations Food and Agriculture Organization; 2010.
11. Intergovernmental Panel on Climate Change: **Summary for policy makers.** In *Managing the Risks of Extreme Events and Disasters to Advance Climate Change Adaptation. A Special Report of Working Groups I and II of the Intergovernmental Panel on Climate Change*. Edited by Field CB, Barros V, Stocker TF, Qin D, Dokken DJ, Ebi KL, Mastrandrea MD, Mach KJ, Plattner G-K, Allen SK, Tignor M, Midgley PM. Cambridge: Cambridge University Press; 2012:1–19.
12. Foresight: *The future of food and farming. Final Project Report. Futures*. London: Government Office for Science; 2011.
13. Royal Society: *Reaping the Benefits: Science and the Sustainable Intensification of Global Agriculture*. London: The Royal Society; 2009.
14. **UN Climate Change Conference, Durban, November–December 2011.**, [http://unfccc.int/2860.php].
15. Intergovernmental Panel on Climate Change: *IPCC Fourth Assessment Report: Climate Change 2007. Contribution of Working Group III to the Fourth Assessment Report of the Intergovernmental Panel on Climate Change*. Cambridge: Cambridge University Press; 2007:2007.
16. **The Impact of Global Temperature Rise of 4°C.**, [http://www.metoffice.gov.uk/climate-change/guide/impacts/high-end/map].
17. Ministry of Economic Affairs, Agriculture and Innovation of the Netherlands: **Chair's summary.** In *Proceedings of The Hague Conference on Agriculture, Food Security and Climate Change; The Hague, Netherlands, 31 October to 5 November2010. 28 pages*. Edited by. :; [http://www.afcconference.com/the-first-conference/135-final-roadmap-for-action].
18. Foley JA, Ramankutty N, Brauman KA, Cassidy ES, Gerber JS, Johnston M, Mueller ND, O'Connell C, Ray DK, West PC, Balzer C, Bennett EM, Carpenter SR, Hill J, Monfreda C, Polasky S, Rockström J, Sheehan J, Siebert S, Tilman D, Zaks DPM: **Solutions for a cultivated planet.** *Nature* 2011, **478**:337–342.
19. Bai ZG, Dent DL, Olsson L, Schaepman ME: *Global Assessment of Land Degradation and Improvement Identification by Remote Sensing*. International Soil Reference and Information Centre: Wageningen; 2008.
20. United Nations Convention to Combat Desertification: *Desertification: A Visual Synthesis*. Bonn: United Nations Convention to Combat Desertification Secretariat; 2011 [www.unccd.int/knowledge/docs/Desertification-EN.pdf].
21. Gustavsson J, Cederberg C, Sonesson U: *Global Food Losses and Food Waste*. Rome, Italy: Food and Agriculture Organization of the United Nations; 2011 [www.fao.org/fileadmin/user_upload/ags/publications/GFL_web.pdf].
22. Beddington J, Asaduzzaman M, Clark M, Fernandez A, Guillou M, Jahn M, Erda L, Mamo T, Van Bo N, Nobre CA, Scholes R, Sharma R, Wakhungu J: *Achieving Food Security in the Face of Climate Change: Final Report from the Commission on Sustainable Agriculture and Climate Change. CGIAR Research Program on Climate Change, Agriculture and Food Security (CCAFS)*. Copenhagen: Commission on Sustainable Agriculture and Climate Change; 2012 [www.ccafs.cgiar.org/commission].
23. Tilman D, Balzerb C, Hill J, Beforta BL: **Global food demand and the sustainable intensification of agriculture.** *Proc Natl Acad Sci U S A* 2011, **108**:20260–20264.
24. French National Institute for Agricultural Research/Centre de coopérationinternationale en rechercheagronomique pour le développement: *Agrimonde: Scenarios and Challenges for Feeding the World in 2050*. Versailles: Editions Quae; 2011.
25. Rogelj J, Chen C, Nabel J, Macey K, Hare W, Schaeffer M, Markmann K, Höhne N, Krogh Andersen K, Meinshausen M: **Analysis of the Copenhagen Accord pledges and its global climatic impacts–a snapshot of dissonant ambitions.** *Environ Res Lett* 2010, **5**(3):1–9.
26. **Case Study Summary: BIOTECH, India.** (2007 Ashden Award)[http://www.ashdenawards.org/files/BIOTECH%20full_0.pdf].
27. Vermeulen SJ, Aggarwal PK, Ainslie A, Angelone C, Campbell BM, Challinor AJ, Hansen JW, Ingram JSI, Jarvis A, Kristjanson P, Lau C, Nelson GC, Thornton PK, Wollenberg E: **Options for support to agriculture and food security under climate change.** *Environ Sci Policy* 2012, **15**:136–144.
28. Nguyen VT, Nguyen QT, Nguyen VA: *Influence of On Farm Water Management to the Methane Emissions in the Red River Delta Area*. Vietnam: Vietnam National Commission on Large Dams; 2007.

29. Worldwatch Institute: *State of the World: Innovations that Nourish the Planet*. New York: W.W. Norton & Company; 2011.
30. **Groupon Earth Observations 2005.**, [www.earthobservations.org].
31. Bangladesh Country Investment Plan: **A Road Map Towards Investment in Agriculture, Food Security and Nutrition.** Updated version[www.nfpcsp.org/agridrupal/sites/default/files/Bangladesh%20Food%20security%20CIP%202011%20Final.pdf].
32. Smith P, Martino D, Cai Z, Gwary D, Janzen H, Kumar P, McCarl B, Ogle S, O'Mara F, Rice C, Scholes B, Irotenko O: **Agriculture.** In *Climate Change 2007: Mitigation. Contribution of Working Group III to the Fourth Assessment Report of the Intergovernmental Panel on Climate Change*. Edited by Metz B, Davidson OR, Bosch PR, Dave R, Meyer LA. Cambridge: Cambridge University Press; 2007:497–540.
33. National Development and Reform Commission: *China's Policies and Actions for Addressing Climate Change –The Progress Report 2009*. Beijing: National Development and Reform Commission; 2009 [www.ccchina.gov.cn/WebSite/CCChina/UpFile/File571.pdf].
34. Yan Z, Liu Y, Ni H, Zeng Z, Men Y: *Effect of Ecological Restoration Techniques on Vegetation and Soil CO_2Discharge of Songnen Saline-alkali Grassland*. New York, NY Institute of Electrical and Electronics Engineers. China: Water Resource and Environmental Protection Symposium; 2011 [http://ieeexplore.ieee.org/xpl/freeabs_all.jsp?arnumber=5893712].
35. Pretty J, Toulmin C, Williams S: **Sustainable intensification in African agriculture.** *Int J Agric Sustain* 2011, **9**:5–24.
36. Pereira PAA, Martha GB, Santana CAM, Alves E: **The development of Brazilian agriculture: future technological challenges and opportunities.** *Agric Food Secur* 2012, **1**:4.
37. Committee on World Food Security: *Land Tenure and International Investments in Agriculture: Summary and Recommendations for Policymakers. Extract from the Report by the High Level Panel of Experts on Food Security and Nutrition of the Committee on World Food Security*. Rome: Food and Agriculture Organization of the United Nations; 2011 [www.fao.org/cfs/cfs-hlpe].
38. International Fund for Agricultural Development: *Rural Poverty Report: New Realities, New Challenges: New Opportunities for tomorrow's Generation*. Rome: International Fund for Agricultural Development; 2011.
39. **Agricultural Production and Soil Nutrient Mining in Africa.**, [http://www.eurekalert.org/africasoil/report/Soil_Nutrient_Mining_in_Africa_Report_Final.pdf].
40. Barrett CB, Barnett BJ, Carter MR, Chantarat S, Hansen JW, Mude AG, Osgood D, Skees JR, Turvey CG, Ward MN: *Poverty Traps and Climate Risk: Limitations and Opportunities of Index-based Risk Financing*, IRI Technical Report No. 07–02. New York, NY: Columbia University International Research Institute for Climate and Society; 2007.
41. Parry M, Evans A, Rosegrant MW, Wheeler T: *Climate Change and Hunger: Responding to the Challenge*. Rome: World Food Programme; 2009.
42. Etiévant P, Bellisle F, Dallongeville J, Etilé F, Guichard E, Padilla M, Romon-Rousseaux M: *Dietary Behaviours. What Factors Come into Play? What Action, for What Result? Collective Scientific Expertise*. Paris: INRA; 2010 [www.international.inra.fr/the_institute/scientific_expertise/expert_reports/dietary_behaviours].
43. Guyomard H, Darcy-Vrillon B, Esnouf C, Marin M, Momot A, Russel M, Guillou M: *Eating Patterns and Food Systems: Critical Knowledge Requirements for Policy Design and Implementation*. Paris: INRA; 2012 [http://ccafs.cgiar.org/sites/default/files/assets/docs/guyomard_et_al_eating_patterns_and_food_systems.pdf].
44. The Cannes Summit Final Declaration: *Building our Common Future: Renewed Collective Action for the Benefit of All*. 2011 [http://www.g20-g8.com/g8-g20/g20/english/for-the-press/news-releases/cannes-summit-final-declaration.1557.html].
45. Food and Agriculture Organization/Organization for Economic Co-operation and Development: *Price Volatility in Food and Agricultural Markets: Policy Responses. Policy Report Coordinated by FAO and OECD Including Contributions by IFAD, International Monetary Fund, UNCTAD, WFP, the World Bank, the WTO, International Food Policy Research Institute and the UN High-Level Task Force*. Rome: Food and Agriculture Organization of the United Nations; 2011.

15

Alternative mechanisms for achieving food security in Oman

Msafiri Daudi Mbaga

Abstract

Food security exists when every person has physical and economic access at all times to healthy and nutritious food in sufficient quantity. There are three fundamental pillars in achieving food security. The first is food availability, second is access to food – economic and physical access – and the third pillar is food utilization. In terms of food availability (the first pillar) today, the world is food secure and statistics shows that global grain prices are less costly in real terms than at any time in recent decades. The second pillar is critical at the national and household levels. At the national level, Oman produces only a fraction of the food it consumes – most of the food, especially grains and red meat, are imported. Land and water scarcity are among the leading constraints to agricultural production such that by 2050 Oman is expected to depend solely on imports to meet food security needs. The third pillar is the utilization of food through adequate diet, clean water, sanitation and healthcare to reach a state of nutritional well-being where all physiological needs are met. Focusing on the two food security pillars (food availability and access to food), this paper proposes three alternative mechanisms for achieving food security in Oman. The three mechanisms proposed are the formation of a National Food Trading Company, setting up an efficient Strategic Grain Reserve System along the lines suggested by the World Bank, and the promotion of private-sector participation in the importation of grain and other critical food commodities. These proposed mechanisms are general and therefore can be easily applied to other Gulf Cooperation Council countries with minor modifications.

Keywords: Market mechanisms, Strategic grain reserve system, Food availability, Price shocks, Food access

Background

Food security exists when all people at all times have both physical and economic access to sufficient food to meet their dietary needs for a productive and healthy life [1,2]. Achieving food security requires that the aggregate availability of physical supplies of food is sufficient, that households have adequate access to those food supplies through their own production, through the market or through other sources, and that the utilization of those food supplies is appropriate to meet the specific dietary needs of individuals. Based on this widely accepted definition of food security, there are three fundamental pillars to food security. The first is food availability, second is access to food – economic and physical access – and the third pillar is food utilization.

In terms of food availability (the first pillar), recent US Department of Agriculture (USDA) statistics indicate that overall the world is food secure [3]; for example, based on USDA statistics, global wheat and coarse grain supplies for 2009/10 had been respectively projected at 2.5 and 4.7 million tons higher in January 2010. In terms of world market prices, notwithstanding the recent agricultural commodity and food prices surge in 2007 and the first half of 2008, the global grain prices are still less costly in real terms than at any time in recent decades [4].

The second pillar is critical at the national and household levels. At the national level, Oman produces only a fraction of the food it consumes – most of the food, especially grains, are imported (Table 1). Land and water scarcity are among the leading constraints to agricultural production in Oman. Average annual

Correspondence: msafiri@squ.edu.om
Department of Natural Resource Economics, Sultan Qaboos University College of Agriculture and Marine Science, P.O. BOX 34Postal Code 123, Muscat, Oman

Table 1 Oman self-sufficiency ratio for important food groups for 2010

Food group	Domestic production (×1,000 tons)	Imports (×1,000 tons)	Exports (×1,000 tons)	Available for consumption (×1,000 tons)	Self-sufficiency ratio[a]
Grains[b]	4.60	630.60	64.90	570.30	0.80
Vegetables	271.47	162.97	61.45	372.98	72.80
Fruits	365.51	147.04	18.39	494.16	74.00
Red meat	24.00	60.63	14.10	70.53	34.00
Chicken meat	42.10	74.56	13.35	103.31	40.80
Eggs	10.06	16.39	6.77	19.68	51.10
Milk	69.60	147.40	61.29	155.71	44.70

Data from Ministry of Agriculture, Department of Statistics; and Royal Oman Police, Directorate General of Customs. [a]Domestic production as a percentage of the amount available for consumption. [b]Statistics for 2008.

rainfall in Oman ranges from a few millimeters to 300 mm and the evaporation rate varies from 1,600 mm to over 2,000 mm. As a result, water loss exceeds water gain and therefore significant and sustainable local food production expansion to meet food security needs is realistically impossible.

Food access at a household level (the second pillar) is also a problem because a segment of households in Oman have insufficient income to afford their food needs in times of crisis, as evidenced by the recent food price surge.

This paper therefore analyzes recent global events such as the recent food price surge and trends in the global food commodity supply and demand situation. Based on the analysis and the two food security pillars, this paper proposes alternative strategies for achieving food security in Oman. The proposed strategies are practically applicable to the other Gulf Cooperation Council (GCC) countries with some minor modifications. The three strategies are proposed in light of land and water shortage in the country as well as the overall world market supply and demand situation as indicated in Tables 2, 3, 4, 5, 6, and 7. This paper is therefore arranged as follows. The second section looks at the recent food price shock and tries to answer the question of whether this is a new phenomenon or whether there is an underlying cyclical cause to it. The third section discusses the prospects for Oman to expand domestic production as a strategy to achieve food security. The fourth section presents an overview of world supply and demand for leading agricultural food commodities. The fifth section presents the three mechanisms for achieving food security in Oman. The sixth section presents the conclusions of this paper.

Table 2 Global total supply, use and ending stocks for major grains[a] 2010/11 to 2012/13

World		Output	Total supply	Trade[b]	Total use[c]	Ending stock
Total grains[d]	2010/11	2,197.05	2,687.72	283.61	2,229.86	457.87
	2011/12 (estimate)	2,300.38	2,758.25	303.91	2,295.91	462.34
	2012/13 (projection) April	NA	NA	NA	NA	NA
	May	2,372.05	2,834.39	303.46	2,356.46	477.93
Wheat	2010/11	651.14	851.20	132.43	654.46	196.74
	2011/12 (estimate)	694.64	891.38	146.74	694.35	197.03
	2012/13 (projection) April	NA	NA	NA	NA	NA
	May	677.56	874.59	137.02	686.47	188.13
Coarse grain[e]	2010/11	1,096.61	1,292.05	116.34	1,129.62	162.43
	2011/12 (estimate)	1,142.44	1,304.86	123.09	1,143.71	161.16
	2012/13 (projection) April	NA	NA	NA	NA	NA
	May	1,228.04	1,389.20	131.31	1,204.27	184.92
Rice milled	2010/11	449.30	544.47	34.83	445.78	98.70
	2011/12 (estimate)	463.31	562.01	34.07	457.85	104.15
	2012/13 (projection) April	NA	NA	NA	NA	NA
	May	466.45	570.60	35.13	465.73	104.88

Data presented in millions of metric tons. NA, not available. Data from [15]. [a]Aggregate of local marketing years. [b]Based on export estimate. See individual commodity tables for treatment of export/import imbalances. [c]Total use for the United States is equal to domestic consumption only (excludes exports). [d]Wheat, coarse grains and milled rice. [e]Corn, sorghum, barley, oats, rye, millet and mixed grains (for the United States, excludes millet and mixed grains).

Table 3 Global supply, use and ending stocks for wheat 1990/91 to 2012/13

Year	Area harvested	Yield	Production	Exports	Feed domestic consumption	Domestic consumption	Ending stock
1990/91	231.0	2.5	588.8	101.3	131.5	554.3	171.6
1991/92	222.8	2.4	543.5	111.4	115.1	551.8	163.2
1992/93	222.1	2.5	562.6	113.2	111.6	547.7	177.9
1993/94	221.0	2.5	558.5	101.7	109.6	553.0	183.4
1994/95	213.3	2.5	523.0	101.5	101.4	542.1	164.2
1995/96	216.7	2.5	537.5	99.2	95.2	546.1	155.7
1996/97	227.1	2.6	581.5	104.0	99.6	573.4	163.8
1997/98	226.4	2.7	610.2	104.5	103.8	576.7	197.2
1998/99	219.2	2.7	590.4	102.0	106.8	578.3	209.4
1999/00	212.8	2.8	586.8	111.9	102.7	565.5	210.7
2000/01	215.7	2.7	583.1	102.4	108.8	586.7	207.1
2001/02	214.5	2.7	583.6	108.0	109.4	586.5	204.2
2002/03	213.8	2.7	569.6	106.8	113.7	604.5	169.2
2003/04	207.8	2.7	555.3	103.6	99.1	588.9	135.6
2004/05	216.1	2.9	626.7	113.9	108.7	606.8	155.4
2005/06	218.7	2.8	618.8	114.1	114.8	621.3	151.2
2006/07	212.2	2.8	596.1	115.6	110.1	615.8	131.6
2007/08	217.1	2.8	611.9	116.8	102.3	617.8	125.6
2008/09	224.6	3.0	682.8	143.5	119.8	641.1	167.3
2009/10	225.8	3.0	686.2	135.6	119.8	652.4	200.1
2010/11	218.3	3.0	651.1	133.7	116.3	654.5	196.7
2011/12	221.7	3.1	694.6	145.1	147.6	694.4	197.0
2012/13	221.2	3.1	677.6	137.4	133.7	686.5	188.1

Data presented in millions of metric tons. Data from [15].

The recent food price shock, is this a new phenomenon?
The surge in the world commodity prices in 2007 and the first half of 2008 took the world by surprise. A lot has been written so far regarding the surge in commodity prices, and overall there appears to be a common consensus that even though in nominal terms commodity prices increased significantly, in real terms the recent increases are more or less like the minor price fluctuations of the 1980s and 1990s [4]. Figure 1, adapted from Piesse and Thirtle [4], tracks prices of four leading food commodities from 1954 to 2008.

Figure 1 shows that in real terms (constant 2007 US $) the current price spikes are more or less like the price fluctuations of the 1980s and 1990s compared with the 1970s crisis. In real terms, prices of all the four commodities were lower in 2008 than they were in 1974. Therefore, in real terms, food was far cheaper in 2008 than it was back in the 1970s. Figure 2 shows results of a 2008 USDA study tracking the price index for food commodities along with the index for the average of all commodities and an index for crude oil [5].

Figure 2 shows that even though the price index for food commodity rose in the last 2 years leading up to 2008, the price index for all commodities and the price index for oil rose even faster.

Is the recent global food crisis unique? According to Timmer [6], global food crises are relatively rare events and they seem to occur roughly three times a century; that is, every three decades or so. Timmer states that the recurring nature of food crises implies there is a long-run cycle of decision-making that derives investment in agricultural technology and productivity [6]. If these investments are induced, as suggested by Hayami and Ruttan [7], then their cyclical nature could be explained by periodic high food prices. If these crises are cyclical as suggested, they are therefore predictable; and if so, we could put in place strategies to reduce their effects.

Prospects for expanding domestic production as a strategy for achieving food security in Oman
>As highlighted in the introductory section, land and water scarcity seriously limits agricultural production in Oman. With respect to water availability, which is critical to increasing local production, Oman is among countries that will face serious water deficit by 2050, as indicated in Figure 3.

Figure 3 illustrates (in green) the number of countries in 2050 that will face freshwater surpluses (>1,300 cubic meters per capita per year in water for food availability

Table 4 Global supply, use and ending stocks for coarse grains 1990/91 to 2012/13

Year	Area harvested	Yield	Production	Exports	Feed domestic consumption	Domestic consumption	Ending stock
1990/91	315.3	2.6	828.8	88.9	537.3	815.2	197.1
1991/92	321.3	2.5	812.2	96.2	536.2	813.1	196.3
1992/93	324.4	2.7	868.9	91.9	551.3	841.4	221.3
1993/94	315.6	2.5	797.6	86.5	546.1	836.2	182.7
1994/95	321.3	2.7	869.4	98.6	567.4	853.8	198.3
1995/96	311.0	2.6	800.9	88.3	544.7	835.8	163.4
1996/97	319.5	2.8	909.1	94.4	573.0	869.5	203.0
1997/98	309.4	2.8	881.4	85.8	579.2	867.9	216.4
1998/99	301.8	3.0	891.5	96.7	575.7	869.7	238.1
1999/00	294.2	3.0	878.1	102.4	585.2	882.6	233.1
2000/01	294.7	2.9	863.7	102.7	588.7	854.7	212.0
2001/02	299.3	3.0	896.6	100.5	601.6	909.0	199.6
2002/03	289.8	3.0	873.6	102.8	595.1	901.2	172.1
2003/04	304.1	3.0	916.0	102.6	615.3	944.8	143.3
2004/05	299.6	3.4	1,015.6	101.3	645.6	978.8	180.1
2005/06	300.2	3.3	980.4	108.6	639.7	994.2	166.2
2006/07	304.1	3.3	988.7	114.7	635.1	1,013.7	141.3
2007/08	316.1	3.4	1,080.8	128.9	656.9	1,057.7	164.3
2008/09	312.3	3.6	1,112.8	110.7	648.9	1,082.1	195.0
2009/10	305.9	3.6	1,110.9	119.1	656.5	1,110.5	195.4
2010/11	303.1	3.6	1,096.6	116.2	655.1	1,129.6	162.4
2011/12	311.6	3.7	1,142.4	122.3	659.3	1,143.7	161.2
2012/13	319.4	3.8	1,228.0	128.1	705.0	1,204.3	184.9

Data presented in millions of metric tons. Data from [15].

over current agricultural land). Also illustrated (in orange and red) are countries facing freshwater deficits (<1,300 cubic meters per capita per year). Based on Food and Agricultural Organization data [10], by 2030 developing countries (such as Oman) would require an average production of 3,000 kcal per capita per day. Assuming that 20% of this is animal protein, an estimated 1,300 cubic meters per capita per year of freshwater would be needed [11]. As indicated in Figure 3, by 2050 Oman will have between 0 and 500 cubic meters per capita per year of freshwater available, which is a serious water deficit since the amount required is 1,300 cubic meter per capita per year. Land suitable for agriculture is also seriously scarce. The arable land in Oman is about 2.2 million hectares [12], which is equivalent to 7% of the total area of Oman (31.4 million hectares). However, the actual cropped area in the Sultanate is 62,000 hectares or 2.8% of the total arable land (due to an irrigation water shortage) and 0.2% of the total area of the country. This indicates that irrigated land in the Sultanate of Oman is below 1%. If only 2.8% of the 2.2 million hectares of arable land can be supported by available water, it is therefore impossible for Oman to achieve self-sufficiency from local production. The Oman self-sufficiency ratio for important food groups as of 2010 is presented in Table 1.

Table 1 shows that Oman is less than 1% self-sufficient in grain production. To achieve self-sufficiency from local production Oman will have to increase local grain production by 99%, which is impossible given that only 2.8% (or 62,000 hectares) of the total arable land is cropped. In addition, to produce a ton of wheat you need 1,000 tons of water (1,000 cubic meters) and for a ton of rice you need 3,400 cubic meters [13] – Oman does not have water to sustainably produce these grains locally. Table 1 also shows that Oman is 34% self-sufficient in red meat. To achieve full self-sufficiency from own production, Oman needs to increase current production by 66%; this is also next to impossible given that 16,000 cubic meters of water are needed to produce a ton of red meat.

As highlighted herein in relation to land and water availability in Oman, it is clear that the prospects for Oman to achieve self-sufficiency from local production are not good. Oman is expected to continue as a food-importing country, as indicated in Figure 4.

Significant expansion of local production to meet food security needs is realistically impossible. Any deliberate efforts to increase local production will put more pressure on the scarce water resources and is therefore not sustainable. Domestic production is therefore not

Table 5 Global supply, use and ending stocks for milled rice 1990/91 to 2012/13

Year	Area harvested	Yield	Production	Exports	Feed domestic consumption	Domestic consumption	Ending stock
1990/91	147.0	3.5	519.4	351.4	12.3	345.3	126.7
1991/92	147.5	3.5	522.8	353.2	14.4	353.2	126.7
1992/93	146.5	3.6	524.2	354.0	14.9	357.5	123.2
1993/94	145.3	3.6	526.1	354.7	16.6	358.9	119.0
1994/95	147.3	3.7	539.9	364.1	20.8	365.5	117.6
1995/96	148.4	3.7	547.3	368.8	19.7	368.3	118.1
1996/97	150.1	3.8	565.3	381.4	18.9	379.2	120.3
1997/98	151.7	3.8	575.2	387.4	27.6	380.0	127.7
1998/99	153.1	3.8	587.0	394.9	24.8	388.6	134.0
1999/00	155.9	3.9	608.6	409.2	22.8	400.1	143.1
2000/01	152.4	3.9	594.1	399.3	24.4	395.6	146.7
2001/02	151.4	3.9	594.3	399.5	27.8	413.3	132.9
2002/03	146.9	3.8	563.1	378.2	27.5	408.1	103.0
2003/04	149.3	3.9	585.1	392.3	27.3	413.7	81.6
2004/05	151.8	3.9	596.6	400.9	29.0	408.5	74.0
2005/06	153.8	4.0	621.4	417.3	29.1	414.6	76.7
2006/07	154.5	4.0	624.7	419.9	31.9	421.3	75.4
2007/08	155.1	4.1	642.5	433.0	29.6	427.5	80.8
2008/09	158.2	4.2	667.7	448.7	29.4	437.1	92.4
2009/10	156.1	4.2	657.1	441.2	31.5	438.5	95.2
2010/11	157.7	4.3	670.4	449.3	35.9	445.8	98.7
2011/12	158.9	4.3	690.7	463.3	34.2	457.9	104.2
2012/13	159.8	4.4	695.6	466.4	35.2	465.7	104.9

Data presented in millions of metric tons. Data from [15].

sustainable. Marginally, however, it might be possible to increase local production through improvements in land and water productivity. Otherwise there may be other sustainable ways to attain food security than attempting to increase local production.

Global supply and demand of important agricultural food commodities

The global food supply in general looks promising [3,14,15]. Based on the USDA's World Agricultural Supply and Demand Estimates, the global wheat and coarse grain supply outlook is good [3]. The total supply, use and ending stocks for major grains are presented in Table 2.

The stocks for grains at the close of the 2011/12 season were estimated at 462.34 million metric tons and are projected to have increased to 477.93 million metric tons in May 2012. Even much more revealing is what we see in Tables 3, 4, 5, 6.

Tables 3, 4, 5, 6 show that for the last 22 years (1990/91 to 2011/12) global ending stocks for all important grains have always been substantial. There is therefore clearly enough global supply of the major grains in the world market. Countries such as Oman that have petrodollars should not have problems procuring enough food to feed its people.

The global total supply, use and ending stocks for oilseeds are as presented in Table 7. The ending stocks for oil seeds are projected to reach 65.63 million metric tons in May 2012/13, slightly lower than the 81.41 million metric tons recorded in the 2010/11 season. The ending stocks for vegetable oils are also projected to reach 12.32 million metric tons in May 2012/13, slightly lower than the 13.37 million metric tons in the 2010/11 season. From Table 7 it is therefore clear that there is enough supply of oilseeds in the global market.

Mechanisms for achieving food security in Oman

The Sultanate government, and the other GCC governments for that matter, has always emphasized the need to undertake agriculture in a sustainable fashion. Sustainable and ecologically sound agriculture guarantees the ability of farmers to continue producing food indefinitely in harmony with the biodiversity. Achieving sustainable agriculture in Oman would require policymakers (the government) to make choices – whether to expand domestic production while knowing that such a choice may put strain on scarce resources (Land and water) and is not sustainable, or whether to invest in

Table 6 Global supply, use and ending stocks for corn 1990/91 to 2012/13

Year	Area harvested	Yield	Production	Exports	Feed domestic consumption	Domestic consumption	Ending stock
1990/91	129.1	3.7	482.0	58.8	318.5	473.4	141.4
1991/92	132.5	3.7	492.0	63.0	334.4	493.3	141.0
1992/93	133.1	4.0	535.6	62.6	348.7	512.1	162.7
1993/94	130.7	3.6	475.8	57.4	342.0	509.0	129.4
1994/95	135.2	4.1	559.3	71.9	371.5	535.6	153.2
1995/96	135.0	3.8	516.4	64.8	365.1	536.6	133.0
1996/97	141.5	4.2	593.0	66.7	387.7	560.1	165.8
1997/98	136.2	4.2	574.4	62.1	400.4	573.8	166.4
1998/99	138.9	4.4	606.0	68.7	404.3	581.5	191.0
1999/00	138.9	4.4	608.1	72.3	422.4	605.3	193.8
2000/01	137.2	4.3	591.4	75.9	427.1	610.1	175.1
2001/02	137.5	4.4	601.0	72.8	436.4	624.8	151.4
2002/03	137.3	4.4	603.2	77.0	433.2	627.5	127.0
2003/04	142.0	4.4	627.4	79.1	445.8	649.2	105.3
2004/05	145.4	4.9	715.8	76.0	476.0	689.4	131.7
2005/06	145.7	4.8	699.7	82.7	478.5	706.7	124.7
2006/07	149.5	4.8	714.0	91.5	478.8	728.5	110.2
2007/08	160.5	5.0	794.3	98.3	498.4	772.9	131.6
2008/09	158.5	5.1	801.2	84.0	431.8	784.7	148.1
2009/10	157.4	5.2	819.4	92.0	490.1	823.5	143.9
2010/11	162.6	5.1	829.1	91.9	501.1	848.6	124.4
2011/12	168.9	5.2	870.5	97.0	508.5	867.3	127.6
2012/13	174.5	5.4	945.8	101.4	548.5	921.0	152.3

Data presented in millions of metric tons. Data from [15].

improving land and water productivity in addition to taking advantage of opportunities presented by the world market.

There are issues that the government might need to address in combination with the three mechanisms/ strategies proposed in this paper. First, the government may need to increase investment in new technologies and innovations to increase water availability, including water desalinization and wastewater recycling.

Second, the government may need to create an environment conducive for foreign investors so as to attract more foreign direct investments. More foreign direct investments will help create jobs and improve the

Table 7 Global supply, use and ending stocks for oilseeds[a] 2010/11 to 2012/13

World		Output	Total supply	Trade	Total use[b]	Ending stock
Oilseeds	2010/11	456.47	529.86	109.23	377.43	81.41
	2011/12 (estimate)	437.26	518.67	106.52	388.29	62.77
	2012/13 (projection) April	NA	NA	NA	NA	NA
	May	471.48	534.25	115.10	399.38	65.63
Oil meals	2010/11	257.04	265.10	77.34	251.33	10.08
	2011/12 (estimate)	263.82	273.90	80.00	260.87	9.53
	2012/13 (projection) April	NA	NA	NA	NA	NA
	May	271.61	281.14	80.39	268.30	10.17
Vegetable oils	2010/11	147.50	161.21	60.13	144.84	13.37
	2011/12 (estimate)	153.47	166.83	62.60	151.01	13.16
	2012/13 (projection) April	NA	NA	NA	NA	NA
	May	157.64	170.80	64.76	155.84	12.32

Data presented in millions of metric tons. NA, not available. Data from [15]. [a]Aggregate of local marketing years with Brazil and Argentina based on an October to September year. [b]Crush only for oilseeds.

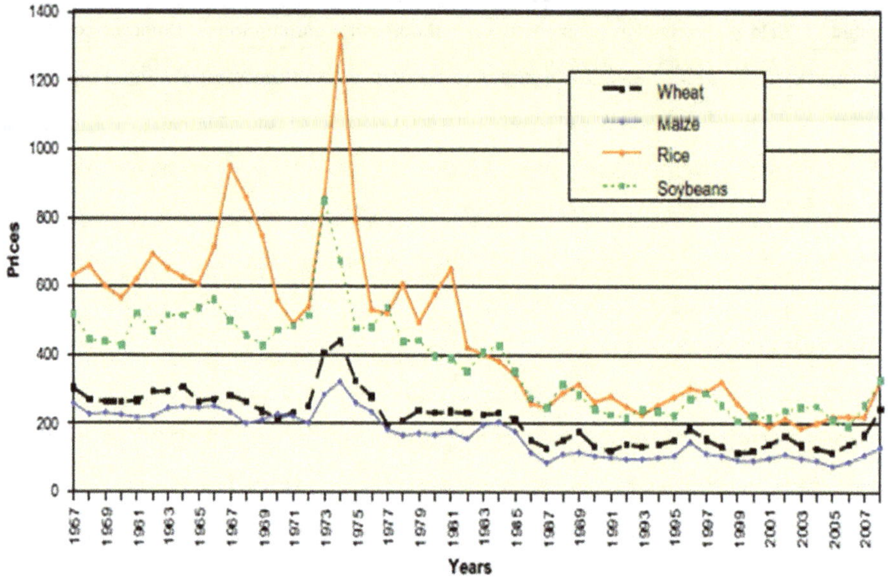

Figure 1 Constant prices (constant 2007 US$ per ton) for bulk commodities (1957 to 2008). Adapted from [4] and Food and Agricultural Organization internal data.

incomes of the people, and hence the capacity to afford/access imported food commodities.

Third, as highlighted in the third section on prospects for expanding domestic production, food production in Oman and other GCC countries is limited by land and water scarcity. The emphasis should be directed towards increasing land and water productivity in terms of yields and value per unit of land and water. This would require government investment in research to improve/increase land and water productivity; developing drought resistance and salinity tolerance crops; and improving crop husbandry practices through effective and focused extension services.

A fourth issue is that, since land and water resources are scarce, there is a need to come up with a national strategy stipulating priority crops that should be grown in the country in order to conserve water, instead of the current situation where everything is grown with no clear focus. For example, the list could be established based on water and land-use efficiency and profitability. The focus should also be on crops that can be grown sustainably without compromising the country's scarce water and land resources. This approach will lead to diversification out of regular staple crops into high-value crops such as fruits and vegetables – these are crops for which the country has already registered high self-sufficiency ratios as indicated in Table 1.

Another issue is that the government may need to invest in buying or leasing land in land-abundant countries to secure food supply. This strategy has already

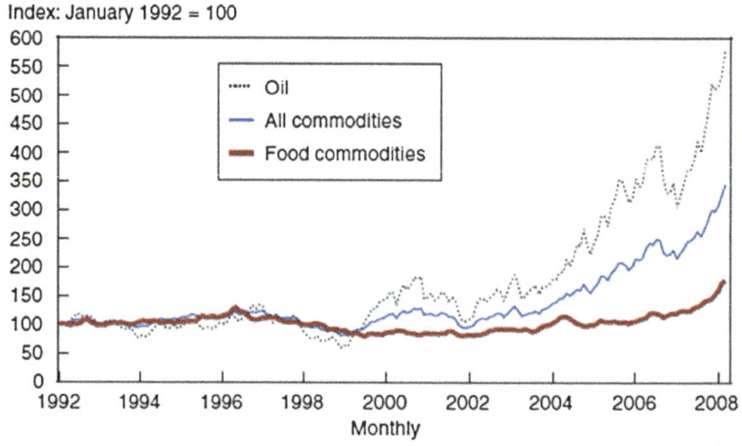

Figure 2 Index of oil, all commodities and food commodities. January 1992 = 100. Adapted from [5].

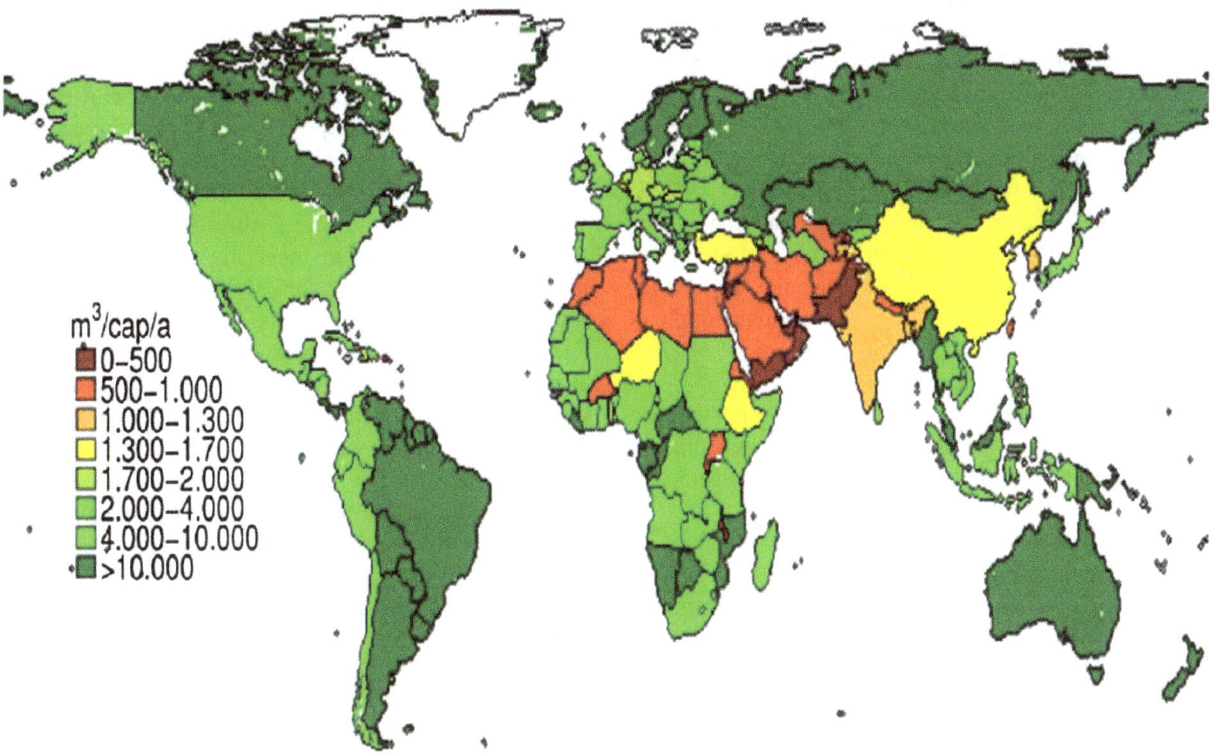

Figure 3 **Number of countries in 2050 facing freshwater surpluses and deficits.** Adapted from [8,9].

been adopted by other GCC countries such as Saudi Arabia and United Arab Emirates. The two countries currently hold more than 2.8 million hectares, mostly in Indonesia, Pakistan and Sudan [16].

A final issue is to provide Oman people with family planning services and promote nutrition education. Promoting nutrition education will go a long way towards increasing food security by reducing demand for cereals. Currently, for example, while the average global per-capita consumption for wheat is 100 kg per annum, the average per-capita consumption in Oman is 115 kg. Reducing this to the global average could be possible.

These issues highlighted above need to be addressed in combination with the following three mechanisms for ensuring food security.

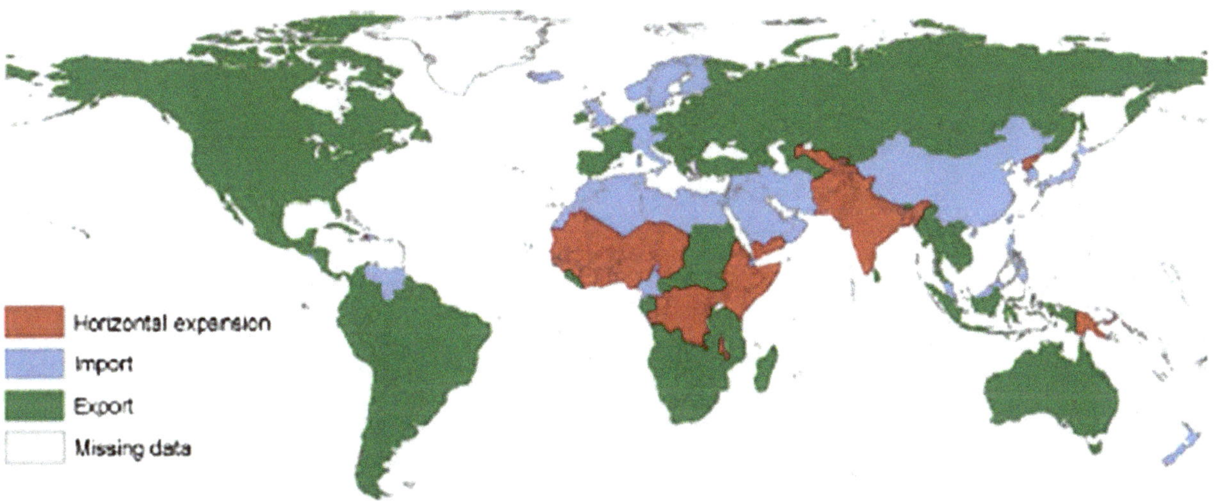

Figure 4 **Food importing and exporting countries from a water availability perspective in 2050.** Adapted from [8].

Formation of a National Food Trading Company

Land and water scarcity seriously limits agricultural production in Oman and other GCC countries. As indicated in Figures 3 and 4, Oman has no other credible and compelling options but to continue importing its food needs, especially grains and red meat (Table 1). One possibility is to establish a National Food Trading Company (NFTC) that will operate independently with minimal government oversight. The NFTC will deal with food commodity trading internationally and domestically, with the mandate or responsibility to make sure that there is enough supply of food commodities in the country during a time of crisis such as that experienced recently.

The advantage of setting-up a NFTC is that such a company can employ financial risk-hedging instruments such as futures contracts and options as cost-effective ways to run its affair.

Futures contracts are one way to manage food commodity price risk. Commodity futures contracts are agreements to purchase or sell a given quantity of a commodity at a predetermined price, with settlement expected to take place at a future date as a result reducing future price risk. Furthermore, futures contracts are standardized in terms of quality and quantity, and place and date of delivery of the commodity. Futures contract transactions are usually made in formal exchanges through clearing houses and are generally closed out before delivery [17].

Setting up and running a NFTC will not impose significant financial burden on the country because, aside from initial set-up costs, thereafter the company will be required to make a profit and survive on its own.

Setting up an efficient strategic grain reserve system

Oman has what is known as the Public Authority for Stores and Food Reserves (PASFR). Other GCC countries have similar kinds of public authorities. The PASFR is responsible for maintaining reserves of rice, sugar, milk powder and edible oil. The PASFR was established in 1980 to maintain strategic food reserves to meet the essential requirements of the population in an emergency.

The PASFR deals only in four items: rice, sugar, milk powder and edible oil. The authority maintains reserve stocks of these items to meet the emergency requirements for 4 to 6 months. The stocks are regularly rotated; that is, old stocks are sold in the market and these are replenished by new purchases. The PASFR purchases milk powder and edible oil from importers/ traders in the Omani market and imports rice and sugar.

The PASFR could be transformed into a strategic grain reserve (SGR) system and restructured to operate much more efficiently, along the lines suggested by the World Bank in relation to a wheat import supply chain [18]. This will, however, depend on the government objectives. A SGR can be used as an instrument to stabilize price at times of crisis like the recent one in 2008. At its simplest, a SGR that seeks to stabilize prices builds up substantial reserves of food grains during periods of surplus and releases them when prices start to rise.

Maintaining a SGR can be quite expensive, but understanding the relationship between grain prices and stocks [18,19] can go a long way towards helping to reduce substantially the cost of maintaining a SGR. Figures 5 and 6 show the relationship between wheat and corn price spikes and the world stock-to-use ratios. For both commodities, price spikes occurred when world stock-to-use ratios were low.

Making sure that the SGR is well stocked around the time when the stock-to-use ratios are low and to draw

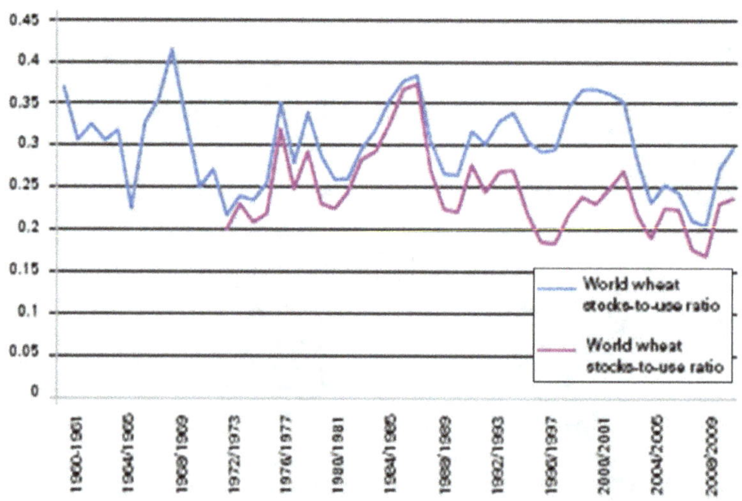

Figure 5 Relationship between wheat price spikes and stock-to-use ratio 1900/01 to 2008/09. Adapted from [19].

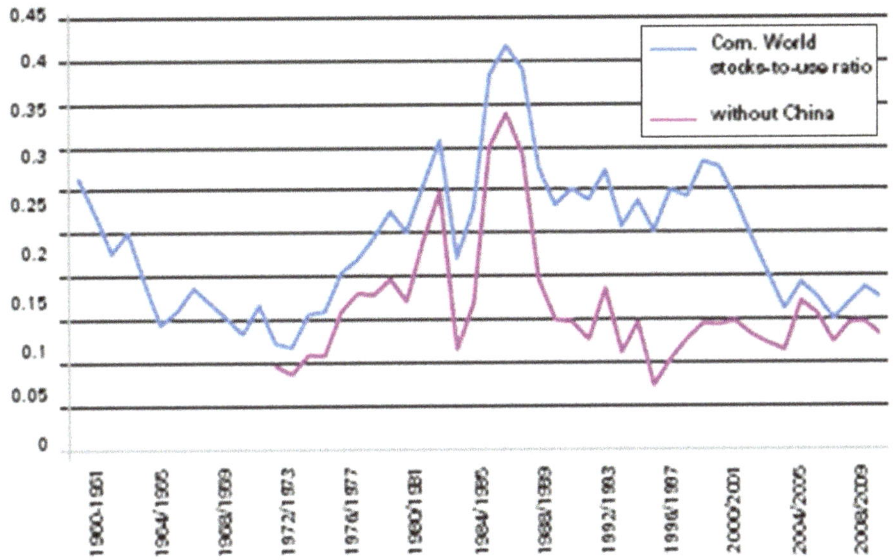

Figure 6 Relationship between corn price spikes and stock-to-use ratio 1900/01 to 2008/09. Adapted from [19].

down stocks when the stock-to-use ratios are high will go a long way towards reducing the costs of maintaining SGR.

Promotion of private-sector participation

Promotion and strengthening of the private-sector capacity to participate in the importation of food commodities is another way to ensure availability of food for food security. To achieve this, the government needs to create an enabling environment to attract private-sector participation. This may include availability of lines of credit through commercial banks with government guarantees, and government investment in modern storage facilities (silos) to be leased to the private food importers. Private companies could also be encouraged to invest in agricultural and food production abroad to contribute towards the country's food needs. In Saudi Arabia, for example, Almarai and Savola Al-Azizia United Company have increased their investments in agricultural and food production abroad to meet growing demand in the kingdom [20].

Conclusion

With land and water scarcity seriously limiting agricultural production in Oman, as well as in other GCC countries, increasing agricultural production through expansion is not feasible. A sustainable marginal increase in output might be achievable through improvements in land and water productivity. In this condition, therefore, the alternative available for Oman is food imports to meet domestic food need, especially grains and red meat (Table 1).

A number of issues need to be addressed in combination with the three alternative mechanisms/strategies proposed in this paper: increase investment in new technologies and innovations to increase water availability, including water desalinization and wastewater recycling; create a conducive environment for foreign investors so as to attract more foreign direct investments to help create jobs and improve the incomes of the people; increase land and water productivity in terms of yields and value per unit of land and water, through investment in research to improve/increase land and water productivity; come up with a national strategy stipulating priority crops that should be grown in the country in order to conserve water instead of the current situation where everything is grown with no clear focus – the focus should, for example, be on crops that can be grown in a sustainable way without compromising the country's scarce water and land resources; invest in buying or leasing land in land-abundant countries to secure food supply; and provide the Oman people with family planning services and promote nutrition education.

The three alternative mechanisms proposed in this paper are the formation of a NFTC, setting up an efficient SGR system, and the promotion of private-sector participation in the importation of food commodities.

Abbreviations

GCC: Gulf Cooperation Council; NFTC: National Food Trading Company; PASFR: Public Authority for Stores and Food Reserves; USDA: US Department of Agriculture.

Competing interests

The author declares that he has no competing interests.

Acknowledgements

The author would like to acknowledge the contribution of the two anonymous reviewers whose comments have helped to shape this paper.

References

1. USDA: **Food Security Indicators and Framework for Use in the Monitoring and Evaluation of Food Aid Programs.** 1999, http://www.fantaproject.org/downloads/pdfs/fsindctr.pdf.
2. **World Food Summit.** http://www.fao.org/wfs/index_en.htm.
3. **World Agricultural Supply and Demand Estimates WASDE-506.** http://www.usda.gov/oce/commodity/wasde/latest.pdf.
4. Piesse J, Thirtle C: **Three bubbles and a panic: an explanatory review of recent food commodity price events.** *Food Policy* 2009, **34:**119–112.
5. Trostle R: **Global Agricultural Supply and Demand: Factors Contributing to the Recent Increase in Food Commodity Prices.** 2008, WRS-0801. http://www.ers.usda.gov/publications/wrs-international-agriculture-and-trade-outlook/wrs-0801.aspx.
6. Timmer PC: **Reflections on food crises past.** *Food Policy* 2010, **35:**1–11.
7. Hayami Y, Ruttan VW, Vernon W: *Agricultural Development: An International Perspective (Revised and Expanded).* Johns Hopkins University Press; 1985.
8. Falkenmark M, Rockström J, Karlberg L: **Present and future water requirements for feeding humanity.** *Food Sec* 2009, **1:**59–69.
9. Rockström J, Falkenmark M, Karlberg L, Hoff H, Rost S, Gerten D: **Future Water Availability for Global Food Production: The Potential of Green Water for Increasing Resilience to Global Change, Water Resource Research.** 2009, W00A12. doi:10.1029/2007WR006767.45. http://www.agu.org/pubs/crossref/2009/2007WR006767.shtml.
10. FAO: *World Agriculture: Towards 2015/30. A FAO Perspective.* London: Earthscan; 2003.
11. Rockström J, Lannerstad M, Falkenmark M: **Assessing the water challenge of a new green revolution in developing countries.** *Proc Natl Acad Sci U S A* 2007, **104:**6253–6260.
12. Ministry of Agriculture, Department of Statistics: 2010. http://www.maf.gov.om/Pages/index.aspx?CMSId=8&lang=EN.
13. Allan JA: **Virtual Water – Economically Invisible and Politically Silent – A Way to Solve Strategic Water Problems.** *International Water & Irrigation* 2001, **21**(4):39–41.
14. OECD-FAO: *Agricultural Outlook. 2008–2017.* 2008. http://www.fao.org/es/esc/common/ecg/550/en/AgOut2017E.pdf.
15. **Grain: World Markets and Trade.** http://www.fas.usda.gov/psdonline/circulars/grain.pdf.
16. World Bank, FAO and IFAD: *Improving Food Security in Arab Countries.* 2009. http://siteresources.worldbank.org/INTMENA/Resources/FoodSecfinal.pdf.
17. Tomek WG, Peterson HH: **Risk management in agricultural markets: a review.** *J Futures Markets* 2001, **21:**953–985.
18. World Bank: 2012, http://www.fao.org/fileadmin/user_upload/tci/docs/The%20Grain%20Chain_ENG.pdf.
19. Wright BD: **International grain reserves and other instruments to address volatility in grain markets: issues and options.** In *Technical Background Paper for the World Grain Forum, May 2009; Russia.* Berkeley: University of California; 2009. http://www-wds.worldbank.org/servlet/WDSContentServer/WDSP/IB/2009/08/25/000158349_20090825154655/Rendered/PDF/WPS5028.pdf.
20. **Khaleej Times.** http://www.menafn.com/qn_print.asp?storyID=1093291209&subl=true.

16

Climate-smart soil water and nutrient management options in semiarid West Africa: a review of evidence and analysis of stone bunds and *zaï* techniques

Robert Zougmoré[1*], Abdulai Jalloh[2] and Andre Tioro[3]

Abstract

There is an increasing need for West African Sahel farmers to be availed with appropriate technologies and management practices that sustainably increase productivity and resilience, while reducing greenhouse gas emissions where possible. In this paper, we analysed prevailing key water and nutrient interaction practices namely stone bund, *zaï* and half-moon techniques in order to explore how their outcomes could be linked to climate change adaptation and mitigation. Data were derived from previous publications but were reinterpreted in the angle of technologies' alignment with the three pillars of climate-smart agriculture (CSA). We found that the most successful systems are those that provide water, nutrients and a supportive soil structure in a synergistic manner. Indeed, technologies such as *zaï*, half-moons and stone bunds, combined with an application of organic/inorganic sources of nutrients, are promising climate-smart agricultural practices that could be widely used by smallholder farmers to maintain food production and secure farmers' livelihoods, while contributing to ecosystem services. However, accompanying incentive measures are needed to empower and capacitate rural farmers to adopt the considered techniques. These are successful examples that can form the basis for informing farmers and agricultural extensionists, as well as future agricultural development plans, policies and investments that pursue sustainable food and nutritional security in the West African Sahel.

Keywords: Climate change, Climate-smart agriculture, Soil and water management, Food security, Sahel

Introduction

The population of West Africa has quadrupled over 50 years (from 90 million in 1960 to 342 million in 2011) [1]. This rapid population growth has had a huge impact on the food demand. Many smallholder farmers must deal with low and unpredictable crop yields and incomes, as well as chronic food insecurity. These challenges are particularly acute in the dry lands, where land degradation, depleted soil fertility, water stress and high costs of fertilizers contribute to low crop yields [2]. Nearly 3.3% of agricultural GDP in sub-Saharan Africa is lost annually because of soil and nutrient losses, estimated at over 30 kg/ha/year [3].

Climate change brings additional threats arising from stresses and shocks caused by higher temperatures and a lack of rainfall. According to Lobell et al. [4], the potential yield loss due to the climate change impact is about 5% for each degree Celsius of global warming.

The Sahel region in West Africa covers approximately 5.7 million km^2 and is home to about 58 million people. Due to its geographic location at the southern edge of the Sahara desert, the region is highly vulnerable to climate change as its population livelihood depends on rain-fed agriculture and livestock. Over the years, crop–livestock production systems in the Sahel have been adapting their practices and ways of life to various risks: climatic variability, economic risks and livestock diseases. To cope with shocks and crisis but also to support changes, various strategies based on the mobility of livestock and/or families, reorganization and diversification of activities, reciprocity

* Correspondence: r.zougmore@cgiar.org
[1]CGIAR Research Program on Climate Change, Agriculture and Food Security (CCAFS), ICRISAT Bamako, BP 320 Bamako, Mali

and social networks have been developed and used successfully. Integrating climate change adaptation and mitigation may ensure food security and also reduce agriculture ecological footprint. Adaptation is a priority for smallholder farmers, who will pursue mitigation when it brings benefits without increasing costs and risks [5].

Climate-smart agriculture (CSA) is an approach that can greatly help us achieve this target. In a recent article, Neufeldt et al. [6] provided substantial information on the concept and its implications. It embraces multiple objectives, aims to increase agricultural productivity and farmers' income, strengthens the resilience of ecosystems and livelihoods to climate change and reduces greenhouse gas emissions. As there are huge variations between geographic locations in terms of risks to be faced and capacities to face them, CSA takes into consideration context-specific and locally adapted actions and interventions, along the whole agricultural value chain.

In semiarid West Africa, where major constraints that impinge on agricultural development are droughts, soil acidity and nutrient depleted and degraded soils [7], most successful CSA options will require a synergistic provision of water, nutrients and a supportive soil structure. These are the key elements needed to support effective plant growth and to reduce crop failure. Water and nutrient interaction practices such as *zaï* and half-moons techniques, fertilizer micro-dosing, stone bunds or vegetation strips combined with an application of organic/inorganic sources of nutrients, cereal-legume intercropping, crop-livestock systems appear to be promising CSA options in the Sahel. Indeed, constructing stone bunds along the contours is an effective way to harvest rain water and decrease run-off erosion in the field [8]. Stone bunding was primarily an indigenous practice that has been later on improved through action research on various parameters (slope, spacing, number of stone lines per bund, etc.) [9]. Also initiated by farmers in West Africa, the *zaï* and half-moon are indigenous practices that consist in concentrating run-off water and organic matter in small pits (*zaï*) or semicircular basins (half-moons). Stone bunds, *zaï* and half-moons are improved land management techniques that often lead to improved soil fertility and soil moisture. Applied at a watershed scale, they also increase ground water level and tree cover [10]. This paper analysed the ability of the above existing technologies and practices and explore how their outcomes are linked to climate change adaptation and mitigation and to increased agricultural resilience in West Africa. This may help in clarifying the concept of CSA through concrete examples while paving appropriate approaches to promote CSA amongst a wide variety of stakeholders in agricultural development.

Climate change projections for semiarid West Africa

Recorded variability in the climate of West Africa during the last century concluded that 1930–1960 was a wet period and 1970–1980 was a drought slot and the good rainfall seasons returned in the 1990s and 2000s [11]. The 1980s have been declared the driest years in the twentieth century, though an increasing trend in precipitation has been observed over the past 15 years [11]. The complex nature of West Africa's climate, dominated by the monsoon, makes it challenging for reanalysis and GCMs to accurately capture the temporal and spatial variability observed in the region [12]. In recent decades, decreased precipitation and drought in the Sahel has become a particular focus for research as it represents one of the largest recent observed climate changes of any region [13]. According to Washington and Hawcroft [14], the projections for changes in crop cultivation limits are variable over space and time, and so the outlook for agriculture is highly uncertain, particularly in the vulnerable Sahelian region. Insufficient observational records constrain the accuracy of reanalysis and gridded data, making the identification of local trends and mechanisms difficult. In addition, there is wide divergence in model projections for the region's climate by the end of this century. This poses a significant challenge to designing agricultural adaptation strategies.

Two physically consistent scenarios for the Sahel have been put forward; one drying and one wetting [15]. Simplistically, the drying interpretation centers on overall warming of the oceans and associated oceanic convection, leading to decreased convection at the continental margins and resultant drying of the interior. Alternatively, a wetter Sahel could be caused by an enhanced land-sea temperature gradient, either due to greater warming of the land and a stronger monsoon flow or due to a reversal in the north–south Atlantic sea surface temperature gradient [16]. Washington and Hawcroft [14] concluded that the evidence therefore remains inconclusive as to whether a climatological increase or decrease in precipitation is the more plausible projection of change for the Sahel. A study by Ericksen et al. [17] indicated that the length of growing period will decline by 5% or more across a broad area of the global tropics, including heavily cropped areas of West Africa. The IPCC [18] predicts an approximate 50% decrease in yields from rain-fed agriculture by 2020 in some countries. The recent study by Jalloh et al. [7], covering 11 West African countries, reported detailed projection results for major crops such as millet, sorghum, maize, rice, groundnut, beans, cassava, cocoyam and cotton. The models predict significant decline in crop yields between 2000 and 2050 if no adaptation measures are undertaken. As an example, CSIRO and MIROC models predict a general decrease in maize yields of 5%–25% over a baseline in most parts of countries which lie on the southern coast of West Africa and a yield increase of 5%–25% in the Sahel zone [7].

Also, heavy and persistent rainfall in hitherto dry areas of the Sahel could cause an increase in diseases and pests to which livestock in those areas are not adapted. On the other hand, a marked decrease in rainfall in hitherto wet regions like Liberia could cause significant changes in the growing conditions that may require changes in the farming system with regard to crops and livestock composition and management [7]. The real issue is the inability of resource-poor farmers to react appropriately and fast enough. Unless strong adaptation measures are taken, projected changes in rainfall and temperatures may cause significant declines in crop yields in semi-arid, tropical and sub-tropical regions such as West Africa.

It is worthy to note that against this background, farmers have been finding ways and means to adapt to the particularly drying conditions. These efforts have been often supported by research in order to provide a scientific evidence of the effectiveness of these adopted climate change adaptation options. This adoption is a reflection of the realities of climate change and variability on the ground and the ability of farmers to make informed choices from a range of available options. Prevalent technologies that have been used by farmers in the Sahel include stone bunds, *zaï* and half-moons.

Are contour stone bunds successful CSA options in the Sahel?

Background information

Two major factors characterizing agriculture in the Sahel are (i) erratic climatic conditions with frequent periods of water shortages and (ii) the presence of large areas of inherently low fertility and crust-prone soils [19]. What is responsible for water deficiency (i.e. more and/or longer periods of water stress) and low water use efficiency is not primarily water shortage but loss of water through runoff, soil evaporation and drainage below the root zone [8]. The loss through run-off is caused by the high intensity of the rainfall, the low organic matter content of the soils and the extent of soils with surface crusts and seals [20]. These have resulted in severe human-induced land degradation in the Sahel. Indeed, Oldeman et al. [21] indicated some years ago that in Africa, 40% of agricultural lands were affected by human-induced land degradation. Because of the degradation phenomenon, crops and animal production are at risk. To solve the problem, farmers have developed a range of measures, including run-off control, soil structure improvement and nutrient management [22].

Soil and water conservation extensionists have put an emphasis on the implementation of the stone bund technique to check run-off and to control erosion. Thus, laying stone bunds in the fields is now well known and is widely practiced by farmers in sub-Saharan West Africa [8]. Indeed, constructing stone bunds is the most widely practiced technique to combat run-off and erosion by farmers. As a result, various government and non-government programmes are promoting the large-scale introduction of the technique and providing technical and logistical backup for collecting and transporting stones. Contour stone bunds are erosion control structures built with quarry rock or stones in a series of two or three. They are constructed in lines along the natural contour of the land after 10–15 cm of the soil has been removed from the line where they are to be built. They should be built to a height of 20–30 cm from the ground and spaced 20–50 m apart depending on the inclination of the terrain.

On-farm research has shown that stone bunds are efficient in increasing the soil water status and in reducing soil erosion and downward particle transport. The technique is particularly efficient in reducing run-off and improving rainwater infiltration [8]; however, as shown in Tables 1 and 2 [23], the best results are achieved when contour stone bunds are used in combination with organic fertilizer application. Indeed, stone bunds with compost application showed the least water run-off during 3 years (Table 1). Plots with stone bunds maintained more moisture than plots with grass strips and plots without soil and water conservation (SWC) measures (Table 2).

Constructing stone bunds is the most widely practiced technique to combat run-off and erosion by farmers [25]. Indeed, plant nutrient use efficiency in cereal-based farming systems is often very low because of limited soil moisture conditions [26]. The low soil quality combined with the harsh Sahelian climate leads to a low efficiency of fertilizers [19]. The beneficial effect of SWC measures such as stone bunds on soil productivity was limited under continuous non-fertilized cereal cropping. However, interactions of SWC measures with organic or mineral source of nutrients optimize water and nutrient use efficiency [27], which can boost crop production and result in economic benefits for resource-poor farmers.

Beneficial impacts for food production, adaptation and/or mitigation

Under water-limiting conditions, the stone bunds are efficient measures in improving soil water content through run-off control, which can reach 59% in plots with barriers alone and even 84% in plots with barriers combined with organic matter [23]. The stone bunds form a barrier that slows down run-off and spreads it more evenly over the land. By slowing the flow of water over the land, it can seep into the soil and prevents the loss of rainwater. Rainwater that filters through the bunds infiltrates into the soil. When rainfall is erratic, the stone bunds contribute to conserving more moisture in the soil thereby helping to alleviate water stress during dry spells. The

Table 1 Treatment effects on run-off during 2000, 2001 and 2002 rainy seasons at Saria, Burkina Faso

	Annual run-off (% of rainfall)			Run-off reduction (%)		
	2000	2001	2002	2000	2001	2002
T_0	15.9 (2.1)	12.2 (1.1)	17.6 (1.9)	0	0	0
T_{SR}	7.1 (2.0)	3.5 (1.3)	5.0 (0.9)	55	71	71
T_{SRU}	8.3 (2.7)	4.2 (1.6)	5.3 (0.87)	48	65	70
T_{SRC}	6.8 (1.6)	3.2 (1.6)	1.0 (0.6)	58	74	94
T_{GS}	8.3 (1.1)	5.9 (1.0)	8.2 (1.3)	48	51	53
T_{GSU}	11.4 (0.9)	9.5 (1.1)	7.6 (2.2)	29	22	57
T_{GSC}	7.1 (1.6)	4.5 (1.8)	2.8 (1.2)	56	63	84
T_{SRM}/T_U	7.5 (2.2)	6.6 (0.9)	9.0 (1.0)	53	46	49
T_{GSM}/T_C	8.2 (1.8)	8.2 (0.7)	2.4 (0.9)	48	32	87
Number of rain events	10	09	16	10	09	16

Values in brackets: ±standard deviation between run-off volumes measured in pits and recorded values of run-off. T_0 no SWC measures, no nutrient supply (control plot); T_{SR} stone rows, no nutrient supply; T_{SRC} stone rows + compost; T_{SRM} stone rows + manure; T_{SRU} stone rows + urea; T_{GS} grass strips, no nutrient supply; T_{GSC} grass strips + compost; T_{GSM} grass strips + manure; T_{GSU} grass strips + urea. Table adapted from Zougmoré et al. [24].

effect of stone bunds on soil water moisture depends on the space between the bunds. The larger the spacing, the less their effects [28]. In wet years however, the bunds may cause waterlogging in some parts of the field, which can adversely affect crop production. Contour stone bunds protect the land against sheet erosion caused by run-off. Some studies reported that the application of compost led to the reduction of total soil loss by 79% in plots with stone bunds as compared to the losses in plots without barriers [24]. By slowing down the run-off speed, the bunds also induce sedimentation of fine waterborne particles of soil and manure, resulting in a build-up of a layer of sediments rich in nutrients. The seeds of grasses and shrubs are also trapped by the bunds, favouring the establishment of natural vegetation along the structure. This further stabilizes the soil and the bunds and contributes to conserving the biodiversity of plants and small wild animals (monitor lizards, birds, snakes and other reptiles). From the perspective of climate change adaptation, contour stone bunds are useful for a number of reasons. In years with high rainfall, they protect the land from rapid flow and associated erosion in the event of a heavy rain, a phenomenon that tends to increase with climate change. In years with a decline in rainfall, they contribute to more effective rainwater harvesting. They improve water retention and infiltration into the soil, increasing the amount of water available to plants and guaranteeing the harvest. Such an increase in available water in the soil profile will help to mitigate the predicted decrease in the length of growing period (LGP) in the Sahel due to global warming [17]. In addition, IPCC [18] projects that heavy rainfall events in the Sahel are likely to increase in frequency and intensity. And with the development of an important tree cover along the stone bunds, they also lower soil temperature and provide protection against wind erosion. In addition to increasing the adaptive capacity, the better management of these lands may lead to increased productivity.

Table 2 Treatment effects on volumetric soil water content in the root zone (%) at Saria, Burkina Faso

	06 August 2000	27 August 2001	15 July 2002
T_{SRU}	18.9 (1.7)^{a}	14.4 (0.7)^{a}	12.4 (0.4)^{a}
T_{SR}	16.3 (1.6)^{b}	12.8 (1.0)^{b}	11.4 (1.1)^{b}
T_{GSC}	16.5 (1.6)^{b}	13.4 (1.9)^{b}	09.3 (0.6)^{e}
T_{GSU}	16.1 (1.3)^{b}	10.5 (0.7)^{d}	10.0 (0.4)^{cd}
T_{SRM}/T_U	15.8 (1.8)^{b}	10.4 (0.6)^{d}	10.3 (0.3)^{c}
T_{SRC}	15.2 (1.3)^{bc}	10.9 (0.5)^{d}	08.6 (0.5)^{f}
T_{GS}	14.4 (1.0)^{c}	12.1 (0.6)^{c}	10.3 (0.6)^{c}
T_{GSM}/T_C	14.2 (1.4)^{c}	11.6 (1.0)^{c}	09.8 (0.7)^{d}
T_0	14.0 (1.1)^{c}	10.1 (1.3)^{d}	10.1 (0.6)^{cd}
Probability	<0.001	<0.001	<0.001

$^{a, b, c, d, e, f}$express statistical differences. Treatments with the same letter are not statistically different at $P = 0.05$. Values in brackets: ±standard deviation. Treatment keys in Table 1. Table adapted from Zougmoré et al. [24].

During dry years, crops in plots with stone bunds could yield two to three times more than crops in control plots [25]. The increase in sorghum yields varies between 33% and 55% in Burkina Faso's Central Plateau area while grain yields increase by more than 40% for millet up to 15 years after the bunds were established in Niger [29]. A similar picture also emerged in Niger where farm families with SWC produced an estimated grain surplus of 70% in years of good rainfall [30]. Zougmoré et al. [27] reported that the yield increase did not cover annual costs of single SWC measures while application of single compost or urea was cost effective. Combining stone bunds with application of compost increased sorghum grain yield by about 142% and induced positive interaction effects (mean added effects of 185 kg ha^{-1} for stone bunds combined with compost), which resulted in financial gains of

145,000 to 180,000 FCFA ha^{-1} $year^{-1}$ under adequate rainfall conditions (Table 3). These results indicate that despite the harsh Sahelian environment, opportunities do exist for making a more efficient use of the limited local sources in a way to develop CSA options such as contour stone bunds.

Landolt [29] identified a range of both poverty and hunger reduction benefits (increased yields, introduction of cash crops, greater food security and income) and environmental benefits (raised water tables, increased vegetation cover, increased stock of trees, reduced pressure on nearby savannahs, increased species diversity) when contour stone bunds were constructed and organic fertilizer was used. Higher crop production improves household food security in proportion to the area of a farm improved with bunds. Under the Niger project for integrated protection of agro-sylvo-pastoral resources, an average of 16% of the area of a farm was improved with stone bunds, resulting in an increase of between 8% and 33% in annual output with no other additional measures [29]. In some areas, a reduction in temporary migration to pursue off-farm income-generating activities was also observed. The above results show that applying the stone bund technique can strengthen the adaptive capacity of farmers while empowering them to invest in soil fertilization for increased productivity and food security.

Thanks to the stone bunds, the increased vegetation cover and diversity on the rehabilitated areas can also increase the supply of fuel wood. As a result, more organic matter is being applied to fields instead of being used as fuel, further increasing soil fertility and crop yields. This could be an opportunity for soil carbon sequestration if the crop biomass and other organic resources are maintained in the soil while preventing erosion. Also, the vegetation that grows also contributes towards the mitigation by absorbing carbon dioxide from the atmosphere. As groundwater levels rise, farmers started growing vegetables on small plots near wells, thereby increasing both their income and the diversity of their diets. Health benefits from this are likely to be significant, although have yet to be measured. In that respect, durable and effective soil erosion control structures such as stone bunds will assume even greater importance and constitute an important adaptation measure with possible mitigation co-benefits.

Zaï and half-moons: two indigenous land rehabilitation practices to adapt to the changing climate and land use in the Sahel

Background information

In the West African Sahel, the combined effects of climatic conditions, inherent poor soil quality and human activities have resulted in soil degradation. The latter is due to crusting, sealing, erosion by water and wind. Cultivated lands are particularly characterized by a gradual loss of structure, hardpan formation, reduced permeability, compaction, inadequate aeration and limited plant root development. On these soils, increasing erosion has ultimately resulted into the development of totally bare, sealed and crusted soils locally called *zipellé* in Burkina Faso, or *harde* soils in Chad [8]. No one single measure is sufficient to adapt to climate change and variability. Rather, a mix of measures is needed to target the various farm variables—water, soil, micro-climate, seeds and crops as well as labour and capital.

This example shows how traditional integrated soil and water management practices called *zaï* and half-moons can combat land degradation and improve productivity of these previously abandoned bared soils. The *zaï* method (also called tassa in Niger or towalen in Mali) is a soil rehabilitation system that concentrates run-off water and organic matter in small pits. With an average of 20–40 cm in diameter and 10–15 cm deep, the holes are dug manually during the dry season. A handful (≈0.3 kg) of animal manure or compost is supplied per pit, i.e. 9.5 t ha^{-1}. Like *zaï*, the half-moon technique (originating from Niger) is another method for the rehabilitation of sealed and crusted bare soils consisting of a basin of 2 m in diameter. The digging is done with a hoe or a pick so as to break the crusted layer on the soil surface and to collect the run-off water. The cultivated area is 6.3 m^2 for each half-moon. A barrowful (35 kg) of animal manure or compost is supplied in each half-moon, i.e. 14.6 t ha^{-1}. Stone bunds are usually laid along contours in order to slow down run-off and allow better water retention and infiltration in the *zaï* and half-moon basins.

Table 3 Economic benefits of combining stone rows or grass strips with compost-N or urea-N at Saria

	2001				2002			
	T_{SRU}	T_{GSU}	T_{SRC}	T_{GSC}	T_{SRU}	T_{GSU}	T_{SRC}	T_{GSC}
Minimum yield for N inputs (kg ha^{-1})	195	195	271	271	195	195	271	271
Minimum yield for SWC measures (kg ha^{-1})	345	187	345	187	345	187	345	187
Minimum yield for SWC + N input (kg ha^{-1})	540	382	615	457	540	382	615	457
Excess yield (kg ha^{-1})	158	54	821	782	−193	−135	987	915
Economic benefit (FCFA ha^{-1})	22,120	7,560	114,940	109,480	−27,020	−18,900	138,180	128,100

T_{SRC} stone rows + compost-N T_{GSC} grass strips + compost-N, T_{SRU} stone rows + urea-N, T_{GSU} grass strips + urea-N. Table adapted from Zougmoré et al. [24].

Beneficial impacts for food production, adaptation and/or mitigation

The two soil rehabilitation practices are efficient in improving soil productivity mainly through biophysical and biological processes. Indeed, by breaking the soil crust, pit digging facilitates more water infiltration; also, run-off water is harvested due to the earthen bund formed downslope of the pits. The applied organic matter attracts termites which develop nesting and foraging activities. This results in improved soil structure as they open up large and numerous macropores throughout the entire soil profile. The soil structure enhancement following soil fauna activity leads to increased water infiltration and drainage, lower run-off and reduced soil resistance to root penetration. Moreover, the application of organic inputs not only enhances soil nutrient availability but also improves crop nutrient uptake from soil reserves. The improvement of water status in the soil and the increased decomposition and nutrient release result into a beneficial impact of the *zaï* and half-moon systems on crop performance under semiarid conditions. Several studies [20,22] in the Sahel region reported that applying compost with mineral fertilizers allowed substantial gain in sorghum grain yields from the *zaï* pits without any organic or mineral supply (Table 4).

It is a practical solution to reclaim these degraded lands but also to rehabilitate the agroforestry cover in the Soudano-Sahelian semiarid area. Indeed, the plant seeds included in the manure allow the regeneration of shrubs and trees in the *zaï* pits. In line with their objective to restore vegetation in their lands and to expect beneficial by-products from trees, farmers generally proceed to a selection of regenerated young trees; then they are assisted until they become mature plants. Several studies reported the re-establishment on formerly bare soil of over 20 herbaceous species and 15 woody species following two consecutive years of *zaï* in the central part of Burkina Faso [10,31]. The *zaï* method at present is still labour-intensive, about 60 working days for 1 ha. Scientists from the National Institute for Environment and Agricultural Research in Burkina Faso recommended a so-called 'mechanical *zaï*' that consists of making appropriate holes mechanically with animal-drawn tools (Dent IR12 for sandy soils or Dent RS8 for other types of soils). This reduces by more than 90% the amount of time required for making the pits. Indeed, it takes only 11 to 22 h ha^{-1} to construct these pits with oxen that are well fed with crop residues [32]. This also resulted into an economic benefit of 165,000 CFA ha^{-1} compared to only 17,000 CFA ha^{-1} with the manually dug *zaï* (Table 5).

However, it is worth mentioning that the widespread use of the mechanical *zaï* technique is still constrained by the non-affordable cost of the animal-drawn equipment and tools [32]. Thus, a long-term trade-off may consist in developing local capacities within farmer communities in order to empowering them to produce and sell locally the basic agricultural equipment (e.g. plough, cart, pick, shovel). This should be accompanied by widespread promotion of compost production in order to contribute to solving the limited availability of animal manure.

The above results demonstrate that *zaï* and half-moon practices can be considered as CSA options as they contribute effectively to rehabilitate previously abandoned and degraded bare lands. Therefore, they improve the

Table 4 Grain and biomass production on deep, brown Eutropept soil, at Taonsogo, Burkina Faso

Treatment	1992			1993		
	Kg ha^{-1}	±S.D.	Test[1]	Kg ha^{-1}	± S.D.	Test[1]
Grain production						
Control	150	±154	a	3	±0.6	a
Pit	200	±63	a	13	±4.2	a
Pit + neem leaves	395	±151	ab	24	±7.3	a
Pit + compost[2]	654	±145	abc	123	±82.5	a
Pit + mineral fertilizer	1,383	±236	bc	667	±256.3	b
Pit + compost + mineral fertilizer	1,704	±305	bc	924	±346.8	b
Biomass production						
Control	946	±529	a	167	± 75	a
Pit	1,329	±549	a	292	± 49	a
Pit + neem leaves	1,990	±207	ab	875	± 172	ab
Pit + compost[2]	2,843	±945	abc	1,417	± 511	bc
Pit + mineral fertilizer	4,839	±1,105	bc	2,375	± 706	bcd
Pit + compost + mineral fertilizer	5,333	±1,490	bc	3,250	± 857	cd

[1]Tukey-Kramer test (P <0.05). a, b, c, d express statistical differences. Treatments with the same letter are not statistically different at P = 0.05.
[2]Compost: 3 Mg ha^{-1} of a mixture of dry manure, straw and various crop residues composted during three dry months. Table adapted from Roose et al. [20].

Table 5 Economic evaluation of the treatments at Pougyango (CFA/ha)

	Control	Soil scarifying	Manual *zaï*	Mechanized *zaï* with excavation	Mechanized *zaï*
Soil tillage cost	0	3,000	55,000	26,900	15,000
Total cost	58,333	69,003	166,698	93,659	76,790
Income grains	29,820	49,560	146,160	201,740	191,380
Income straw	12,668	14,220	37,808	56,970	52,748
Total income	42,488	63,780	183,968	258,710	244,128
Benefit (income-cost)	−15 846	−5,223	17,270	165,051	167,337
Additional costs	-	0	52,000	23,900	12,000
Total benefits	-	0	120,188	194,930	180,348
Ratio cost/benefit	-	-	2,31	8.16	15.03

1 euro = 655.95 CFA (Table adapted from Barro et al. [32]).

resilience of smallholder farmers; in addition, these practices substantially increase crop productivity and allow farmers to adapt to climate variability. Also, according to Bayala et al. [10], *zaï* and half-moon techniques also favour local species regeneration through their seeds contained in the manure applied. A sound application of these techniques can potentially contribute to GHG reduction through the subsequent impact of the regenerated trees and their effect on soil carbon and crop production. However, these aspects still need further investigation, especially in relation to how trees are managed by farmers.

Conclusion

In this paper, we reviewed data and information on three integrated soil water and nutrient management practices from the semiarid West African region to analyse their contribution to the three pillars of CSA. The goal is to simultaneously achieve increased agricultural productivity and incomes (food security), improved resilience to climate change (adaptation) and reduction of greenhouse gas emissions (mitigation), where possible. We found that stone bunds, *zaï* and half-moon techniques combined with the application of organic and/or mineral fertilizers are sustainable land management practices that have increased agricultural productivity, vegetative cover and carbon sequestration. They also reduce water erosion. Thus, in the context of the West African Sahel, these techniques can be qualified as climate-smart since in various ways, they contribute to the CSA criteria. However, constraints that may limit their adoption and sustainability are mainly the specific requirement of rocks in order to construct the contour stone bunds and the corollary need for considerable cost and labour to collect and transport the rocks from their original locations to the fields. Vegetation barriers can also be considered as an alternative. Planting locally adapted grass strips (e.g. of *Andropogon gayanus*) or live hedges (e.g. of *Guiera Senegalensis, Piliostigma reticulatum*) may be sustainable solutions if farmers ensure their maintenance (pruning, protection). Also, in view of the limited availability of organic matter to be applied with the *zaï*, half-moon and stone bund techniques, compost production should be encouraged in order to valorize the scarce organic resources in the region. Therefore, developing local capacities and empowering rural communities to produce good quality compost and to manufacture basic agricultural equipment could contribute to reducing the labour demand while enabling the widespread implementation of the technique. These successful examples show the many ways CSA can take shape and should serve as inspiration for future policies and investments. It seems therefore important to recognize the opportunities and limitations of specific climate-smart techniques as they can be applied in specific locations when designing local, national and sub-regional agricultural and food security investment plans. There is a growing support for CSA promotion through the formation of global, continental and regional alliances around the concept. Hence, categorizing these technologies according to their climate-smartness will contribute to giving practical evidence of CSA concept to farmers. This may facilitate the widespread adoption and out scaling of these CSA technologies and practices amongst farmers in West Africa.

Competing interests

The authors declare that they have no competing interests.

Authors' contributions

RZ developed the outline and drafted the manuscript. AJ and AT contributed to the evolving outline and writing of the manuscript. All authors read and approved the final manuscript.

Acknowledgements

The authors acknowledge support provided by the CGIAR Research Program on Climate Change, Agriculture and Food Security (CCAFS) with funding from the CGIAR Fund and the European Union (EU) and with technical support from the International Fund for Agricultural Development (IFAD). They also thank CORAF/WECARD and ROPPA for their collaboration and contribution to this paper.

Author details
[1]CGIAR Research Program on Climate Change, Agriculture and Food Security (CCAFS), ICRISAT Bamako, BP 320 Bamako, Mali. [2]West and Central African Council for Agricultural Research and Development (CORAF/WECARD), BP 48 Dakar, RP, Senegal. [3]Réseau des Organisations Paysannes et de Producteurs Agricoles de l'Afrique de l'Ouest (ROPPA), 09 BP 884 Ouagadougou 09, Burkina Faso.

References

1. Elbehri A, Kaminski J, Koroma S, Iafrate M, Benali M: **West Africa food systems: an overview of trends and indicators of demand, supply, and competitiveness of staple food value chains.** In *Rebuilding West Africa's Food Potential: Policies and Private Initiatives to Promote Smallholder-Inclusive Staple Food Value Chains.* Edited by FAO. Rome: Inter-Departmental Working Group; 2013:1–42.
2. Winterbottom R, Reij C, Garrity D, Glover J, Hellums D, Mcgahuey M, Scherr S: *Improving Land and Water Management. Working Paper, Installment 4 of Creating a Sustainable Food Future.* Washington, DC: World Resources Institute; 2013 [http://www.worldresourcesreport.org]
3. The Montpellier Panel: *Sustainable Intensification: A New Paradigm for African Agriculture.* London: Agriculture for Impact, Imperial College; 2013:1–36.
4. Lobell DB, Schlenker WS, Costa-Roberts J: **Climate trends and global crop production since 1980.** *Science* 2011, **333:**616–620. doi:10.1126/science.1204531.
5. Jarvis A, Lau C, Cook S, Wollenberg E, Hansen J, Bonilla O, Challinor A: **An integrated adaptation and mitigation framework for developing agricultural research: synergies and trade-offs.** *Exp Agric* 2011, **47:**185–203.
6. Neufeldt H, Jahn M, Campbell BM, Beddington JR, DeClerck F, De Pinto A, Gulledge J, Hellin J, Herrero M, Jarvis A, LeZaks D, Meinke H, Rosenstock T, Scholes M, Scholes R, Vermeulen S, Wollenberg E, Zougmoré R: **Beyond climate-smart agriculture: toward safe operating spaces for global food systems.** *Agric Food Secur* 2013, **2:**12.
7. Jalloh A, Nelson GC, Thomas TS, Zougmoré R, Roy-Macauley H: *West African agriculture and climate change: a comprehensive analysis.* Washington DC: IFPRI Books and research monographs; 2013. http://dx.doi.org/10.2499/9780896292048. ISBN 978-0-89629-204-8.
8. Zougmoré R, Ouattara K, Mando A, Ouattara B: **Rôle des nutriments dans le succès des techniques de conservation des eaux et des sols (cordons pierreux, bandes enherbées, *zaï* et demi-lunes) au Burkina Faso.** *Sécheresse* 2004, **15:**41–48.
9. Zougmoré R: *Integrated water and nutrient management for sorghum production in semi-arid Burkina Faso. Tropical Resources Management Papers, Volume 45.* The Netherlands: Wageningen University; 2003.
10. Bayala J, Sileshi GW, Kalinganire A, Tchoundjeu Z, Sinclair F, Garrity D: **Cereal yield response to conservation agriculture practices in drylands of West Africa: a quantitative synthesis.** *J Arid Environment* 2012, **78:**13–25.
11. Nicholson S, Selato J: **The influence of La Niña on African rainfall.** *Int J Climatol* 2000, **20:**1761–1776.
12. Cook K, Vizy E: **Coupled model simulations of the West African monsoon system: twentieth and twenty first century simulations.** *J Clim* 2008, **19:**3681–3703.
13. Dai A: **Precipitation characteristics in eighteen coupled climate models.** *J Clim* 2006, **19:**4605–4630.
14. Washington R, Hawcroft M: *Climate Change in West African Agriculture: Recent Trends, Current Projections, Crop-Climate Suitability, and Prospects for Improved Climate Model Information.* Denmark, Copenhagen: CCAFS; 2012 [www.ccafs.cgiar.org]
15. Giannini A: **Mechanisms of climate change in the semiarid African Sahel: the local view.** *J Clim* 2010, **23:**743–756.
16. Haarsma RJ, Selten FM, Weber SL, Kliphuis M: **Sahel rainfall variability and response to greenhouse warming.** *Geophysical Res Lett* 2005, **32:**L17702.
17. Ericksen P, Thornton P, Notenbaert A, Cramer L, Jones P, Herrero M: *Mapping hotspots of climate change and food insecurity in the global tropics, CCAFS Report 5.* Denmark, Copenhagen: CCAFS; 2011. Available online at: [www.ccafs.cgiar.org]
18. IPCC: Climate Change: **The Physical Science Basis.** In *Contribution of Working Group I to the Fifth Assessment Report of the Intergovernmental Panel on Climate Change.* Edited by Stocker TF, Qin D, Plattner GK, Tignor M, Allen SK, Boschung J, Nauels A, Xia Y, Bex V, Midgley PM. Cambridge, United Kingdom and New York, NY, USA: Cambridge University Press; 2013:1–1535.
19. Breman H, Groot JJR, van Keulen H: **Resource limitations in Sahelian agriculture.** *Glob Environ Changes* 2001, **11:**59–68.
20. Roose E, Kaboré V, Guenat C: ***Zaï* practice: A West African traditional rehabilitation system for semiarid degraded lands: a case study in Burkina Faso.** *Arid Soil Res Rehabil* 1999, **13:**343–355.
21. Oldeman LR, Hakkeling RTA, Sombroek WG: *World map of the status of human-induced soil degradation. An explanatory note.* 2nd edition. ISRIC, UNEP: Wageningen, Nairobi; 1991.
22. Reij C, Tappan G, Smale M: **Re-greening the Sahel: farmer-led innovation in Burkina Faso and Niger.** In *Agro-Environmental Transformation in the Sahel: Another Kind of "Green Revolution".* Washington DC: IFPRI Discussion Paper; 2009:1–52. 914.
23. Zougmoré R, Mando A, Ringersma J, Stroosnijder L: **Effect of combined water and nutrient management on runoff and sorghum performance in semiarid Burkina Faso.** *Soil Use Manag* 2003, **19:**257–264.
24. Zougmoré R, Mando A, Stroosnijder L: **Soil nutrient and sediment loss as affected by erosion barriers and nutrient source in semi-arid Burkina Faso.** *Arid Land Res Manage* 2009, **23:**85–101.
25. Kaboré PD, Reij C: *The emergence and spreading of an improved traditional soil and water conservation practice in Burkina Faso. IFPRI discussion paper 114.* Washington DC: Environment and production technology division; 2004.
26. Buerkert A, Piepho HP, Bationo A: **Multi-site time-trend analysis of soil fertility management effects on crop production in sub-Saharan West Africa.** *Exp Agric* 2002, **38:**163–183.
27. Zougmoré R, Mando A, Stroosnijder L: **Economic benefits of combining soil and water conservation measures with nutrient management in semiarid Burkina Faso.** *Nutr Cycl Agroecosyst* 2004, **70:**261–269.
28. Zougmoré R, Guillobez S, Kambou NF, Son G: **Runoff and sorghum performance as affected by the spacing of stone lines in the semiarid Sahelian zone.** *Soil Tillage Res* 2000, **56:**175–183.
29. Landolt M: *Stone lines against desertification.* Rural 21 International Platform; 2011:36–37. 1.
30. Hassan A: **Improved traditional planting pits in the Tahoua Dep. (Niger): an example of rapid adoption by farmers.** In *Sustaining The Soil – Indigenous Soil and Water Conservation in Africa.* Edited by Reij C, Scoones I, Earthscan TC; 1996:56–61.
31. Mando A, Zougmoré R, Zombré NP, Hien V: **Réhabilitation des sols dégradés dans les zones semi-arides de l'Afrique subsaharienne.** In *La jachère en Afrique Tropicale; de la jachère naturelle à la jachère améliorée. Le point des connaissances.* Edited by Floret C, Pontanier R. Paris: John Libbey Eurotext; 2001:311–339. Volume II.
32. Barro A, Zougmoré R, Taonda SJB: **Mécanisation de la technique du *zaï* manuel en zone semi-aride.** *Cahiers Agric* 2005, **14:**549–559.

17

Household adoption of soil-improving practices and food insecurity in Ghana

Jifar T Nata[1], James W Mjelde[2*] and Frederick O Boadu[2]

Abstract

Background: The persistent problem of poor agricultural practices and technology use leads to food insecurity for some farmers in Ghana. Studies show that the adoption of improved agricultural practices and technology may help stabilize production and lessen food insecurity. This study addresses the link between food insecurity and the adoption of soil-improving practices. To address this link, the objectives of this study are to examine factors associated with the adoption of soil-improving practices by Ghanaian farmers and how this adoption impacts the probability of increased food security. Using survey data, two logit models are estimated to determine 1) the likelihood of adopting soil-improving practices including how food security may influence adoption and 2) the relationship between technology adoption and food security.

Results: In the adoption model, being a food secure household, the use of insecticides and seasonal lease land tenure increase the probability of adopting such soil-improving practices. Farming on better soil decreases the probability of adopting soil-improving practices. Higher incomes along with the use of chemical fertilizers, farming on better soil, and the use of commercial seeds increase the probability of the household being food secure. The results on fertilizer, soil, and seeds are most likely associated with increased production. The use of insecticides (which may be an indicator of an insect infestation) lowers the probability of being food secure. Household characteristics such as income, age, education level, and household size are not significant in influencing the adoption decision or improving household food security.

Conclusions: The results lend support to a need to fine-tune the fertilizer subsidy policy implemented by the Government of Ghana. The government needs to consider that the use of fertilizer may have conflicting influences on adoption of soil-improving practices and food security. Improved seed research and distribution would improve the households' food security. Inferences also suggest that it is important to consider land contract markets in policy decisions.

Keywords: Ghana, Food security, Ordered logit, Survey, Soil-improving practices adoption

Background

Agriculture is a major source of livelihood in Ghana, supporting 46% of the total households and accounting for 30% of Ghana's gross domestic product in 2010 [1]. Although agriculture is a growing sector in Ghana's economy with a mean annual growth rate of approximately 5.3% [1], food insecurity remains an issue. Food security is composed of three basic elements: availability, access, and utilization [2,3]. Availability is the condition where there is physical presence of food in the form of domestic production, commercial imports, and aid. Access is the ability to acquire food through a combination of home production and stocks, purchasing, borrowing, personal gifts, and in-kind or food aid. Utilization is the household's use of acquired food including the ability to absorb and metabolize nutrients. Using these elements, food security is generally defined as "Secure access by all people at all times to enough food for a healthy life" [3], p. 2.

Wolter [4] notes that for a proper perspective of agriculture and food security in Ghana, one needs to consider the two faces of Ghana agriculture. Ghana is facing stagnating production in the food crop sector contributing to food insecurity, but increasing exports

* Correspondence: j-mjelde@tamu.edu
[2]Department of Agricultural Economics, Texas A&M University, College Station, TX, USA

in horticulture crops [4]. A 2009 study for the World Food Programme (WFP) estimates that 1.2 million people in Ghana (approximately 5% of the population) have limited access to sufficient and nutritious food [5]. Out of these 1.2 million people, almost 55% are from households that are primarily food and cash crop farmers, agro-pastoralist, food processors, or unskilled laborers. The basic underlying factors noted by the Biederlack and Rivers [5] for the food insecure households are high dependency on agriculture, lack of education, lack of access to output markets, and poverty. Other studies such as [6] and [7] attribute ineffective production techniques, inadequate extension access, and inadequate input supplies as causes of food insecurity.

To improve food security, Ghana has implemented several programs through accelerated agriculture growth and development strategies [8]. Stabilizing production in a sustainable way through soil-improving practices may be one way of ensuring food security for the growing Ghanaian population. Haggblade and Tembo [9], for example, argue that farming with minimum tillage, cover crops, and crop rotation holds promise for stabilizing production and ensuring food security from their experiments in Zambia. There, however, is little information concerning the relationship between food security and adoption of soil-improving practices in Ghana. Adoption of recommended soil management practices may improve soil quality, help adaptation to climate change, and is essential to advancing food security [10].

The agriculture technology adoption literature, which includes the adoption of soil conservation/soil-improving practices, presents diverse results depending on location of the study, type of technology, and other factors considered. Pender and Gebremedhin [11] argue that as a result of population pressure, more land is being cultivated causing arable land to become increasingly scarce. Landholdings per household may become smaller, reducing per farm production and subsequently smaller investments in the land. As a result, a vicious circle of low productivity, inadequate investment capacity, and continued degradation exert further pressure on the existing cultivated land [11]. Solutions to this cycle call for the need to apply sustainable land practices to resolve the long-term impact of degradation on productivity and food production. Reviews of this literature [12-14] suggest that there is a general consensus pointing to a positive relationship between agriculture technology use and output, hence on households' food security status. There are potential location-specific effects [14], including both socio-economic households' characteristics and bio-physical conditions of agricultural practices that need to be understood if relevant information for policy purposes is to be developed. This study starts to fill in the gap by providing information on farmers' decision-making regarding the use of soil-improving practices and food security.

The overall objectives of this study are to examine factors associated with the adoption of soil-improving practices by Ghanaian farmers and how this adoption impacts the probability of increased food security. Factors associated with the probability of adoption of soil-improving practices of households in the GA West District in Greater Accra Region of Ghana are examined. This study also explores if and how food insecure agricultural households differ from food secure agricultural households in terms of agricultural practices, household characteristics, and adoption of soil-improving technologies. To achieve these objectives, a survey of Ghanaian farmers was conducted. Logit models are used to determine likelihood of adopting soil-improving practices and the association between adoption of soil-improving practices and food security.

Methods

Study area

The study area is the GA West District in Greater Accra Region of Ghana (Figure 1). The major agricultural crops in the Greater Accra region are cassava, rice, and maize. Cassava covers one-half of the cropping area, while rice and maize comprise approximately one-fourth of the cropping area in 2010 [15]. Production of cassava was 68,170 tons, while rice and maize production were 12,741 tons and 3,584 tons in 2010 [15].

Description of survey

Survey data was collected during the fall of 2012. Approval of the survey was obtained from the Institutional Review Board at Texas A&M (IRB2013-0513). Informed consent was obtained from each participant before they answered the survey. Information on food security, farming practices, socio-economic characteristics, food consumption, asset ownership, soil-improving practices, technological adoption, and income for 116 households was collected. The detailed survey consisted of 14 pages of questions [16]. The questionnaire was pretested using households within the study region. The survey was part of a larger study on soil fertility improvement in Ghana conducted by the Norman Borlaug Institute at Texas A&M University.

Face-to-face interviews were conducted by extension agents from the Ministry of Food and Agriculture in Ghana, working for the district agriculture office. Respondents were chosen to account for diversity in farming systems and socio-economic conditions within the district. The interviews took place in a cluster of 14 rural villages within Ga West District. Even though all the villages lie in the same region of Ghana, there is considerable diversity in farming practices, especially in terms of

Figure 1 Location of the study area in Ghana, the Greater Accra region and GA West. (Source: Wikomedia Commons).

practices used, crops grown, and education level. Unfortunately, there is no official information on the number of households (to provide a sampling frame); as such, respondents were identified with the help of the extension agents. These agents have considerable knowledge concerning the specific localities of villages. A village consists of two to approximately ten dwellings. Given the dispersion of the villages, the extension agent teams split into two groups to interview the farmers. The data collection process was time-consuming because the data collectors had to walk between the widely dispersed villages to conduct the interviews. Completion of the survey took between 30 min and an hour depending on the complexity of the respondents' household and farming operations. One person per household completed the survey, usually the household head.

Logit model for adoption choice

A logit model is used to provide changes in probability of the adoption of soil-improving practices associated with various household characteristics. A household is categorized as *adopter* if they use (adopt) one or more of the five soil-improving practices: no-till, minimum till, applying mulch, planting cover crops, and using crop rotation. *Non-adopters* are households that do not use any of these practices. The use of logit models in economics is based on random utility theory [17]. Households are assumed to choose the alternative(s) that maximize their utility subject to a set of constraints [17,18]. Indirect utility, the basis for this analysis, measures the maximum utility that a household achieves given the price level and constraints [18]. According to the random utility theory, indirect utility has both a deterministic component and a random (unobservable) component. Assuming two (0 and 1) choices and the commonly used linear utility model, the household's utility associated with choice zero is:

$$U_{i0} = V_{i0} + \varepsilon_{i0}, \text{ where } V_{i0} = X\beta_0 \tag{1}$$

where V_{i0} is the deterministic utility associated with household i and alternative 0, X is a matrix of independent variables with the corresponding parameter vector β_0, and ε_{i0} is the random component. Similarly, for choice one, the household's utility is:

$$U_{i1} = V_{i1} + \varepsilon_{i1}, \text{ where } V_{i1} = X\beta_1. \tag{1a}$$

The utilities are unobservable, but the observed choice reveals which choice provides the larger utility. Letting Y = 1 denote the household's choice of alternative 1, the probability that household will choose 1 [19]:

$$
\begin{aligned}
\Pr[Y=1|X] &= \Pr[U_{i1} > U_{i0}] \\
&= \Pr[X\beta_1 + \varepsilon_{i1} > X\beta_0 + \varepsilon_{i0}|X] \\
&= Pr[X\beta_1 - X\beta_0 + \varepsilon_{i1} - \varepsilon_{i0} > 0|X] \\
&= \Pr[X(\beta_1 - \beta_0) + \varepsilon_{i1} - \varepsilon_{i0} > 0|X] \\
&= \Pr[X\beta + \varepsilon_i > 0|X].
\end{aligned} \tag{2}
$$

One cannot estimate the separate effects of the individual beta coefficients, but can only estimate the differences in the betas. Different assumptions on the error terms give rises to the various quantitative choice models. Here, the error terms are assumed to have the standard logistic distribution which gives rise to the conditional logit model. Here, the predicted household choice is based on the following:

$$
\begin{aligned}
&\text{Choice zero if } X\hat{\beta} < 0, \text{ and} \\
&\text{Choice one if } X\hat{\beta} > 0
\end{aligned} \tag{3}
$$

where the "cutoff" point between the two possible choices is zero. In the model, $Y = 0$ is designated as a household that does not adopt soil-improving practices and $Y = 1$ is a household adopting soil-improving practices. For estimation propose, the base group is *non-adopters*.

Description of the independent variables and expected signs with respect to the base variables are given in Table 1. A positive sign indicates that the characteristic is expected to increase the probability of adoption of soil-improving practices. A "?" indicates no prior expectation of the effect of the variable on adoption. The main variable of interest is food secure vs. insecure households. Chronic, seasonal, vulnerable, and food secure households are as described in the next section. Based on these four categories, food insecure households are those that fall in any one of the three food insecure categories: chronic, seasonal, and vulnerable. Food security is expected to lead to a higher probability of adoption of soil-improving technologies because the household may have been able to forgo short-run food production to increase the productivity of the soil without compromising the hunger of the household.

Ordered logit for food insecurity

An ordered logit model is estimated to investigate the factors influencing a household's food security category. Different scales or categories have been developed to classify levels of food insecurity. The classification of households' food insecurity used here is based on the duration of time a household goes without sufficient food: chronic, transitory, and vulnerable [2,20,21]:

Chronic food insecure—occurs if the household was persistently unable to meet the minimum daily requirement over the past 12 months;
Seasonal food insecure—occurs if there was a cyclical pattern of inadequate access to food over the past 12 months. Typically, the period of food shortage is between planting and harvesting. It is commonly associated with climate shocks, natural disasters, economic crises, or conflicts;
Vulnerable food insecure—occurs if the household is at risk of transitory or chronically food insecure, but currently has a minimum acceptable food intake; and
Food secure household—is a household where all members have not lived in hunger over the past 5 years.

Ordered logit is used to take advantage of the ordinal nature of the food security categories. For example, a household in the seasonal food insecure category is "worse off" than a household in the vulnerable category, but they are "better off" than a chronic food insecure household.

The ordered logit is an extension of the previous conditional logit model by adding multiple cutoff points. For the four food secure categories model, the model predicts that a household is in the four food security categories based on the following [19]:

$$
\begin{aligned}
&\text{Chronic food insecure } X\hat{\beta} < \hat{u}_a, \\
&\text{seasonal food insecure } \hat{u}_a < X\hat{\beta} < \hat{u}_b, \\
&\text{vulnerable food insecure } \hat{u}_b < X\hat{\beta} < \hat{u}_c, \text{ and} \\
&\text{food secure } X\hat{\beta} > \hat{u}_c.
\end{aligned} \tag{4}
$$

With four categories, three cutoff points are estimated, the $\hat{\mu}'s$. The logistic model is estimated for the adoption and food category models using maximum likelihood techniques programmed in STATA version 11 [22].

The adoption and food security model include the same independent variables with two exceptions. In the adoption model, an independent variable indicating whether the household food secure or not is included. In the food security model, a variable indicating the adoption of soil-improving practices is included. In both models, the marginal effects are not equal to the coefficients. Marginal effects, calculated at the mean of the continuous independent variables, are provided by STATA based on either continuous or discrete variables.

Results and discussion

Adoption of soil-improving practices

The chi-squared test p value is 0.00, indicating that the coefficients of independent variables are not jointly equal to zero (Table 2). Moreover, the model fit is within the range expected for cross-sectional data with a pseudo R^2

Table 1 Description of variables and expected signs used in the conditional and ordered logit models

Variables	Expected sign		Values	Details
	Adoption	Food security		
Dependent variables				
Application of soil-improving practices			No adoption = 0	Mulching, cover crops, crop rotation, no-till, and minimum till
			Adoption = 1	
Food insecurity groups			Chronic = 1	
			Seasonal = 2	
			Vulnerable = 3	
			Food secure = 4	
Independent variables				
Application of soil-improving practices		+	No adoption = 0	
			Adoption = 1	
Food insecure/secure	+	?	Food insecure = 0	
			Food secure = 1	
Income	+	+	Continuous	100 cedi currency/year
Chemical fertilizer application	?	+	No application = 0	12-month time frame
			Apply = 1	
Commercial seed application	?	+	No application = 0	12-month time frame
			Apply = 1	
Insecticide application	?	?	No application = 0	12-month time frame
			Apply = 1	
Herbicide application	?	?	No application = 1	12-month time frame
			Apply = 1	
Household head age	+	?		Years
Household head education level	+	+	No or pre-school = 0	
			junior and above school = 1	
Household size	?	?		Number
Medium-quality soil	?	+	Poor = 0	
			Medium = 1	
Seasonal lease	?	?	No seasonal lease = 0	12-month time frame
			Seasonal lease = 1	
Long term proprietorship—dropped to avoid collinearity issues	+	?	No long lease = 0	12-month time frame
			Long lease = 1	
Sharecropping	?	?	No sharecropping = 0	12-month time frame
			Share cropping = 1	
Information and training (access) from extension services	+	+	No access = 0	
			Access = 1	

of 0.32. Estimated coefficients are presented in Table 2 and marginal effects in Table 3. The independent variable signs are interpreted with respect to the base, non-adopters. In the following discussion, a significant level of 0.15 is assumed[a].

The coefficients associated with food security (whether the household is food secure or not), use of chemical fertilizers, use of insecticides, soil quality (medium-quality soil), and seasonal lease arrangement are significant. Food secure households are more likely to adopt soil-improving practices than food insecure households. Food secure households may be willing to sacrifice some production for increased soil quality because they are not hungry. Kristjanson et al. [23] found a similar relationship between farming practices and number of months with food deficit for five African countries. Their study observed that households that experience a larger number of months with a food deficit per year make less

Table 2 Estimated coefficients for adoption of soil-improving practices model

Variable	Coefficient	Standard error	z	$p > \|z\|$	95% confidence interval	
Food insecure/secure	1.746	0.782	2.230	0.026	0.213	3.279
Chemical fertilizer	−1.443	0.785	−1.840	0.066	−2.981	0.095
Commercial seed	0.166	0.733	0.230	0.821	−1.272	1.603
Herbicides	−0.752	0.758	−0.990	0.321	−2.237	0.733
Insecticide	0.949	0.657	1.450	0.148	−0.338	2.236
Medium-quality soil	−1.208	0.723	−1.670	0.095	−2.626	0.210
Off-farm activity	−0.075	0.695	−0.110	0.914	−1.437	1.288
Seasonal lease	2.280	0.655	3.480	0.001	0.995	3.565
Share cropping	−0.295	0.857	−0.340	0.731	−1.975	1.385
Income	0.005	0.015	0.350	0.728	−0.023	0.034
Education level	−0.229	0.580	−0.390	0.693	−1.366	0.909
Household size	0.033	0.147	0.220	0.824	−0.256	0.322
Age of house head	−0.001	0.023	−0.030	0.980	−0.045	0.044
Extension service	0.029	0.655	0.040	0.965	−1.254	1.312
Intercept	1.205	2.145	0.560	0.574	−2.999	5.410

Base group: non-adopters.
Number of observation = 116.
Chi squared = 45.65, probability > chi squared = 0.0000.
Pseudo R^2 = 0.32.

changes to their farming practices (such as introducing micro-catchments, ridges, rotations, improved pastures, and trees) compared to households with fewer months with food deficits. Ajayi [24] notes that farmers have a good understanding of food security and soil fertility issues. This understanding may contribute to the positive association between adoption of soil-improving practices and food insecure/secure category.

The use of chemical fertilizers decreases the probability of adopting soil-improving practices. Given the nature of the survey questions, one can only speculate as to why this is the case. It may be the case that households are substituting fertilizers for soil-improving practices to maintain yields. The use of insecticides increases the likelihood of adoption of soil-improving practices. Application of insecticides may indicate pest infestations, and households are adopting soil conservative practices to help control insects. Soil quality is categorized as either poor or medium quality according to the subjective opinion of the respondent. This coefficient is negative indicating that respondents

Table 3 Marginal effects for the adoption of soil-improving practices model

Variable	Marginal effect	Standard error	z	$p > \|z\|$	95% confidence interval	
Food insecure/secure	0.235	0.099	2.390	0.017	0.042	0.429
Chemical fertilizer	−0.195	0.102	−1.900	0.057	−0.395	0.006
Commercial seed	0.022	0.099	0.230	0.821	−0.171	0.216
Herbicides	−0.101	0.101	−1.010	0.314	−0.299	0.096
Insecticide	0.128	0.086	1.490	0.137	−0.041	0.297
Medium-quality soil	−0.163	0.095	−1.710	0.087	−0.350	0.024
Off-farm activity	−0.010	0.094	−0.110	0.914	−0.194	0.174
Seasonal lease	0.308	0.071	4.360	0.000	0.169	0.446
Share cropping	−0.040	0.115	−0.340	0.730	−0.266	0.186
Income	0.001	0.002	0.350	0.728	−0.003	0.005
Education level	−0.031	0.078	−0.400	0.693	−0.184	0.122
Household size	0.004	0.020	0.220	0.824	−0.035	0.043
Age of house head	0.000	0.003	−0.030	0.980	−0.006	0.006
Extension service	0.004	0.088	0.040	0.965	−0.169	0.177

with higher quality soil are less likely to adopt soil-improving practices than those farming poorer quality soil. It appears that households with better quality soil may have less of an incentive to adopt land-conserving practices than those households with poorer soils. Households may be adopting soil-improving practices as their soil quality deteriorates. As their soil deteriorates and yields decrease, households may only then perceive the benefits of adopting soil-improving practices. These results are in line with Mbaga-Semgalawe and Folmer [25], who found that the perception of soil erosion as a priority issue increases the probability of adoption of improved soil conservation practices.

Seasonal leases tend to increase the probability of adoption over long-term proprietorship. Again, one can only speculate as to why this is the case. Adoption may be because of the short-term lease arrangements. Landowners may require producers to use conserving practices, or leaseholders may be afraid of losing their leases if they do not use conserving practices. It may be the case that the lessee has an incentive to take care of the land to be able to obtain a contract for next season.

Household characteristics such as age, income, household size, and education are not significant in influencing the likelihood of adopting soil-improving practices. Although variables included in previous studies vary, studies such as [26-28] also found that characteristics such as age or experience, type of land tenure, or education are not significant determinates of adoption of soil management practices. Household characteristics do not influence the adoption of soil-improving practices despite the results of the previous studies [29,30] that suggest these factors may influence adoption behavior. Knowles and Bradshaw [14] note that there are few if any universal factors influencing adoption; thereby, concluding efforts to promote conservation agricultural need to be tailor made for individual locations. Although variability exists in the survey data, the study region is fairly homogenous. One unique characteristic of this region is the nearness to the capital city and the benefits this brings in terms of educational and infrastructure facilities. Households in the rural areas of the GA West District in Greater Accra Region are uniformly more educated compared to other regions and have higher incomes. This lack of sufficient variability in these variables may be the cause of their insignificance.

Food security categorization

In the food security model, the chi-squared test p value is 0.001, indicating that the coefficients of independent variables are not jointly equal to zero (Table 4). Chi-squared tests of the equality of the three cutoff points are rejected at p values of 0.00 or less. The significant differences in the cutoff points indicates that the four categories differ; all four categories should be included in the model. Estimated coefficients from an ordered logit model are difficult to interpret because they are in

Table 4 Ordered logit coefficient estimates for food security

Independent variable	Coefficient	Standard error	z	$p > \lvert z \rvert$	95% confidence interval	
Adoption of soil imp.	0.187	0.493	0.380	0.704	−0.778	1.152
Chemical fertilizer	0.979	0.489	2.000	0.045	0.021	1.937
Commercial seed	0.766	0.536	1.430	0.153	−0.285	1.817
Herbicides	0.596	0.560	1.060	0.287	−0.502	1.694
Insecticide	−0.983	0.468	−2.100	0.036	−1.900	−0.066
Medium-quality soil	2.439	0.541	4.510	0.000	1.379	3.498
Off-farm activity	0.396	0.481	0.820	0.410	−0.546	1.338
Seasonal lease	−0.113	0.553	−0.200	0.838	−1.198	0.972
Share cropping	−0.497	0.713	−0.700	0.486	−1.896	0.901
Income	0.017	0.010	1.750	0.080	−0.002	0.036
Education level	0.468	0.414	1.130	0.259	−0.344	1.280
Household size	0.009	0.101	0.090	0.929	−0.189	0.207
Age of house head	0.013	0.016	0.780	0.435	−0.019	0.044
Extension service	−0.632	0.513	−1.230	0.218	−1.637	0.373
μ_a	1.281	1.448			−1.557	4.119
μ_b	4.946	1.525			1.957	7.934
μ_c	5.976	1.554			2.930	9.021

Number of observation = 116.
Chi squared = 35.00, probability >chi squared = 0.002.
Pseudo R^2 = 0.13.

log-odds units; as such, the marginal effects (Table 5) are discussed.

Marginal effects are interpreted relative to the category and sign. A positive coefficient for a category indicates an increase in that variable increases the probability of being in that category, whereas, a negative coefficient indicates a decrease in probability of being in that category. Four variables have significant marginal effects (p values ≤0.15) in all four equations: use of chemical fertilizers, having medium-quality soil, income, and use of insecticides. The use of chemical fertilizers decreases the probability of being in either the chronic or seasonal insecure categories and increases the probability of being in either the vulnerable insecure and secure categories. Farming medium-quality soil over poor soil has a similar effect, decreasing the probability of being in either chronic or seasonal and increasing the probability of being in either vulnerable or secure categories. These results are most likely associated with increased production from the use of chemical fertilizers and/or farming better soils.

Previous studies have shown that the adoption of agricultural innovation and sustainable land practices may influence the level of agricultural productivity which in turn helps determine farm income and household food insecurity [31,32]. Minten and Barrett [33] show that Madagascarian farmers regard productivity-enhancing innovations, such as improved access to agricultural equipment, irrigation, and cattle, as key elements to increase their food security. Asfaw and Shiferaw [34] show that small holder farmers in Tanzania and Ethiopia are able to produce more and create a market surplus by planting improved varieties of chickpea and pigeon pea. Wollini et al. [35] note that the use of improved agricultural practices can have a positive effect on food security and biodiversity through crop rotation and intercropping. Farmers may be able to grow crops that can be harvested at different times and have different climate and environmental stress response characteristics. Such practices are a way of hedging the risks of drought, irregular temperatures, and rainfall variability. Specifically, for Ghana, Diao and Sarpong [36] identify soil fertility depletion as the main bio-physical factor that deters improvements in productivity and per capita food production.

Increasing income decreases the probability of being food insecure and increases the probability of the household being in either the vulnerable or food secure category. As incomes increase, households have more funds to purchase inputs or food; the marginal effects reflect this idea. Studies such as [37,38] also show positive associations between income and food security.

The use of insecticides has an opposite effect of the previous two variables, namely increases the probability of being in the chronic or seasonal food insecure categories and decreases the probability of being in vulnerable insecure or secure category. Again, one can only speculate as to why, but insecticide use may be an indicator that there is a pest attack on the crop. Hence, households that experienced pest attacks may have decreased production and reduced food security. Failure to apply insecticides would most likely cause even smaller yields. Conway and Toenniessen [39] suggest that low soil fertility and crop losses from pests have reduced harvests and food security in Africa.

The marginal effect of the use of commercial seeds is significant in two categories: seasonal and secure. Commercial seed use is very close to significant (p = 0.153) for the vulnerability category and is not significant for the chronic category. The use of commercial seeds increases the probability of being in the vulnerable or secure categories while decreasing the probability of being seasonal food insecure. As with the use of chemical fertilizers, commercial seed marginal effects are most likely related to increased yields. Commercial seeds are expected to increase yields providing higher yields for personal consumption or sale.

Household characteristics, such as education level of the household, age of the household head and household size, and sharecropping, and training and information from the Ghana extension service are not significantly associated with the probability of households improving their food security position. As in the adoption model, their insignificance may be because of the households in the survey area are relatively homogenous in those factors.

Three variables of particular interest to agricultural policy-makers are adoption of soil improvement techniques, education level, and access to extension service. Adoption of soil-improving practices and increased education levels have the expected effects, namely decreasing the probability of being in chronic and seasonal food insecure categories and increasing the probability of being in vulnerable and food secure categories. Adoption, however, is not significant. It may be the case (given the conditional logit results) that soil quality is related to adoption of soil-enhancing practices, as such soil quality is picking up any adoption behavior in the food security model. Training and information from the extension service has an unexpected effect but the effect is insignificant; extension services increases the probability of being in chronic or seasonal categories and decreases the probability of being in vulnerable or food secure categories. Several reasons may exist for the extension results. Similar to Rubas' [12] finding that outreach is not a universal factor influencing technology adoption; extension services may not influence food secure. Rubas [12] concludes that outreach agents may want to alter their approaches. Another reason may be the respondent's social desirability bias behavior [40]. The survey

Table 5 Marginal effects associated with the food security model

Independent variables	Marginal effect	Standard error	z	$p > \|z\|$	95% confidence interval	
Chronic food insecure						
Adoption of soil imp.	−0.011	0.031	−0.370	0.714	−0.073	0.050
Chemical fertilizer	−0.076	0.050	−1.520	0.128	−0.173	0.022
Commercial seed	−0.051	0.042	−1.220	0.223	−0.134	0.031
Herbicides	−0.043	0.049	−0.880	0.381	−0.138	0.053
Insecticide	0.065	0.036	1.780	0.076	−0.007	0.136
Medium-quality soil	−0.200	0.065	−3.080	0.002	−0.327	−0.073
Off-farm activity	−0.022	0.024	−0.880	0.377	−0.069	0.026
Seasonal lease	0.007	0.032	0.210	0.836	−0.056	0.069
Share cropping	0.035	0.058	0.600	0.551	−0.079	0.149
Income	−0.001	0.001	−1.620	0.105	−0.002	0.000
Education level	−0.030	0.029	−1.020	0.307	−0.087	0.027
Household size	−0.001	0.006	−0.090	0.929	−0.012	0.011
Age of house head	−0.001	0.001	−0.760	0.446	−0.003	0.001
Extension service	0.033	0.025	1.310	0.189	−0.016	0.082
Seasonal food insecure						
Adoption of soil imp.	−0.025	0.064	−0.390	0.694	−0.150	0.100
Chemical fertilizer	−0.093	0.040	−2.340	0.019	−0.171	-0.015
Commercial seed	−0.092	0.058	−1.590	0.112	−0.206	0.022
Herbicides	−0.065	0.046	−1.400	0.161	−0.155	0.026
Insecticide	0.122	0.059	2.080	0.037	0.007	0.237
Medium-quality soil	−0.222	0.071	−3.130	0.002	−0.361	−0.083
Off-farm activity	−0.061	0.081	−0.750	0.451	−0.219	0.097
Seasonal lease	0.016	0.080	0.200	0.842	−0.141	0.173
Share cropping	0.056	0.063	0.900	0.371	−0.067	0.179
Income	−0.002	0.001	−1.630	0.104	−0.005	0.000
Education level	−0.060	0.050	−1.200	0.229	−0.158	0.038
Household size	−0.001	0.014	−0.090	0.929	−0.029	0.026
Age of house head	−0.002	0.002	−0.770	0.440	−0.006	0.003
Extension service	0.101	0.092	1.100	0.272	−0.079	0.281
Vulnerable food insecure						
Adoption of soil imp.	0.018	0.046	0.380	0.703	−0.073	0.108
Chemical fertilizer	0.085	0.042	2.020	0.043	0.003	0.168
Commercial seed	0.070	0.049	1.430	0.153	−0.026	0.165
Herbicides	0.054	0.048	1.110	0.268	−0.041	0.149
Insecticide	−0.089	0.044	−2.030	0.042	−0.175	−0.003
Medium-quality soil	0.187	0.049	3.840	0.000	0.092	0.282
Off-farm activity	0.037	0.046	0.820	0.413	−0.052	0.127
Seasonal lease	−0.011	0.052	−0.200	0.839	−0.113	0.092
Share cropping	−0.045	0.062	−0.720	0.470	−0.168	0.077
Income	0.002	0.001	1.630	0.104	0.000	0.003
Education level	0.043	0.039	1.130	0.260	−0.032	0.119
Household size	0.001	0.010	0.090	0.929	−0.018	0.019
Age of house head	0.001	0.002	0.770	0.443	−0.002	0.004

Table 5 Marginal effects associated with the food security model *(Continued)*

Extension service	−0.059	0.048	−1.230	0.220	−0.153	0.035
Food secure						
Adoption of soil imp.	0.019	0.049	0.390	0.698	−0.077	0.115
Chemical fertilizer	0.084	0.037	2.250	0.025	0.011	0.157
Commercial seed	0.074	0.049	1.510	0.130	−0.022	0.169
Herbicides	0.054	0.044	1.230	0.220	−0.032	0.140
Insecticide	−0.098	0.047	−2.060	0.039	−0.191	−0.005
Medium-quality soil	0.235	0.056	4.200	0.000	0.125	0.345
Off-farm activity	0.045	0.059	0.760	0.447	−0.071	0.161
Seasonal lease	−0.012	0.060	−0.200	0.841	−0.129	0.105
Share cropping	−0.046	0.057	−0.800	0.425	−0.158	0.066
Income	0.002	0.001	1.680	0.092	0.000	0.004
Education level	0.046	0.040	1.170	0.240	−0.031	0.124
Household size	0.001	0.011	0.090	0.929	−0.020	0.022
Age of house head	0.001	0.002	0.780	0.437	−0.002	0.005
Extension service	−0.075	0.069	−1.090	0.274	−0.209	0.059

The marginal effects are for the discrete change in the qualitative 0–1 variables.

was conducted by extension agents who have worked with farmers in the region over a long period of time. It is possible as Legget et al. [40] has suggested that the respondents presented themselves in a socially desirable frame (respond that they received training/information) to look good in the eyes of the survey personnel, in this case extension agents.

Conclusions

It appears that soil quality should be the center of the focus in addressing the issues of adopting soil-improving practices and food insecurity in Ghana. Households with higher quality soil are more likely to be associated with the use of soil-improving practices and are more likely to be in an improved food insecure category. Given the high priority, the government of Ghana has placed on food security, policies that encourage the use of yield enhancing inputs, chemical fertilizers, and commercial seeds that may be beneficial in improving farm households' food security status. Adoption of soil-improving practices and yield enhancing inputs, however, may not always be compatible.

Priorities may differ between food secure and food insecure households in terms of adopting soil-improving practices. It is estimated that food secure households are more likely to be associated with the use of soil-improving practices than food insecure households. Food insecure households may be weighing the immediate need for increased production from the limited land they farm against longer term needs. Food secure households may have been able to afford to adopt soil-improving practices even if there is temporary drop in yields. Hence, policies may have to be customized for the different categories of food secure households to increase adoption of soil-improving practices.

Household characteristics such as income, age, education level, and household size emerge as not associated with improving household food security or the adoption decision in all food insecurity groups. There might be other factors that are outside the control of households to determine the food security position of households. Exogenous factors such as weather and crop prices are beyond the farmer control but may determine food security "better" than the influence of household characteristics. Further research into why households appear to differ little between the categories should be conducted or is it entirely related to soil quality.

Four recommendations emerge to improve the food security status of households based on their soil-improving practice choice decisions. First, the results from the food security model lend support for a need for the government to eliminate the problems that saddled the distribution of fertilizers in response to the fertilizer subsidy program announced during the 2008 planting season. Results suggest that the use of chemical fertilizers improves the food security status of households. The adoption model, however, suggests a catch-22 may exist. Although increasing use of chemical fertilizers reduces food insecurity, it also reduces the probability of adopting soil-improving practices. This is an unexpected result. Households may be using the application of chemical fertilizer as a short-term solution to restore soil fertility. This practice, however, may be a short-sighted substitute activity for a longer term solution of adoption of soil-improving practices. This is just speculation; more research work is necessary in this area. At this time, the government needs to approach these two different issues cautiously.

Second, the results suggest a need to improve the seed research and distribution. A recent study has indicated a woeful adoption of improved seed in Ghana since 1997 [41]. Banful [41] notes that approximately 3% of Ghana's maize area is planted to hybrids which have twice the yield potential of the currently planted seeds. She notes that the system is either not solving the binding constraints on Ghana's farmers to adopt the variety or the extension service is unable to disseminate the net benefits of the new varieties. Third, the results point to a need to improve the institutional framework, especially land markets to improve land use contracting. Farmers leasing land have been shown to increase technology adoption, but there is little contract enforcement mechanisms within the judicial system to promote land lease contracts. Fourth, given the importance of soil quality is in both models, policies that enhance soil quality are important. Information on improving soil quality may need to be delivered to farmers in a way that is understandable and easy to use.

Contrary to *a priori* expectations, both access to agricultural extension services and education are insignificant in both adoption of soil-improving practices and food insecurity categorization. *A priori*, it was felt that these factors should increase adoption of soil-improving practices and improve the food security position of the household. Further, research is necessary to better understand the relationship between these variables.

Endnote

[a]As requested by a reviewer, the level of significance is arbitrary, depends on the sample size, and involves trade-offs between consequences of type I and II errors. A level of 0.15 is used in this study to include in the discussion any potentially important variables that might be nonsignificant because of the sample size and data limitations. Further, given that several of the variables are significant between 0.05 and 0.11 and relatively few variables between 0.15 and 0.25 (a natural break), it was felt that a larger significance level of 0.15 is appropriate. Exact p values are provided in the tables for the interested reader.

Abbreviation

WFP: World Food Programme.

Competing interests

The authors declare that they have no competing interests.

Authors' contributions

JTN, JWM, and FOB are involved in the conception and design, data collection and analysis, manuscript writing, and final approval of the manuscript.

Acknowledgements

This work was supported by the McKnight Foundation Grant No. 09-1170, administered by Norman Borlaug Institute for International Agriculture at Texas A&M University.

Author details

[1]Water Management and Hydrological Science, Texas A&M University, College Station, TX, USA. [2]Department of Agricultural Economics, Texas A&M University, College Station, TX, USA.

References

1. Institute of Statistical, Social, and Economic Research: *The State of the Ghanaian Economy in 2010.* Legon: University of Ghana; 2010.
2. Devereux S: *Desk Review: Distinguishing Between Chronic and Transitory Food Insecurity in Emergency Needs Assessments.* World Food Programme; 2006. http://www.livestock-emergency.net/userfiles/file/assessment-review/Institue-Development-Studies-2006.pdf. Accessed 20 October 2012.
3. Maxwell S, Frankenberger T: *Household Food Security: Concepts, Indicators, and Measurements, a Technical Review.* Rome: International Fund for Agricultural Development; 1992. http://www.ifad.org/hfs/tools/hfs/hfspub/index.htm. Accessed 4 November 2013.
4. Wolter D: **Ghana: Agriculture is Becoming a Business.** *OECD Journal General Papers* 2009, **5**(2):9–32. http://dx.doi.org/10.1787/gen_papers-2009-5ks9zs5gt1d2. Accessed 18 September 2013.
5. Biederlack L, Rivers J: *Comprehensive Food Security & Vulnerability Analysis (CFSVA): Ghana. United Nations World Food Programme.* 2009. http://documents.wfp.org/stellent/groups/public/documents/ena/wfp201820.pdf. Accessed 12 November 2014.
6. Mulugeta T, Hundie B: *Impacts of Adoption of Improved Wheat Technologies on Households' Food Consumption in Southeastern Ethiopia, Prepared for Presentation at the International Association of Agricultural Economists (IAAE) Triennial Conference.* Foz do Iguaçu, Brazil; 2009. http://ageconsearch.umn.edu/bitstream/126766/2/Mulugeta.pdf. Accessed 29 November 2012.
7. World Bank: *Agriculture for Development, World Development Report 2008.* Washington, DC; 2008. http://web.worldbank.org/WBSITE/EXTERNAL/EXTDEC/EXTRESEARCH/EXTWDRS/0,,contentMDK:23062293~pagePK:478093~piPK:477627~theSitePK:477624,00.html. Accessed 4 November 2013.
8. Ghana Ministry of Food and Agriculture: *Food and Agricultural Sector Development Policy. August 2007.* Accra; 2011. http://mofa.gov.gh/site/?page_id=598. Accessed 19 September 2012.
9. Haggblade S, Tembo G: *Development, Diffusion and Impacts of Conservation Farming in Zambia.* International Food Policy Research Institute, EPTD Discussion Paper 108; 2003. http://ageconsearch.umn.edu/handle/54464. Accessed 4 November 2013.
10. Lal R, Stewart BA: **Sustainable management of soil resources and food security.** In *Advances in Soil Science World Soil Resources and Food Security.* Edited by Lal R, Stewart BA. Boca Raton: CRC Press, Taylor and Francis Group, LLC; 2012:1–10.
11. Pender J, Gebremedhin B: **Determinants of agricultural and land management practices and impacts in crop production and household income in highlands of Tigray, Ethiopia.** *J Afr Econ* 2006, **17**:395–450.
12. Rubas D: *Technology adoption: who is likely to adopt and how does the timing affect the benefits?* PhD thesis. TX, USA: Texas A&M University, Department of Agricultural Economics, College Station; 2004.
13. Lee DR: **Agricultural sustainability and technology adoption: issues and policies for developing countries.** *Am J Agric Econ* 2005, **87**(5):1325–1334.
14. Knowler D, Bradshaw B: **Farmers' adoption of conservation agriculture: a review and synthesis of recent research.** *Food Policy* 2007, **32**(1):25–48.
15. Ghana Statistical, Research and Information Directorate: *Agriculture in Ghana: Facts and Figures.* Accra: Ghana Ministry of Food and Agriculture; 2010. http://mofa.gov.gh/site/wp-content/uploads/2011/10/AGRICULTURE-IN-GHANA-FF-2010.pdf.
16. Nata J: *Household segmentation in food insecurity and soil improving practices in Ghana. M.S. Thesis.* Texas A&M University, College Station; 2013.
17. Train KE: *Discrete Choice Models with Simulation.* New York: Cambridge University Press; 2007.

18. Mas-Colell A, Whinston MD, Green JR: *Microeconomic Theory*. New York: Oxford University Press; 1995.
19. Wiggins S, Maunder N, Carnegie J, Roberts B, Mokoena R, Tregurtha N: *Scoping Study Towards DFIDSA's Regional Hunger and Vulnerability Programme*. Pretoria: Department Fund for International Development, Southern African Regional Poverty Network; 2004. http://www.eldis.org/go/home&id=24431&type=Document#.VGO7yWftiBu. Accessed 12 November 2014.
20. Bickel G, Nord M, Price P, Hamilton W, Cook J: *Guide to Measure Household Food Security*. Washington, DC: U.S. Department of Agriculture, Office of Analysis, Nutrition, and Evaluation; 2000. http://www.fns.usda.gov/sites/default/files/FSGuide_0.pdf. Accessed 12 November 2014.
21. Greene WH: *Econometric Analysis*. 7th edition. Upper Saddle River: Pearson Prentice Hall; 2012.
22. StataCorp: *Stata Statistical Software: Release 11*. College Station: StataCorp LP; 2009.
23. Kristjanson P, Neufeldt H, Gassner A, Mango J, Kyasse FB, Desta S, Sayula G, Thiede B, Furch W, Thornton PK, Coe R: **Are food insecure smallholder households making changes in their farming practices: evidence from East Africa.** *Food Security* 2012, **2012**(4):381–397.
24. Ajayi OC: **User acceptability of sustainable soil fertility technologies: lessons from farmers' knowledge, attitude and practice in Southern Africa.** *J Sustain Agric* 2007, **30**(3):21–40.
25. Mbaga-Semgalawe Z, Folmer H: **Household adoption behaviour of improved soil conservation: the case of the North Pare and West Usambara Mountains of Tanzania.** *Land Use Policy* 2000, **17**:321–336.
26. Anim FDK: **A note on the adoption of soil conservation measures in the northern province of South Africa.** *J Agric Econ* 1999, **50**(2):336–345.
27. Nkamleu GB: **Modeling farmers' decisions on integrated soil nutrient management in sub-Saharan Africa: a multinomial logit analysis in Cameroon.** In *Advances in Integrated Soil Fertility Management in Sub-Saharan Africa: Challenges and Opportunities*. Edited by Bationo A, Waswa B, Kihara J, Kimetu J. Dordrecht: Springer; 2007:891–904.
28. Lapar LA, Pandey S: **Adoption of soil conservation: the case of the Philippine uplands.** *Agric Econ* 1999, **21**:241–256.
29. Franzel S: **Socioeconomic factors affecting the adoption potential of improved tree fallows in Africa.** *Agrofor Syst* 1999, **47**:305–321.
30. Marenya PP, Barrett CB: **Household-level determinants of adoption of improved natural resources management practices among smallholder farmers in western Kenya.** *Food Policy* 2007, **32**:515–536.
31. Savadogo K, Reardon T, Pietola K: **Adoption of improved land use technologies to increase food security in Burkina Faso: relating animal traction, productivity, and non-farm income.** *Agric Syst* 1998, **58**(3):441.464.
32. Branca G, Lipper L, McCarthy N, Jolejloe MC: **Food security, climate change, and sustainable land management: a review.** *Agron Sustain Dev* 2013, **33**(4):635–650.
33. Minten B, Barrett CB: **Agricultural technology, productivity, and poverty in Madagascar.** *World Dev* 2008, **36**(5):797–822.
34. Asfaw S, Shiferaw B: *Agricultural Technology Adoption and Rural Poverty: Application of Endogenous Switching Regression for Selected East African Countries*. Poster Presented at the Joint 3rd African Association of Agricultural Economists (AAAE) 19-23 September; 2010. http://ageconsearch.umn.edu/handle/97049. Accessed 2 December 2012.
35. Wollini D, Lee R, Thies JE: *Effects of Participation in Organic Markets and Farmer-Based Organizations on the Adoption of Soil Conservation Practices Among Small-Scale Farmers in Honduras*. Paper Presented at the American Agricultural Economics Association Annual Meeting, Orlando, FL 27-29 July; 2008. http://ideas.repec.org/p/ags/aaea08/6423.html. Accessed 20 October 2012.
36. Diao X, Sarpong D: *Cost Implication of Agricultural Land Degradation in Ghana: an Economy Wide Multimarket Model Assessment. Discussion Paper 698*. Washington, DC: International Food Policy Research Institute; 2007. http://www.ifpri.org/publication/cost-implications-agricultural-land-degradation-ghana. Accessed 2 November 2012.
37. Owusua V, Abdulaib A, Abdul-Rahmana S: **Non-farm work and food security among farm households in Northern Ghana.** *Food Policy* 2011, **36**(2):108–118.
38. Babatunde RO, Qaim M: **Impact of off-farm income on food security and nutrition in Nigeria.** *Food Policy* 2010, **35**:303–311.
39. Conway G, Toenniessen G: **Science for African food security.** *Science* 2003, **299**(5610):1187–1188.
40. Leggett CG, Kleckner NS, Boyle KJ, Duffield JW, Mitchell RC: **Social desirability bias in contingent valuation surveys administered through in-person interviews.** *Land Econ* 2003, **79**(4):561–575.
41. Banful AB: *Operational Details of the 2008 Fertilizer Subsidy in Ghana - Preliminary Report. IFPRI Ghana Strategy Support Program (GSSP), Background Paper GSSP 18, May 2009*. http://www.ifpri.org/sites/default/files/publications/gsspwp18.pdf. Accessed 3 November 2013.

PERMISSIONS

All chapters in this book were first published in A&FS, by BioMed Central; hereby published with permission under the Creative Commons Attribution License or equivalent. Every chapter published in this book has been scrutinized by our experts. Their significance has been extensively debated. The topics covered herein carry significant findings which will fuel the growth of the discipline. They may even be implemented as practical applications or may be referred to as a beginning point for another development.

The contributors of this book come from diverse backgrounds, making this book a truly international effort. This book will bring forth new frontiers with its revolutionizing research information and detailed analysis of the nascent developments around the world.

We would like to thank all the contributing authors for lending their expertise to make the book truly unique. They have played a crucial role in the development of this book. Without their invaluable contributions this book wouldn't have been possible. They have made vital efforts to compile up to date information on the varied aspects of this subject to make this book a valuable addition to the collection of many professionals and students.

This book was conceptualized with the vision of imparting up-to-date information and advanced data in this field. To ensure the same, a matchless editorial board was set up. Every individual on the board went through rigorous rounds of assessment to prove their worth. After which they invested a large part of their time researching and compiling the most relevant data for our readers.

The editorial board has been involved in producing this book since its inception. They have spent rigorous hours researching and exploring the diverse topics which have resulted in the successful publishing of this book. They have passed on their knowledge of decades through this book. To expedite this challenging task, the publisher supported the team at every step. A small team of assistant editors was also appointed to further simplify the editing procedure and attain best results for the readers.

Apart from the editorial board, the designing team has also invested a significant amount of their time in understanding the subject and creating the most relevant covers. They scrutinized every image to scout for the most suitable representation of the subject and create an appropriate cover for the book.

The publishing team has been an ardent support to the editorial, designing and production team. Their endless efforts to recruit the best for this project, has resulted in the accomplishment of this book. They are a veteran in the field of academics and their pool of knowledge is as vast as their experience in printing. Their expertise and guidance has proved useful at every step. Their uncompromising quality standards have made this book an exceptional effort. Their encouragement from time to time has been an inspiration for everyone.

The publisher and the editorial board hope that this book will prove to be a valuable piece of knowledge for researchers, students, practitioners and scholars across the globe.

LIST OF CONTRIBUTORS

François Collart Dutilleul
MSH Ange Guépin, 5 allée Jacques Berque, Nantes cedex 1, 44021, France

Bart van den Borne
Department of Health Education & Promotion, Faculty of Medicine & Life Sciences, Maastricht University, 6200 MD Maastricht, The Netherlands

Susan Keino
Department of Health Education & Promotion, Faculty of Medicine & Life Sciences, Maastricht University, 6200 MD Maastricht, The Netherlands
Department of Epidemiology & Nutrition, School for Public Health, Moi University, 30100 Eldoret, Kenya

Guy Plasqui
Department of Human Biology, School for Nutrition, Toxicology & Metabolism, Maastricht University University, 6200 MD Maastricht, The Netherlands

Jessica Oh
Sandra Rotman Centre, University Health Network and University of Toronto, Toronto, Ontario, Canada

Obidimma C Ezezika
Sandra Rotman Centre, University Health Network and University of Toronto, Toronto, Ontario, Canada
African Centre for Innovation and Leadership Development, Federal Capital Territory, Abuja, Nigeria
Dalla Lana School of Public Health, University of Toronto, Toronto, Canada

Mary Stella N Wabwoba and Jacob W Wakhungu
Department of Disaster Management and Sustainable Development, MasindeMuliro University of Science and Technology, Kakamega, Kenya

Martine M Rutten
International Policy Department, LEI Wageningen UR, Alexanderveld 5, 2502, The Hague, LS, Netherlands

Hervé Guyomard, Béatrice Darcy-Vrillon, Catherine Esnouf, Marie Russel and Marion Guillou
INRA Paris, 147 rue de l'université, 75 338, Paris Cedex 07 France

Michèle Marin
INRA Toulouse, 24 Chemin de Borde Rouge, CS 52627, 31326 Castanet Tolosan, Cedex, France

Peter R Steward, Gorm Shackelford, Tim G Benton and Steven M Sait
Africa College and School of Biology, University of Leeds, Leeds LS2 9JT, UK

Luísa G Carvalheiro
Africa College and School of Biology, University of Leeds, Leeds LS2 9JT, UK
Naturalis Biodiversity Center, Leiden 2300RA, The Netherlands

Lucas A Garibaldi
Sede Andina, Universidad Nacional de Río Negro (UNRN) and Consejo Nacional de Investigaciones Científicas y Técnicas (CONICET), Mitre 630, San Carlos de Bariloche, CP 8400, Río Negro, Argentina

Ghose Bishwajit, Hui Gao and Daogen Yin
Department of Nutrition and Food Hygiene, School of Public Health, Tong Ji Medical College, Wuhan, China

Marce-Amara Kpoghomou
Department of Epidemiology and Biostatistics, School of Public Health, Tong Ji Medical College, Wuhan, China

Liu Jun
Department of Respiration Medicine, Guangzhou First People's Hospital, Guangzhou, China

Sajeeb Sarker
Department of Journalism, Communication and Media Studies, State University of Bangladesh, Dhaka, Bangladesh

Sharmistha Ghosh
Department of Public Administration, Huazhong University of Science and Technology, Wuhan, China

Bronwyn Isaacs, Laura Ford, Jane Dixon, Martyn Kirk and Sharon Friel
National Centre for Epidemiology and Population Health, The Australian National University, Canberra, ACT 0200, Australia

Gillian Hall
National Centre for Epidemiology and Population Health, The Australian National University, Canberra, ACT 0200, Australia
Medical School, The Australian National University, Canberra, ACT 0200, Australia

Alison Rothwell
School of Science and Health, University of Western Sydney, Locked Bag 1797, Penrith, NSW 1797, Australia

Tim Grant
Life Cycle Strategies, 29 Forest Street, Collingwood, VIC 3066, Australia

Nelson Turyahabwe
Department of Extension and Innovation Studies, College of Agricultural and Environmental Sciences, Makerere University, Kampala, Uganda

Willy Kakuru and Mnason Tweheyo
Department of Forestry, Biodiversity and Tourism, College of Agricultural and Environmental Sciences, Makerere University, Kampala, Uganda

David Mwesigye Tumusiime
Department of Environmental Management, School of Forestry, Environment and Geographical Sciences, Makerere University, Kampala, Uganda
Department of International Environment and Development Studies (Noragric), Norwegian University of Life Sciences (UMB), Aas, Norway

Govindan Parayil
United Nations University-Institute of Advanced Studies (UNU-IAS), 6F International Organisations Centre, Pacifico Yokohama, 1-1-1 Minato Mirai, Nishi-ku, Yokohama 220-8502, Japan

Ademola A Adenle
United Nations University-Institute of Advanced Studies (UNU-IAS), 6F International Organisations Centre, Pacifico Yokohama, 1-1-1 Minato Mirai, Nishi-ku, Yokohama 220-8502, Japan
National Graduate Institute for Policy Studies (GRIPS), 7-22-1 Roppongi, Minato-ku, Tokyo 106-8677, Japan

Ogugua C Aworh
Department of Food Technology, University of Ibadan (UI), Ibadan, Nigeria

Richard Akromah
Kwame Nkrumah University of Science and Technology (KNUST), Faculty of Agriculture, Kumasi, Ghana

Costancia Peter Rugumamu
Department of Zoology and Wildlife Conservation, University of Dar es Salaam, Dar es Salaam, Tanzania

Dilrukshi Hashini Galhena, Russell Freed and Karim M Maredia
College of Agriculture and Natural Resources (CANR), Michigan State University, East Lansing, MI 48824, USA

Norman Uphoff
SRI International Network and Resources Center (SRI-Rice), Cornell University, Ithaca, NY 14853, USA
Cornell International Institute for Food, Agriculture and Development (CIIFAD), Cornell University, Ithaca, NY 14853, USA

Joshua Sikhu Okonya
Global Program of Integrated Crop and Systems Research, International Potato Center (CIP), PO Box 22274, Kampala, Uganda

Jürgen Kroschel
Global Program of Integrated Crop and Systems Research, International Potato Center (CIP), Apartado 1558, Lima 12, Peru

Genesis T Yengoh
Department of Physical Geography and Ecosystem Sciences, Geobiosphere Science Centre, Lund University, Sölvegatan 12, Lund SE-223 62, Sweden

Robert Zougmoré
CGIAR Research Program on Climate Change, Agriculture and Food Security (CCAFS), ICRISAT Bamako, BP 320 Bamako, Mali

Abdulai Jalloh
West and Central African Council for Agricultural Research and Development (CORAF/WECARD), BP 48 Dakar, RP, Senegal

Andre Tioro
Réseau des Organisations Paysannes et de Producteurs Agricoles de l'Afrique de l'Ouest (ROPPA), 09 BP 884 Ouagadougou 09, Burkina Faso

Jifar T Nata
Water Management and Hydrological Science, Texas A&M University, College Station, TX, USA

James W Mjelde and Frederick O Boadu
Department of Agricultural Economics, Texas A&M University, College Station, TX, USA

Index

www.ingramcontent.com/pod-product-compliance
Lightning Source LLC
LaVergne TN
LVHW080647210926
848485LV00003B/340